Nucleic Acids and Molecular Biology

Volume 32

Series editor
Allen W. Nicholson
Philadelphia, PA 19122, USA

More information about this series at http://www.springer.com/series/881

Béatrice Clouet-d'Orval
Editor

RNA Metabolism and Gene Expression in Archaea

Springer

Editor
Béatrice Clouet-d'Orval
Laboratoire de Microbiologie et de Génétique
Moléculaires (LMGM), Centre de Biologie Intégrative (CBI)
Université de Toulouse, CNRS
Toulouse, France

ISSN 0933-1891 ISSN 1869-2486 (electronic)
Nucleic Acids and Molecular Biology
ISBN 978-3-319-65794-3 ISBN 978-3-319-65795-0 (eBook)
DOI 10.1007/978-3-319-65795-0

Library of Congress Control Number: 2017957625

Printed on acid-free paper

This Springer imprint is published by Springer Nature
The registered company is Springer International Publishing AG
The registered company address is: Gewerbestrasse 11, 6330 Cham, Switzerland

Outline and Synopsis

Although many reviews and books survey many aspects of RNA in Bacteria and in Eukarya, no book has so far attempted to provide a panoramic view of enzymes, machines, and pathways that synthetize, modify, and degrade RNA in Archaea, the third domain of life.

The early studies of C.R Woese of microbial phylogeny based on rRNA sequences yielded the remarkable finding that the prokaryotic world was not monophyletic but separated into two distinct domains: Archaea and Bacteria. This major advance during the last part of the twentieth century opened a novel vision of the living world. The universal tree predicted that the Archaea were specific relatives of the Eukarya, to the exclusion of the Bacteria. Astonishingly, genetic information processing systems in the Archaea are often closely related to their counterparts in eukaryotes (translation, transcription, replication, DNA recombination, and repair) even if Archaea have a bacteria-like cellular structure. Archaea are now considered as models with an emphasis on studying the molecular mechanisms that universally conserve the evolutionary history of life and for the emergence of new technologies. With the advent of whole genome sequencing, a promising way to take advantage of these similarities is to analyze the evolution of cellular systems through phylogenomic approaches. Over the years, our vision of archaeal microorganisms has evolved from exotic microbial extremophiles to organisms of general importance that are used to elucidate fundamental biological questions. More recently, a newly discovered deep-sea archaeon named Loki suggests eukaryotes evolved directly from archaea rather than representing a separate branch of life. In consideration of this fundamental finding, Archaea are now renowned to be important model microorganisms for the study of molecular mechanisms that are conserved between Eukarya and Archaea. In particular, Archaea and Eukarya share many RNA biology aspects which are at the center of cellular regulation pathways. Major examples are the 3D structures of the archaeal RNA polymerase and archaeal exosome which resemble eukaryal RNA polymerases and RNA-degrading exosome complexes, respectively. Moreover, most of the archaeal tRNA and rRNA modifications rely on RNP guide machineries orthologous to the eukaryal C/D and H/ACA RNP complexes. This highlights the advantage of an archaeal model to gain further mechanistic and evolutionary information of fundamental processes across the three domains of life. Nevertheless, for over a decade, numerous signaling pathways have been described in Eukarya and Bacteria in which RNA processing regulates gene expression. In Archaea, these processes have been overlooked. Molecular mechanisms of RNA maturation and decay and posttranscriptional control of gene expression are far from understood.

Given that the RNA field is moving very fast, it is time to cover many of the exciting and sometimes overlooked developments in the field that reveal originality of the archaeal system and illuminate the fascinating biology that sets the stage for RNA. Therefore, this volume of Nucleic Acid and Molecular Biology provides a review of our knowledge on different aspects of RNA biology in archaeal cells from synthesis to degradation through modification, maturation, and regulatory function as well as defense system including evolutionary considerations of those processes. However, the field of archaeal virology will not be discussed in this book since to date there is no evidence on the existence of archaeal RNA viruses or any regulatory RNA molecules involved in the virus life cycles.

This task is achieved by 11 chapters that collectively summarize recent advances in our understanding of RNA metabolism at all levels.

Contents

Chapter 1
A Global Characterisation of the Archaeal Transcription Machinery

Katherine Smollett, Fabian Blombach, Thomas Fouqueau, and Finn Werner

Abstract Archaea employ a eukaryote-like transcription apparatus to transcribe a bacteria-like genome; while the RNA polymerase, basal factors and promoter elements mirror the eukaryotic RNA polymerase II system, archaeal genomes are densely packed with genes organised into multicistronic transcription units. The molecular mechanisms of archaeal transcription have been studied and characterised in great detail in vitro, but until recently relatively little was known about its global characteristics. In this chapter we discuss an integrated view of transcription from the molecular to the global level. Systems biology approaches have provided compelling insights into promoter and terminator DNA elements, the genome-wide distribution of transcription initiation- and elongation factors and RNA polymerase, the archaeal transcriptome and chromatin organisation. Overall these analyses illuminate transcription from a genome-wide perspective and serve as a resource for the community. In addition, Big Data can often validate mechanistic models based on biochemical and structural information, and generate new working hypotheses that can be thoroughly tested and dissected in vitro. This is an exciting time to study gene expression in the archaea since we are at the brink of a comprehensive yet detailed understanding of transcription.

1.1 Introduction

Archaea are prokaryotes and as such share many properties with bacteria including circular genomes, densely packed with genes organised into operons. However, their transcription machinery is closely related to that of RNA polymerase II, the enzyme responsible for mRNA transcription in eukaryotes (Fig. 1.1). This similarity extends from the RNA polymerase (RNAP) subunit composition, via general transcription factors required for initiation, to their cognate promoter elements (Fig. 1.1a, b) (Werner and Grohmann 2011). In essence, archaeal transcription

K. Smollett • F. Blombach • T. Fouqueau • F. Werner (✉)
Institute of Structural and Molecular Biology, Division of Biosciences, University College London, Gower Street, London WC1E 6BT, UK
e-mail: f.werner@ucl.ac.uk

B. Clouet-d'Orval (ed.), *RNA Metabolism and Gene Expression in Archaea*,
Nucleic Acids and Molecular Biology 32, DOI 10.1007/978-3-319-65795-0_1

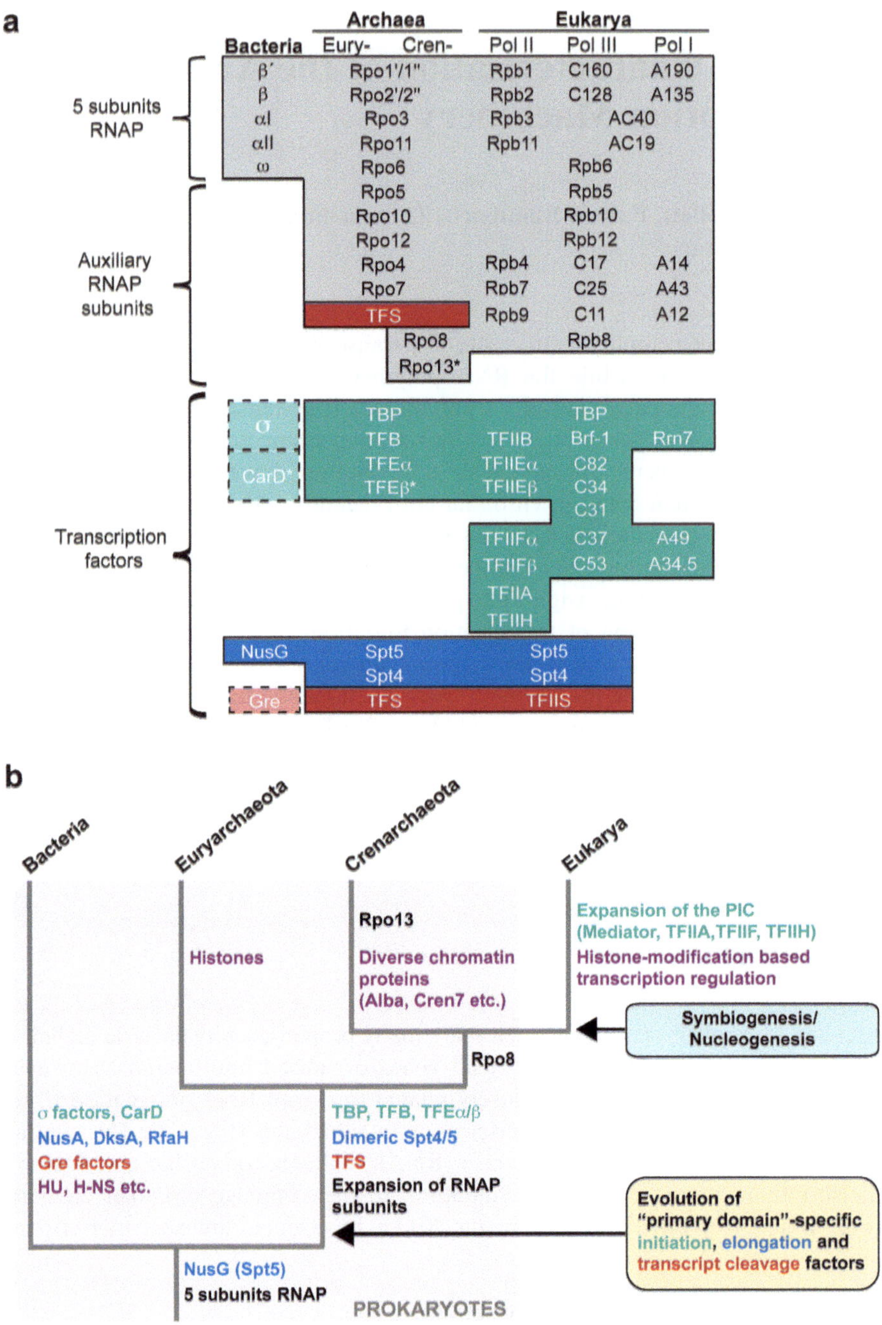

Fig. 1.1 Evolution of the basal transcription machinery in the three domains of life. (**a**) Table of RNAP subunits and general transcription factors in the three domains of life. The columns represent the single RNAP transcription systems in bacteria, eury- and crenarchaea, and the three orthodox RNAPI, II and III systems in eukaryotes. The rows depict homologous factors,

involves a eukaryotic-like machinery acting upon a bacterial-like template, making it an interesting and important subject to study. Archaeal transcription can be considered a simpler, stripped-down version of the RNAPII system, generally consisting of fewer and smaller components that facilitate the basic mechanisms of transcription. These are often obscured by the baroque complexity in eukaryotes—making archaea invaluable tools to dissect them. In vitro studies of archaeal transcription have focused on hyperthermophilic archaea due to their high biochemical tractability including the in vitro assembly of RNAPs from *Methanocaldococcus jannaschii* and *Pyrococcus furiosus* from individual recombinant subunits under defined conditions in the test tube (Naji et al. 2007; Smollett et al. 2015; Werner and Weinzierl 2002). This approach has not been successful with any eukaryotic RNAP thus far, and archaea have therefore provided invaluable model systems to elucidate the molecular mechanisms of RNAPII transcription (Fouqueau et al. 2013; Grohmann et al. 2011; Hirtreiter et al. 2010a, b; Kostrewa et al. 2009; Tan et al. 2008; Werner and Weinzierl 2005). Whilst such recombinant systems are required to carry out a definitive functional dissection of transcription, less attention has been paid to the systems level properties of the basic transcription machinery in archaea, including whole genome occupancy, transcription start site- and transcriptome mapping. As high-throughput sequencing technologies have become more accessible new avenues of research have become possible. In this chapter, we outline how systems biology can complement classical biochemistry/structural biology, and how this enhances our understanding of the different stages of the transcription cycle and the structure and function of chromatin (Fig. 1.2).

1.2 The Basal Transcription Machinery and the Archaeal Transcription Cycle

1.2.1 Promoter Recognition and Recruitment of the RNAP

In all domains of life transcription is initiated by the recruitment of basal, or general, transcription initiation factors to the promoter. Most archaeal promoters rely on three elements: the TATA box, B-recognition element (BRE) and the Initiator (Inr). TATA box and BRE are DNA sequence recognition motifs of the two general transcription factors TBP and TFB, respectively (Bell et al. 1999; Qureshi et al. 1995; Rowlands et al. 1994), both TBP and TFB are necessary and

Fig. 1.1 (continued) while functionally analogous but evolutionarily unrelated factors are shown as separated fields with dashed borders. Factors are colour-coded according to their function in transcription initiation (*green*), elongation (*blue*) and transcript cleavage (*red*). Subunits and factors that are not conserved in all domains are indicated with *asterisks*. (**b**) Schematic representation of key events in the evolution of the basal transcription machinery. General transcription factors are *colored* as in panel **a**, with chromatin proteins added in *purple*

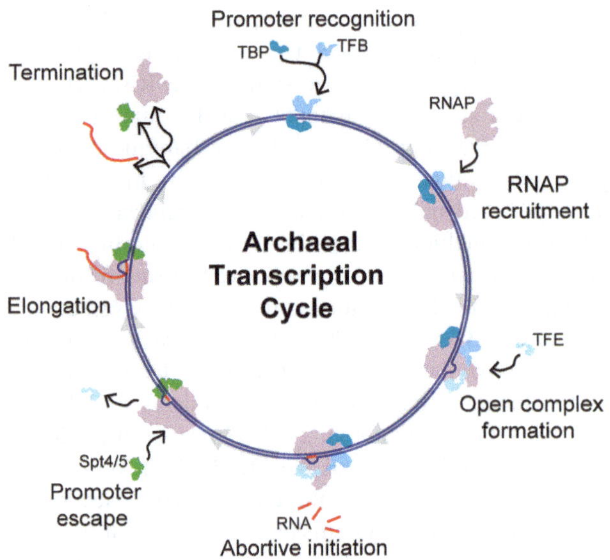

Fig. 1.2 The Archaeal transcription cycle. Transcription initiation is a recruitment cascade, the BRE and TATA promoter motifs sequester TBP and TFB, which in turn recruits RNAP to form the PIC. TFE stimulates DNA strand separation of the promoter IMR region, which stabilises the PIC. The later stages of initiation involved the synthesis of abortive transcripts, and promoter escape, which is likely facilitated by the swapping of TFE for Spt4/5, forming a processive transcription elongation complex. During elongation additional factors including transcript cleavage factors and Spt4/5 ensure highly processive transcription. At the 3′ end of the gene transcription is terminated by short poly-U signatures, and likely by hitherto uncharacterised termination factors

sufficient to facilitate promoter-directed transcription in vitro (Werner and Weinzierl 2002). TBP and TFB are homologous to eukaryotic TBP and TFIIB (Fig. 1.1), respectively, and have identical functions, albeit with a faster DNA-binding dynamics (Gietl et al. 2014) that may reflect different mechanisms of regulation (Blombach and Grohmann 2017). Global mapping of transcription start sites (TSSs) and subsequent promoter sequence analysis confirm in vitro observations in as much as TATA and BRE motifs are dominant elements in most archaeal promoters, with a few notable exceptions including the *M. jannaschii* ribosomal RNA promoter (Figs. 1.3 and 1.4) (Babski et al. 2016; Cho et al. 2017; Jäger et al. 2009, 2014; Li et al. 2015; Smollett et al. 2017; Wurtzel et al. 2010). This is in contrast to eukaryotes where strong TATA motifs (i.e., close to consensus sequence) are absent from the majority of promoters (Yang et al. 2007). Recently we have used Chromatin Immunoprecipitation followed by high-throughput sequencing (ChIP-seq) to characterise how promoter elements direct the recruitment of TBP, TFB and RNAP in vivo in the euryarchaeon *M. jannaschii* (Smollett et al. 2017). While BRE and TATA elements are the main contributors to promoter strength in vitro. There is only a weak correlation between BRE/TATA consensus score and TBP/TFB ChIP signals, and RNA steady-state levels in vivo.

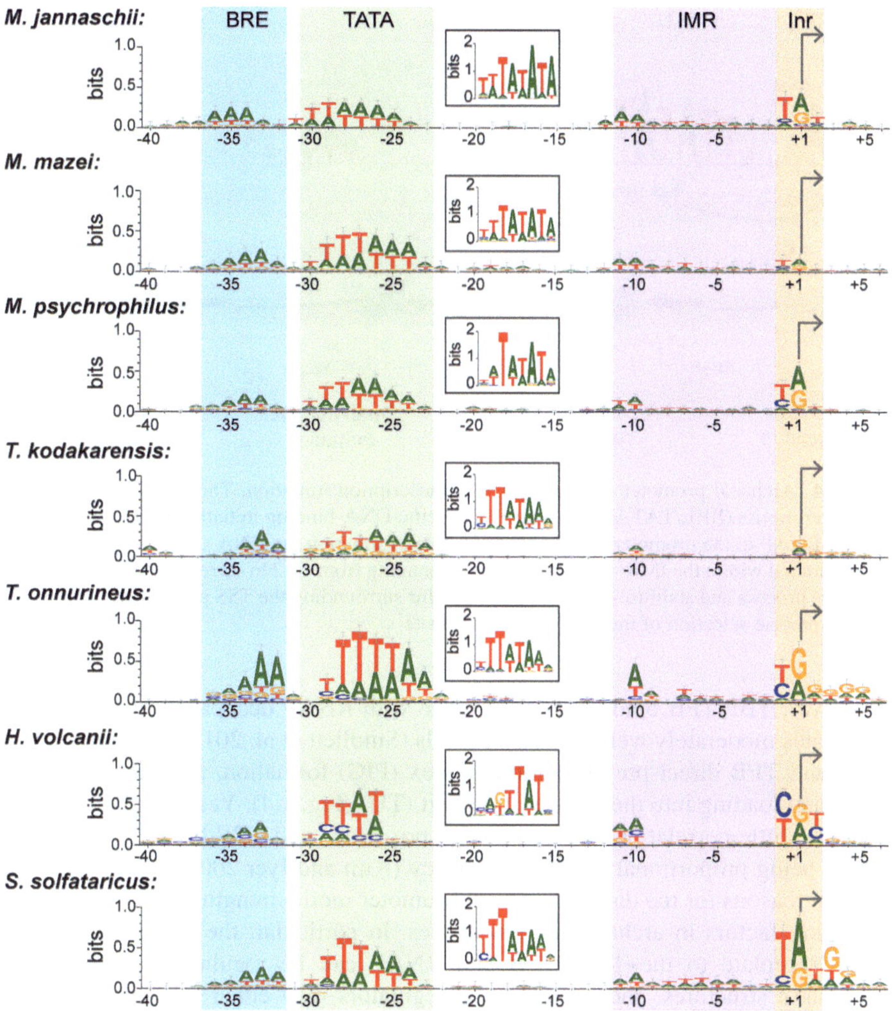

Fig. 1.3 Comparison of promoter consensus motifs in different archaea. Alignment of the DNA sequences upstream of TSS identified on a genome-wide scale identifies individual promoter elements including BRE, TATA box, IMR and Inr elements surrounding the TSS. Alignment of primary TSSs identified by whole genome sequencing of *M. jannaschii* (Smollett et al. 2017), *Methanosarcina mazei* (Jäger et al. 2009), *Methanolobus psychrophilus* (Li et al. 2015), *Thermococcus kodakarensis* (Jäger et al. 2014), *T. onnurineus* (Cho et al. 2017), *Haloferax volcanii* (Babski et al. 2016) and *Solfolobus solfataricus* (Wurtzel et al. 2010). Alignment visualised using WebLogo 3 adjusting to the background GC content for each organism (31.3% *M. jannaschii*, 41.5% *M. mazei*, 44.6% *M. psychrophilus*, 52% *T. kodakarensis*, 51.3% *T. onnurineus*, 65.5% *H. volcanii*, 35.8% *S. solfataricus*, http://weblogo.threeplusone.com/). *Inset* shows TATA box motif determined from same DNA sequences using MEME (http://meme-suite.org/tools/meme-chip). Adapted from Smollett et al. (2017)

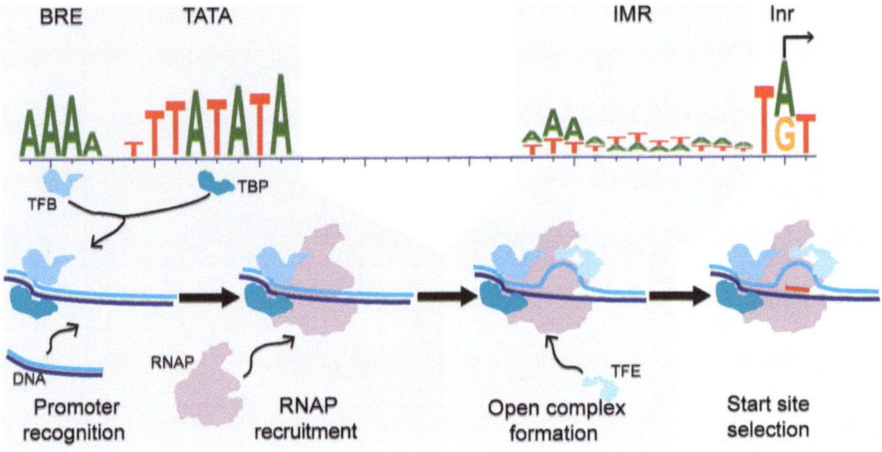

Fig. 1.4 Archaeal promoter elements govern transcription initiation. The interactions between promoter motifs (BRE/TATA), and sequence specific DNA-binding initiation factors (TBP/TFB) recruit RNAP to the promoter. During open complex formation the DNA strands of the promoter are separated within the IMR, an AT-rich region spanning from −12 to +2 relative to the TSS. TFE aids this process and stabilises the open PIC. The Inr surrounding the TSS plays an important role for the precise selection of the transcription start site

However, TBP/TFB binding does correlate with RNAP occupancy, which in turn correlates moderately well with RNA levels (Smollett et al. 2017). This shows that TBP and TFB direct pre-initiation complex (PIC) formation, and RNAP recruitment and loading into the transcription unit (TU) (Fig. 1.4). Yeast promoters show likewise little correlation between TATA box motif and TBP binding, with RNA levels being proportional to TBP occupancy (Kim and Iyer 2004). There could be several reasons for the discord between promoter motif strength and the binding of initiation factors in archaea and eukaryotes. In particular, the availability of the DNA template to the TBP, TFB and RNAP can be regulated by alternative chromatin structures, and gene-specific regulators may either enhance or inhibit PIC assembly (see Sect. 1.3). Several archaea encode multiple variants of TBP and TFB, in particular halophilic species such as *Halobacterium* NRC-1 contain 6 TBP and 7 TFB variants (Baliga et al. 2000); it has been proposed that the combination of TBP and TFB variants can direct a degree of promoter-specific regulation of transcription akin to bacterial sigma factors (Facciotti et al. 2007). A combination of different TBP/TFB deletion strains and ChIP analyses has revealed that only some TBP and TFB variants are essential and that different combinations of TBP/TFB bind to distinct promoters in vivo. Many promoters were associated with multiple TFB variants demonstrating a significant degree of redundancy (Facciotti et al. 2007), while subtle sequence biases in the BREs account for preferential binding of the different TFB variants (Seitzer et al. 2012).

1.2.2 Stabilisation of the PIC by Open Complex Formation

The transcription initiation factor TFE (homologous to eukaryotic TFIIE) enhances the stability of the PIC by aiding DNA strand separation and loading of the template strand into the active site of RNAP (Blombach et al. 2015, 2016; Grohmann et al. 2011). This process is referred to as 'open complex' formation. The regulation of open complex formation is a crucial step in defining transcription output across all domains of life (reviewed in Blombach et al. 2016). The region of DNA to be separated, the initially melted region (IMR), extends from position −12 to +2 relative to the TSS (Bell et al. 1998; Blombach et al. 2015; Nagy et al. 2015). Global sequence analysis reveals that the IMR does not contain a specific sequence motif, but throughout the archaea have a significantly higher A and T content compared to the genome average, particularly at the upstream edge (Fig. 1.3) (Smollett et al. 2017). As A-T basepairs require less energy for DNA strand separation compared to G-C basepairs the AT-bias may have been selected to facilitate open complex formation (Fig. 1.4), while there is no correlation between AT content and promoter strength. This is similar to the bacterial −10 element, which is also AT-rich and forms the upstream edge of the transcription bubble (Sasse-Dwight and Gralla 1989; Zuo and Steitz 2015). Short AT-rich DNA motifs (IMR, −10 element) and factors (TFE, sigma, CarD, TFIIH) that contribute to open complex formation and stability have coevolved in all domains of life (Fig. 1.1). The eukaryotic counterpart of TFE, TFIIE, is a dimeric factor consisting of subunits TFIIEα and TFIIEβ. Many archaea employ monomeric TFE variants (homologous to TFIIEα), whereas crenarchaeal TFE variants are α/β heterodimers (Blombach et al. 2009, 2015, 2016).

1.2.3 Selection of the Transcription Start Site

Genome-wide studies show that mammalian genes can be transcribed from multiple promoters using multiple TSSs (Sandelin et al. 2007). In bacteria, the discriminator promoter element is important for genome-wide start site selection (Winkelman et al. 2016), and the bacterial core recognition element has been shown to influence TSS selection by interactions between a G nucleotide at register +2 in the non-template strand and the core RNAP (Vvedenskaya et al. 2016). The archaeal Inr is comprised of a dinucleotide motif '−1T+1[A/G]' which directs precise start site selection in vivo (Figs. 1.3, 1.4 and 1.5) (Smollett et al. 2017). The Inr is a common feature in archaeal promoters, although the prevalence can vary between closely related species, e.g., it is present in the promoters of *Thermococcus onnurineus*, but not in *T. kodakarensis* (Fig. 1.3) (Cho et al. 2017; Jäger et al. 2014; Smollett et al. 2017). This suggests that TSS precision is not selected for some organisms, or may be compensated for by other factors. The archaeal Inr—essentially a preference for purine at the +1 and pyrimidine at the −1 position—is conserved in bacterial and eukaryotic promoters (Kadonaga 2012; Shultzaberger

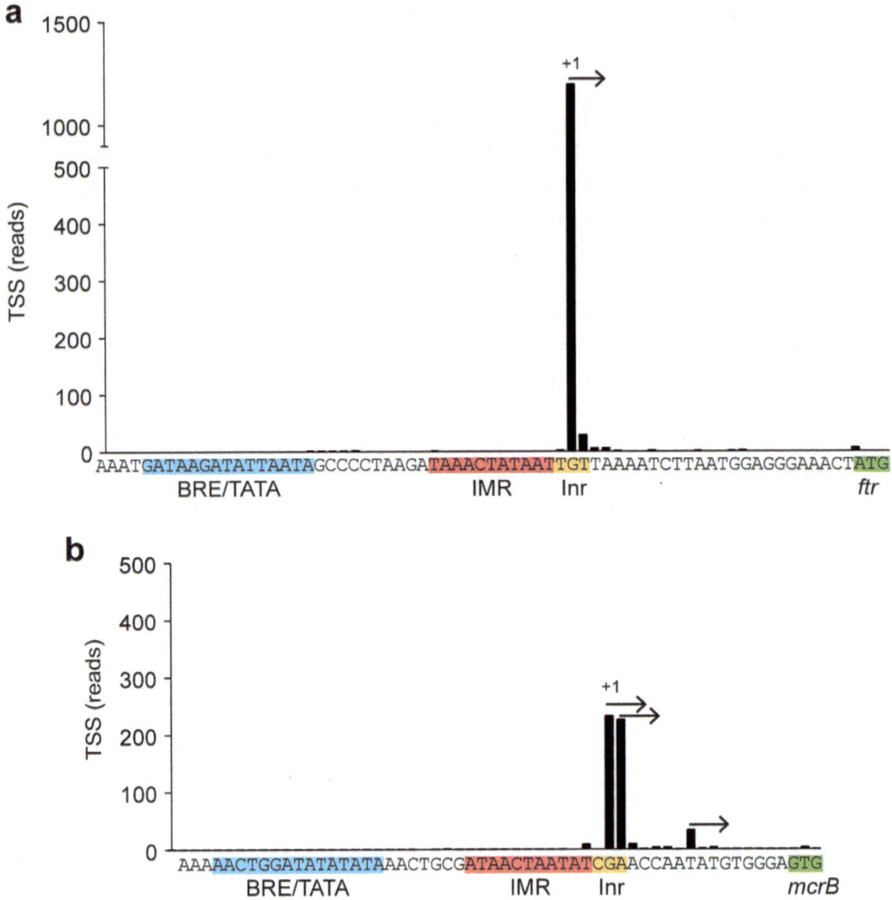

Fig. 1.5 The Inr motif influences TSS selection. Many archaeal promoters include an Inr motif (−1T/+1[A/G]) and utilise single or multiple TSSs (+1). (**a**) A strong Inr motif will direct one specific TSS resulting in transcripts with identical 5′-termini. (**b**) Promoters with a weaker Inr motif will direct transcription from several TSS leading to RNA species with heterogenous 5′-termini (Smollett et al. 2017)

et al. 2007). Structural analyses suggests that base stacking interactions between the −1 nucleotide of the template strand and the initiating NTP plays a role in template DNA strand stabilisation within the PIC (Basu et al. 2014).

1.2.4 Promoter Escape Facilitated by Factor Swapping

All RNAP face a similar mechanical engineering challenge; while a network of high affinity interactions between promoter-bound initiation factors and RNAP is essential to facilitate efficient recruitment and PIC formation, the escape of RNAP from the promoter (i.e., productive transcription) requires that this network is dismantled (Werner 2012). Spt4/5 is homologous to DSIF in humans and NusG in bacteria, it is the only RNAP-associated transcription factor that is universally conserved in all domains of life (Fig. 1.1) (Werner 2012). Spt4/5 is not essential for transcription in vitro, but ChIP-seq profiles demonstrate that it associates with elongating RNAPs throughout the genome, on coding as well as noncoding TUs. As Spt4/5 and the initiation factor TFE bind to the RNAP clamp in a mutually exclusive manner in vitro, we have proposed that this exchange, or swap, between TFE and Spt4/5 occurs every time the RNAP progresses through the transcription cycle, and that the swap could enhance promoter escape (Grohmann et al. 2011; Werner 2012). Spt4/5 is recruited proximal to the promoter in vivo, in agreement with facilitating the transition from initiation to elongation (Figs. 1.6 and 1.7a) (Smollett et al. 2017). This is different from bacterial NusG, which is recruited to TEC in a stochastic fashion, and it is similar to the early recruitment of Spt4/5 in yeast (Mayer et al. 2010). In addition, a similar exchange between TFIIE and Spt4/5 has been shown at RNAPII promoters (Diamant et al. 2016; Larochelle et al. 2012).

1.2.5 An Alternative Mode of Spt4/5 Recruitment

Genome-wide occupancy analysis allows us to not only define the 'norm' but also identify notable exceptions to the promoter-proximal Spt4/5 recruitment model (Fig. 1.7) (Mooney et al. 2009; Smollett et al. 2017). These exceptions include the ribosomal RNA operons and the abundant CRISPR loci where Spt4/5 is recruited during transcription elongation hundreds of base pairs downstream of the TSS. The underlying mechanisms behind this 'delayed' recruitment is currently not known, but likely includes novel gene-specific transcription factors, strong RNA secondary-structure or co-transcriptional processing—all of which are relevant for rRNA and CRISPR transcripts. The *Sulfolobus solfataricus* and *Pyrococcus furiosus* rRNA promoters have well defined BRE/TATA motifs and are very strong in vitro (Blombach et al. 2015; Micorescu et al. 2008; Qureshi et al. 1997), however, the *M. jannaschii* rRNA promoter shows surprisingly poor promoter motifs, and performs weakly in vitro, in apparent contrast with the high RNA levels and RNAP occupancy on the rRNA operons in vivo (Smollett et al. 2017). The lack of strong promoter motifs is akin to bacterial rRNA promoters, which tend to form unstable PICs, making them more amenable to regulation (Jensen and Pedersen 1990). It is possible that unknown transcription factors mask the Spt4/5 binding site on RNAP (clamp coiled coil) and activate *M. jannaschii* rRNA promoters. Alternatively, efficient promoter escape may occur at the weak rRNA promoter without Spt4/5. This is not the case for the CRISPR promoters, which have multiple promoters with strong matches to the consensus sequence (Smollett et al. 2017).

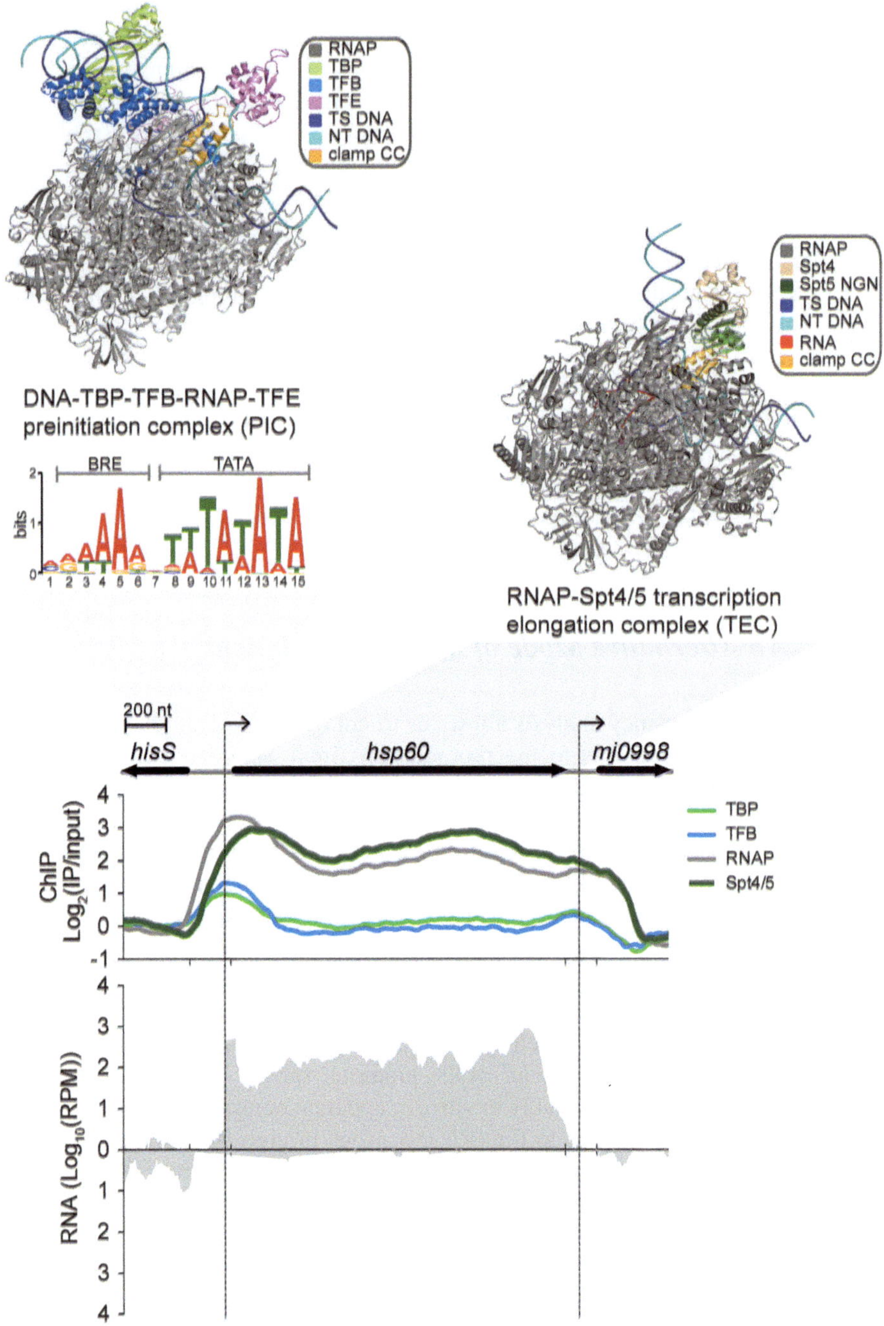

Fig. 1.6 An integrated view of transcription in archaea. ChIP occupancy profiles of the basal transcription machinery reflect the binding of PICs to promoters, and the distribution of RNAP-Spt4/5 TECs within the coding region as shown for the *M. jannashii hsp60* TU. Both plus- and

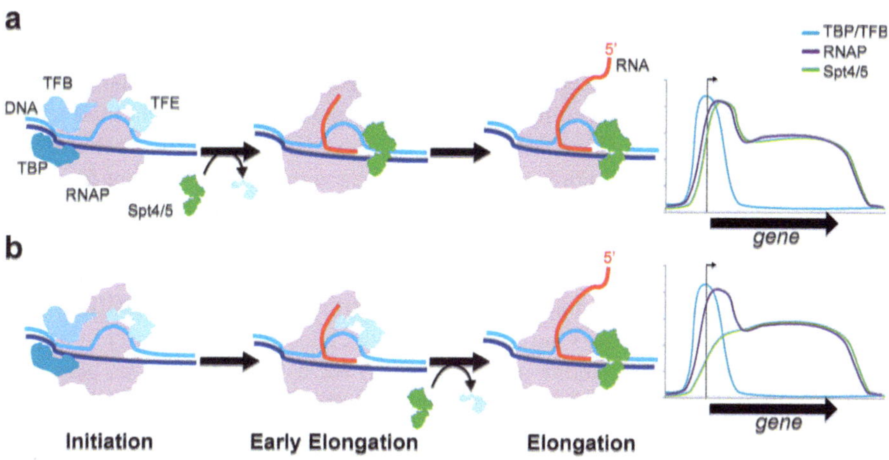

Fig. 1.7 Two modes of Spt4/5 recruitment to RNAP. Global occupancy profiling of *M. jannaschii* RNAP and Spt4/5 reveals two patterns of recruitment. (**a**) Spt4/5 is recruited to RNAP proximal to the promoter at the majority of transcription units. This recruitment profile supports the theory that swapping between initiation factor TFE and elongation factor Spt4/5 aids promoter escape. (**b**) At a small subset of genes including the rRNA and CRISPR loci Spt4/5 is recruited hundreds of base pairs downstream of the transcription start site, during the early elongation phase of the transcription cycle (Smollett et al. 2017)

1.2.6 Termination of Transcription

Transcription termination remains one of the least understood mechanisms of gene expression in archaea. Specific DNA sequences and auxiliary factors can slow down the TEC and trigger dissociation of the TEC into RNAP, transcript and template, but the precise mechanisms and order of events is unclear. While the fundamental process appears conserved in all multisubunit RNAPs, the requirements for DNA sequence motifs and exogenous termination factors differs substantially (Epshtein et al. 2007, 2010; Porrua et al. 2016; Proudfoot 2016). Bacterial intrinsic terminators consist of a short RNA hairpin structure and a poly-U stretch; these terminators induce pausing and enable RNAP to undergo conformational changes such as an opening of the RNAP clamp (Hein et al. 2014), a process likely facilitated by the RNA hairpin that invades the DNA binding channel of RNAP (Epshtein et al. 2007). These allosteric changes lead to the dissociation of the TEC with the last residue of the poly-U stretch forming the RNA 3′ terminus (Ray-Soni et al. 2016). The limited number of archaeal terminators that have been studied

Fig. 1.6 (continued) minus-strand RNA steady-state levels serve as proxy for transcription output of RNAP. Interestingly RNAPs do not strictly require Spt4/5 for transcription elongation in vitro, yet Spt4/5 closely follows RNAP in a genome-wide fashion, behaving as an 'honorary' RNAP subunit (Blombach et al. 2016; Smollett et al. 2017)

in vitro share the requirement for a poly-U stretch (5–8 U-residues), but are not dependent on any RNA secondary structure elements, reminiscent of the eukaryotic RNAPIII system (Hirtreiter et al. 2010a; Santangelo et al. 2009; Santangelo and Reeve 2006; Spitalny and Thomm 2008) (Fig. 1.8a). This suggests that the termination mechanism is conserved across all domains of life, but that the archaeal and RNAPIII TECs dissociate more readily than the bacterial TEC—without the intervention of exogenous factors or RNA hairpins. It is noteworthy that one of the key differences between bacterial and archaeal RNAPs is the Rpo4/7 stalk domain, which enhances transcription termination and has been likened to an 'inbuilt' NusA elongation factor (Belogurov and Artsimovitch 2015; Hirtreiter et al. 2010a).

The genome-wide RNA 3′ termini of a euryarchaeon (*Methanosarcina mazei*) and a crenarchaeon (*Sulfolobus acidocaldarius)* have been mapped at base pair resolution using a systems biology approach (Term-seq) (Dar et al. 2016a, b). In agreement with the mechanisms characterised in vitro, the Term-seq dataset revealed that termination occurred in vivo immediately downstream of a poly-U motif without the need for RNA secondary structure elements (Dar et al. 2016a). In approximately half of convergent (i.e., head-to-head oriented) genes in *S. acidocalaricus* the terminator signal of a given TU was located in the coding region of the other TU, resulting in a potential antisense transcript overlap. This could be due to the high coding density of archaeal genomes and the resulting short intergenic regions, or have regulatory significance. Many TUs were associated with multiple RNA 3′ termini likely due to inefficient termination. Such 'leaky' termination could direct the synthesis of RNA isoforms that differ in the 3′-untranslated region (3′-UTR) targeted by small regulatory RNAs (Fig. 1.8b) (Dar et al. 2016a). However, Term-seq results have two principal caveats. Firstly, Term-seq cannot discriminate between 'native' RNA 3′ ends generated by transcription termination and 'processed' RNA 3′ ends resulting from nucleolytic digestion, either RNA-processing or -degradation. Secondly, termination motifs and RNA 3′ ends could only be identified in 30–39% of TUs, which suggests that alternative- or additional termination mechanisms are at work including template topology (positive supercoiling in hyperthermophiles) and hitherto unidentified termination factors. Strong terminator (poly-U) signals are present in intragenic regions but only 25% of these led to transcription termination. This could be due to transcription antitermination, a well-described phenomenon in bacteria that relies on factors that are conserved between bacteria and archaea including NusG (Spt4/5), NusA, NusE, and co-translating ribosomes (Santangelo and Artsimovitch 2011; Santangelo et al. 2008). Little is known about archaeal termination factors, but we can speculate about their properties. Both bacterial and eukaryotic termination factors use a 'torpedo' mechanism, i.e. they engage with the nascent transcript, translocate along the RNA in the 5′→3′ direction, and ultimately dissociate the TEC upon impact. 5′→3′ RNases (Xrn2 in mammals, Rat1 in yeast) and RNA-helicases (Rho factor in bacteria, Sen1 in eukaryotes) facilitate transcription termination in this fashion (Fig. 1.8c) (Han et al. 2016; Kim et al. 2004; El Hage et al. 2008; West et al. 2004; Epshtein et al. 2010; Porrua and Libri 2013). Archaeal genomes encode several candidates for torpedo-factors but none have been experimentally tested yet (Phung et al. 2013).

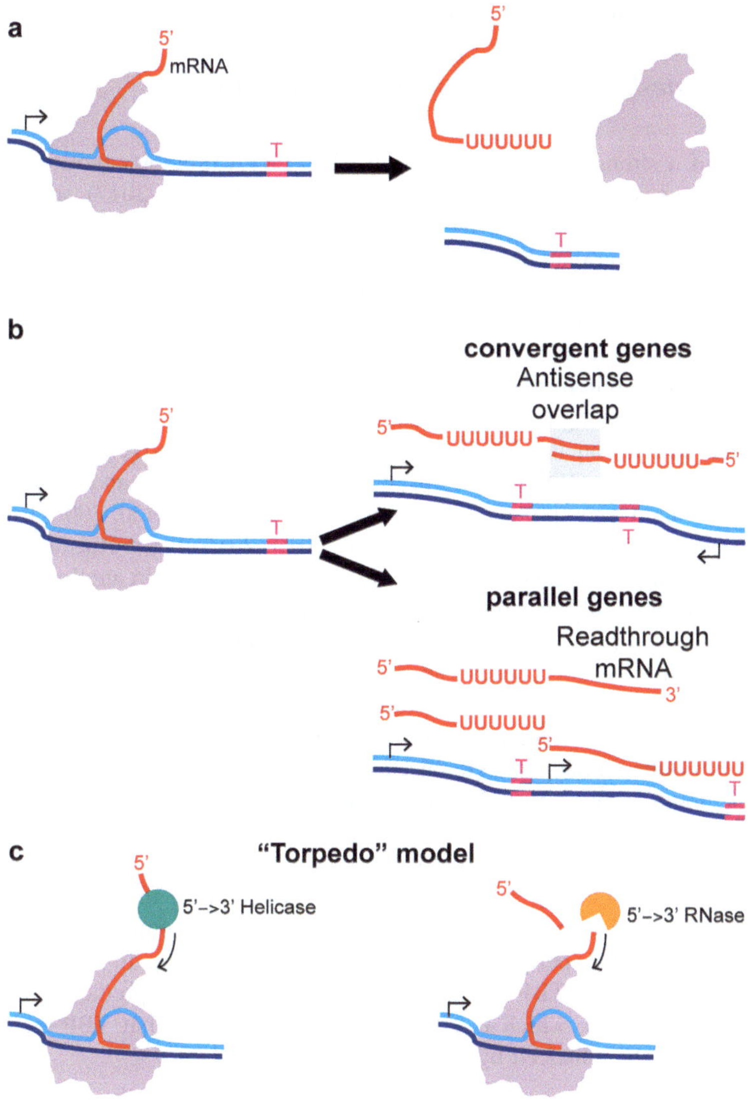

Fig. 1.8 Transcription termination in archaea. (**a**) Short poly-U stretches implicated in triggering transcription termination in vitro and in vivo, and global RNA 3′ mapping demonstrates that transcript 3′ termini consist of U-residues for 30–40% of TU genome-wide. (**b**) Leaky termination can lead to alternative and extended 3′-UTRs, which can result in antisense transcript overlap between two genes organised in a convergent orientation, or read through into downstream TUs. (**c**) Archaeal genomes encode putative termination factors including 5′→3′ RNases and RNA helicases. In eukaryotes and bacteria factors with these activities facilitate transcription termination by 'torpedo' mechanisms

1.3 Additional Factors Affecting Transcriptional Output

1.3.1 Gene-Specific Transcription Regulators

The lack of a strong correlation between BRE/TATA promoter motifs and RNA levels genome wide (Kim and Iyer 2004; Smollett et al. 2017) suggests that additional forces are at work, including gene-specific regulators. The molecular mechanisms of several archaeal metabolic and stress response regulators have been elucidated in vitro, and their regulons characterised by ChIP-chip and ChIP-seq methods (Liu et al. 2016; Nguyen-Duc et al. 2013; Reichelt et al. 2016; Rudrappa et al. 2015; Tonner et al. 2015; Wilbanks et al. 2012). Archaeal regulators operate by a range of different mechanisms including repression by promoter occlusion and activation by enhancing the recruitment of the PIC; the mode of action of the same factor can depend on the location of the binding site relative to the promoter (Aravind and Koonin 1999; Charoensawan et al. 2010; Dahlke and Thomm 2002; Geiduschek and Ouhammouch 2005; Kanai et al. 2007; Lee et al. 2008; Lipscomb et al. 2009; Ochs et al. 2012; Peeters et al. 2013, 2015; Perez-Rueda and Janga 2010). Transcription regulators are described in greater detail in another chapter of this tome, we will only briefly mention example below.

The *M. jannaschii* Lrp-type regulator Ptr2 is an excellent example of how in vitro and in vivo approaches can complement each other. Ptr2 activates transcription from the *rb2* promoter by recruiting TBP to the TATA box, a mechanism that was elucidated by elegant in vitro transcription experiments in the Geiduschek laboratory (Ouhammouch and Geiduschek 2001; Ouhammouch et al. 2003, 2005). Whole genome occupancy studies of *M. jannaschii* TBP validated this mechanism in vivo. By analysis of promoter sequences genome-wide, each TATA motif could be assigned a score that quantified its similarity to the global TATA consensus, i.e. the ideal TBP binding site. Subsequently, TBP binding to specific promoters could be predicted using a linear regression model, and compared to the actual occupancy of TBP experimentally determined by ChIP-seq. In case of the *rb2* promoter the actual TBP occupancy far exceeded the predicted one (0.1 vs. 1 Log_2[IP/input]), which is congruent with the notion that TBP-recruitment in vivo is strongly enhanced by Ptr2 (Smollett et al. 2017).

1.3.2 The Impact of Chromatin Structure on Transcription

All cellular genomes are organised and compacted by DNA-binding proteins that protect the DNA while still allowing the access of molecular machines that facilitate DNA replication, repair and recombination and last but not least transcription (Ammar et al. 2012; Cubonovaa et al. 2012; Peeters et al. 2015; Visone et al. 2014; Xie and Reeve 2004). In eukaryotes, histone-based chromatin has evolved into a major regulatory mechanism, with hundreds of post-translational modifications and remodelling complexes facilitating the precise execution of the genetic programme. Many archaea encode histone homologues, but it remains to be proven to which extent histone-based chromatin regulates gene expression in archaea. In addition to regulatory functions, histones are likely to protect the genomes of hyperthermophiles from thermal denaturation (Visone et al. 2014). Archaea vary in their repertoire of histones and other chromatin proteins (Peeters et al. 2015). Small archaeal chromatin proteins with the ability to bind and condense DNA were first described in *Thermoplasma acidophilum* (DeLange et al. 1981a, b; Searcy 1975; Searcy and Delange 1980), but archaeal histones were first characterised in the hyperthermophile *Methanothermus fervidus* (Sandman et al. 1990). In vitro experiments using a limited number of factors (TBP, TFB and RNAP) have shown that histones inhibit transcription under these conditions, but it remains unknown how additional general factors such as TFE, Spt4/5 and TFS assist RNAP transcribing through chromatin (Wilkinson et al. 2010; Xie and Reeve 2004).

While histones are not essential for cell viability in some archaea, deletion of histones changes the transcriptome by both up- and downregulating genes (Cubonovaa et al. 2012; Heinicke et al. 2004; Nalabothula et al. 2013). In eukaryotes this regulation chiefly occurs via post translational modifications of the histone tails (Bannister and Kouzarides 2011). Archaeal histones generally encompass only the histone fold and lack the tails of their eukaryotic counterparts. In *M. jannaschii,* no histone modifications could be identified in a top-down mass spectrometry approach (Forbes et al. 2004). However, most archaea with histones encode multiple paralogues, enabling different combinations of histone homo- and heterodimers to form alternative chromatin structures, either at specific regulatory sequences, different genomic loci or TU, or under different growth conditions. For example, in *M. fervidus* the expression levels of histone HMfA are higher than HMfB during exponential growth but decrease in stationary phase, a change which may result in more compact chromatin (Sandman et al. 1994).

High-throughput sequencing approaches including nucleosome sequencing have mapped the genome-wide histone occupancy, and identified the optimal archaeal histone DNA binding site, which is near-identical to eukaryotes and reflects a basepair sequence that enables DNA curvature/bending (Ammar et al. 2012; Maruyama et al. 2013; Nalabothula et al. 2013). Generally, archaeal histones dimerise in solution and interact with 30 bp of DNA. Limited MNase digestion of chromatin isolated from *Haloferax volcanii* resulted in nucleosome ladder with 60 bp steps corresponding to histone tetramers (Ammar et al. 2012), while *Thermococcus kodakarensis* and *Methanothermobacter thermautophicus* resulted in a pattern with 30 bp steps, indicative of histone dimers (Maruyama et al. 2013; Nalabothula et al. 2013). Both observations are congruent with a chromatin model where histones polymerise upon DNA binding (Fig. 1.9a). As is seen in eukaryotes, promoter regions, and specific genomic loci including the highly transcribed rRNA operons are apparently devoid of histone binding (nucleosome-free regions or NFR). Moreover, MNase digestion of in vitro reconstituted chromatin reproduces this pattern (Maruyama et al. 2013; Nalabothula et al. 2013). This not only suggests that the DNA sequence alone is sufficient to organise chromatin structure, but also

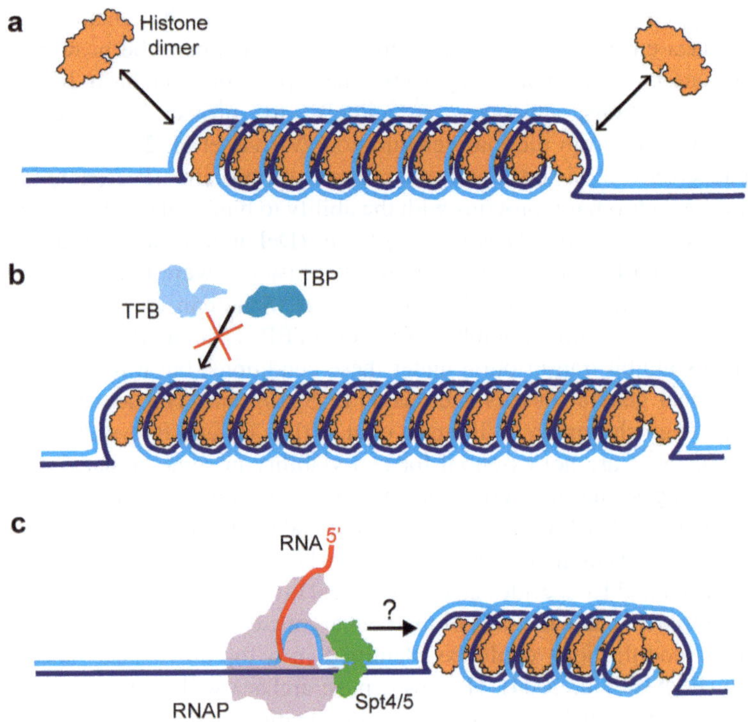

Fig. 1.9 Interference of chromatin and transcription in archaea. (**a**) In euryarchaea histones compact and organise the genome into dynamic chromatin fibres that grow and shrink by association or dissociation of histone dimers at each end. These chromatin structures can interfere with transcription in a number of ways: (**b**) by denying access of initiation factors to promoter or (**c**) by providing a barrier to TEC

implies that on-going transcription has little influence on the deposition of histones across the genome—and altogether emphasises a possible role of histones in transcription regulation (potential mechanisms are shown in Fig. 1.9b, c).

1.4 The Output of the Transcription System

1.4.1 The Archaeal Transcriptomes

The introduction of high-throughput sequencing approaches to determine global RNA levels provide a significant improvement compared to hybridisation-based approaches such as microarrays in terms of the dynamic range and the detection of low abundance transcripts (Zhao et al. 2014). RNA-seq data for several euryarchaeal species (Babski et al. 2016; Cho et al. 2017; Jäger et al. 2009, 2014;

Li et al. 2015; Smollett et al. 2017) and the crenarcheaon *S. solfataricus* (Wurtzel et al. 2010) have provided new insights into archaeal transcriptomes, while other archaeal phyla remain unexplored. Because the RNA-seq approach is independent of prior knowledge about the coding regions and predicted TUs, these data sets can provide us with a wealth of novel non-coding transcripts including small regulator RNAs (discussed in Chap. 10), anti-sense RNA, and newly discovered mRNAs (Fig. 1.10) (Babski et al. 2016; Cho et al. 2017; Jäger et al. 2009, 2014; Li et al. 2015; Smollett et al. 2017; Straub et al. 2009; Tang et al. 2005; Toffano-Nioche et al. 2013; Wurtzel et al. 2010; Dar et al. 2016a). Archaeal ncRNA species with uncharacterized functions include processed fragments of mRNA UTRs. Methanogens (*M. jannaschii, M. mazei* and *Methanolobus psychrophilus*) and Thermococcales (*T. kodakarensis, T. onnurineus* and *P. furiosus*) all contain long 5′UTRs, including ribosome binding sites and potential sites of regulation by riboregulators and riboswitches (Cho et al. 2017; Jäger et al. 2009, 2014; Li et al.

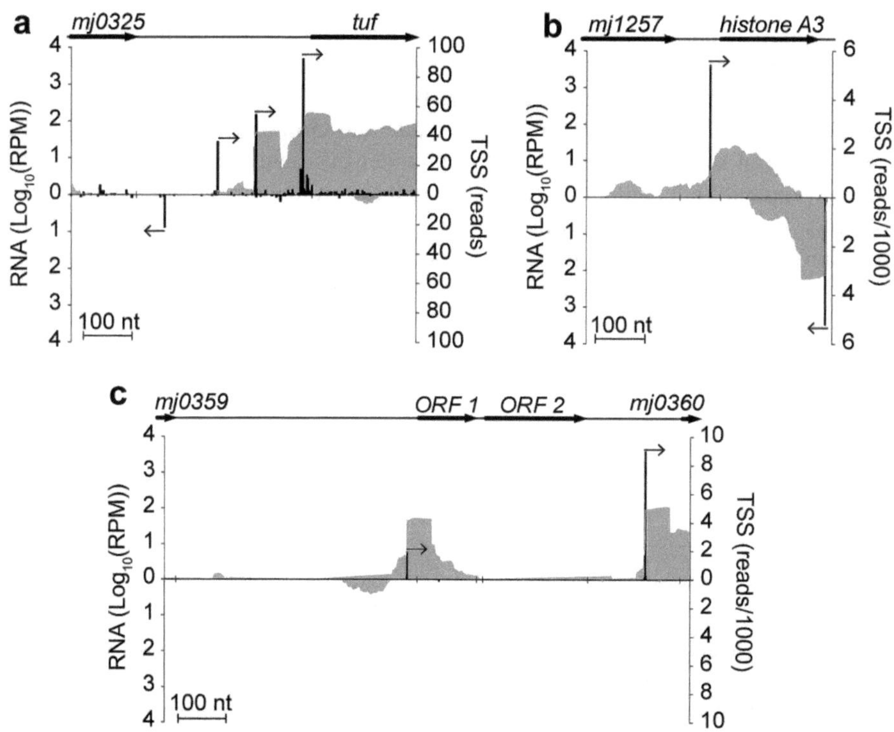

Fig. 1.10 Features of the archaeal transcriptome. Global TSS mapping and RNA-seq highlight the diversity of archaeal transcripts as shown in *M. jannaschii*. (**a**) Transcription initiation using alternative promoters leads to the synthesis of distinct mRNA species with different 5′-UTRs, which provide opportunities for riboregulation by e.g., riboswitches. (**b**) and (**c**) These methods also lead to the discovery of novel transcripts including antisense RNAs (**b**) and small ORFs (**c**) missed in genome sequence-based annotations (Smollett et al. 2017)

2015; Smollett et al. 2017; Toffano-Nioche et al. 2013). In contrast, *Sulfolobus* and halophilic archaea are characterised by leaderless mRNAs where translation is initiated directly from the mRNA 5′-end (Babski et al. 2016; Brenneis et al. 2007; Koide et al. 2009; Torarinsson et al. 2005; Wurtzel et al. 2010). Term-seq has revealed the abundance of 3′-UTRs in archaea, which similar to the 5′-ends are longer in methanogens than in *Sulfolobus* (Dar et al. 2016a). Genes encoding ribosomal proteins tend to have long 5′UTRs in all archaea, even in species predominantly using leaderless mRNAs such as *Sulfolobus* (Li et al. 2015; Toffano-Nioche et al. 2013; Wurtzel et al. 2010), which suggests a common regulatory mechanism for these genes.

1.4.2 Evidence for Pervasive Transcription in Archaea

Pervasive transcription describes the phenomenon of non-coding, often anti-sense transcripts that are not restricted by gene boundaries; it has been implicated in transcription regulation, transcription-coupled repair and genome evolution. RNA-seq demonstrates that pervasive transcription occurs in all domains of life (Clark et al. 2011; Smollett et al. 2017; Wade and Grainger 2014). In *E. coli* the comparison of transcriptome data obtained under different growth conditions, and library preparation techniques has yielded a more genuine and comprehensive map of TSSs. Furthermore the detection of novel transcripts in *E. coli* was aided by the deletion of nucleases including RNase E and RNase III that are involved in RNA turnover (Thomason et al. 2015; Wade 2015). The same approaches will likely enable a more accurate estimation as to the amount of pervasive transcription in archaea.

1.4.3 Deconvoluting RNA Synthesis and RNA Steady-State Levels

There are several limitations one needs to be aware of when analysing archaeal transcriptomes, in particular when attempting to correlate genome occupancy profiles of basal transcription factors and RNAP with RNA levels. Due to its high abundance rRNA is often depleted using standard procedures of RNA isolation and subsequent transcriptomics analyses. In addition RNA isolation methods and library preparation techniques tend to include size selection steps that introduce bias against small RNAs. Most importantly, RNA-seq data represent steady-state RNA levels that reflect RNA synthesis and degradation, and not nascent RNA synthesis. Attempts to determine mRNA half-lives haven been made for *S. solfataricus* and *S. acidocaldarius*. These studies revealed important differences in RNA stability depending on functional category of genes and RNA expression levels (Andersson et al. 2006). Several techniques have been developed to map the nascent transcriptome to obtain a more accurate global picture of ongoing RNA synthesis. In a NET-seq (Native elongating transcript sequencing) approach TEC are purified from biomass, the RNA associated with RNAPs is isolated and sequenced, which provides a snapshot of active transcription at a single-nucleotide resolution (Churchman and Weissman 2012). In transient transcriptome sequencing (TT-seq) approaches nascent RNA is metabolically labeled with uridine base analogues that allow the specific purification of the nascent RNA prior to sequencing (Schwalb et al. 2016). A caveat from an archaeal perspective is that the narrow phylogenetic distribution of the required uridine kinase activity would require the introduction of this enzyme by genetic manipulation to adapt such methods for archaea. A slightly different approach has been recently adapted for in vivo labeling of RNA in archaea for the first time using 4-Thiouracil rather than uridine analogues involving a different biochemical pathway from 4-Thiouracil to UMP via uracil phosphoribosyltransferase (Knüppel et al. 2017). Finally, approaches such as Gro-seq (Global run-on sequencing) that isolate TECs and carry out the metabolic labeling of nascent RNA by transcription elongation in vitro can be adapted to archaeal transcriptomics in a reasonably straightforward fashion (Core et al. 2008).

1.5 Future Directions

High-throughput sequencing approaches have greatly improved our understanding of the mechanisms of transcription in archaea. We can now begin to unravel connections between perturbations at the molecular level and changes of the entire system, between in vitro and in vivo data, aiming to understand transcription in a multiscalar fashion. The current experimental portfolio at our fingertips needs to be expanded by mapping the transcriptome-wide occupancy of RNA-binding transcription factors by techniques such as iCLIP (Konig et al. 2011), mapping of TECs by NET-seq and Gro-seq (see above), and genome-wide mapping of evolutionary pervasive DNA- and RNA-modifications (Huber et al. 2015). Once these aspects of archaeal transcription are described on a systems level, it will be possible to characterise the transcription apparatus in flux—as it changes in response to external stimuli and environmental insults. Future research would benefit from being expanded to include little characterised phyla including Nano- Thaum- and Lokiarchaeota to provide further insights into the evolution of transcription regulation in archaea. Last but not least, the ability to examine features of macromolecular metabolism genome-wide will allow us to correlate transcription with DNA replication, -recombination and -repair, and protein translation.

References

Ammar R, Torti D, Tsui K, Gebbia M, Durbic T, Bader GD, Giaever G, Nislow C (2012) Chromatin is an ancient innovation conserved between archaea and eukarya. Elife 1:e00078. doi:10.7554/eLife.00078

Andersson AF, Lundgren M, Eriksson S, Rosenlund M, Bernander R, Nilsson P (2006) Global analysis of mRNA stability in the archaeon Sulfolobus. Genome Biol 7(10):R99. doi:10.1186/gb-2006-7-10-r99

Aravind L, Koonin EV (1999) DNA-binding proteins and evolution of transcription regulation in the archaea. Nucleic Acids Res 27(23):4658–4670

Babski J, Haas KA, Nather-Schindler D, Pfeiffer F, Forstner KU, Hammelmann M, Hilker R, Becker A, Sharma CM, Marchfelder A, Soppa J (2016) Genome-wide identification of transcriptional start sites in the haloarchaeon *Haloferax volcanii* based on differential RNA-Seq (dRNA-Seq). BMC Genomics 17(1):629. doi:10.1186/s12864-016-2920-y

Baliga NS, Goo YA, Ng WV, Hood L, Daniels CJ, DasSarma S (2000) Is gene expression in Halobacterium NRC-1 regulated by multiple TBP and TFB transcription factors? Mol Microbiol 36(5):1184–1185

Bannister AJ, Kouzarides T (2011) Regulation of chromatin by histone modifications. Cell Res 21 (3):381–395. doi:10.1038/cr.2011.22

Basu RS, Warner BA, Molodtsov V, Pupov D, Esyunina D, Fernandez-Tornero C, Kulbachinskiy A, Murakami KS (2014) Structural basis of transcription initiation by bacterial RNA polymerase holoenzyme. J Biol Chem 289(35):24549–24559. doi:10.1074/jbc.M114.584037

Bell SD, Jaxel C, Nadal M, Kosa PF, Jackson SP (1998) Temperature, template topology, and factor requirements of archaeal transcription. Proc Natl Acad Sci U S A 95(26):15218–15222

Bell SD, Kosa PL, Sigler PB, Jackson SP (1999) Orientation of the transcription preinitiation complex in archaea. Proc Natl Acad Sci U S A 96(24):13662–13667. doi:10.1073/pnas.96.24.13662

Belogurov GA, Artsimovitch I (2015) Regulation of transcript elongation. Annu Rev Microbiol 69:49–69. doi:10.1146/annurev-micro-091014-104047

Blombach F, Grohmann D (2017) Same same but different: the evolution of TBP in archaea and their eukaryotic offspring. Transcription 8(3):162–168. doi:10.1080/21541264.2017.1289879

Blombach F, Makarova KS, Marrero J, Siebers B, Koonin EV, van der Oost J (2009) Identification of an ortholog of the eukaryotic RNA polymerase III subunit RPC34 in crenarchaeota and thaumarchaeota suggests specialization of RNA polymerases for coding and non-coding RNAs in archaea. Biol Direct 4:39. doi:10.1186/1745-6150-4-39

Blombach F, Salvadori E, Fouqueau T, Yan J, Reimann J, Sheppard C, Smollett KL, Albers SV, Kay CW, Thalassinos K, Werner F (2015) Archaeal TFEalpha/beta is a hybrid of TFIIE and the RNA polymerase III subcomplex hRPC62/39. Elife 4:e08378. doi:10.7554/eLife.08378

Blombach F, Smollett KL, Grohmann D, Werner F (2016) Molecular mechanisms of transcription initiation-structure, function, and evolution of TFE/TFIIE-like factors and open complex formation. J Mol Biol 428(12):2592–2606. doi:10.1016/j.jmb.2016.04.016

Brenneis M, Hering O, Lange C, Soppa J (2007) Experimental characterization of cis-acting elements important for translation and transcription in halophilic archaea. PLoS Genet 3(12): e229. doi:10.1371/journal.pgen.0030229

Charoensawan V, Wilson D, Teichmann SA (2010) Genomic repertoires of DNA-binding transcription factors across the tree of life. Nucleic Acids Res 38(21):7364–7377. doi:10.1093/nar/gkq617

Cho S, Kim MS, Jeong Y, Lee BR, Lee JH, Kang SG, Cho BK (2017) Genome-wide primary transcriptome analysis of H2-producing archaeon *Thermococcus onnurineus* NA1. Sci Rep 7:43044. doi:10.1038/srep43044

Churchman LS, Weissman JS (2012) Native elongating transcript sequencing (NET-seq). Curr Protoc Mol Biol Chapter 4:Unit 4 14 11–17. doi:10.1002/0471142727.mb0414s98

Clark MB, Amaral PP, Schlesinger FJ, Dinger ME, Taft RJ, Rinn JL, Ponting CP, Stadler PF, Morris KV, Morillon A, Rozowsky JS, Gerstein MB, Wahlestedt C, Hayashizaki Y, Carninci P, Gingeras TR, Mattick JS (2011) The reality of pervasive transcription. PLoS Biol 9(7):e1000625; discussion e1001102. doi:10.1371/journal.pbio.1000625

Core LJ, Waterfall JJ, Lis JT (2008) Nascent RNA sequencing reveals widespread pausing and divergent initiation at human promoters. Science 322(5909):1845–1848. doi:10.1126/science.1162228

Cubonovaa L, Katano M, Kanai T, Atomi H, Reeve JN, Santangelo TJ (2012) An archaeal histone is required for transformation of *Thermococcus kodakarensis*. J Bacteriol 194(24):6864–6874. doi:10.1128/JB.01523-12

Dahlke I, Thomm M (2002) A Pyrococcus homolog of the leucine-responsive regulatory protein, LrpA, inhibits transcription by abrogating RNA polymerase recruitment. Nucleic Acids Res 30 (3):701–710

Dar D, Prasse D, Schmitz RA, Sorek R (2016a) Widespread formation of alternative 3′ UTR isoforms via transcription termination in archaea. Nat Microbiol 1(10):16143. doi:10.1038/nmicrobiol.2016.143

Dar D, Shamir M, Mellin JR, Koutero M, Stern-Ginossar N, Cossart P, Sorek R (2016b) Term-seq reveals abundant ribo-regulation of antibiotics resistance in bacteria. Science 352(6282): aad9822. doi:10.1126/science.aad9822

DeLange RJ, Green GR, Searcy DG (1981a) A histone-like protein (HTa) from *Thermoplasma acidophilum*. I. Purification and properties. J Biol Chem 256(2):900–904

DeLange RJ, Williams LC, Searcy DG (1981b) A histone-like protein (HTa) from *Thermoplasma acidophilum*. II. Complete amino acid sequence. J Biol Chem 256(2):905–911

Diamant G, Bahat A, Dikstein R (2016) The elongation factor Spt5 facilitates transcription initiation for rapid induction of inflammatory-response genes. Nat Commun 7:11547. doi:10.1038/ncomms11547

El Hage A, Koper M, Kufel J, Tollervey D (2008) Efficient termination of transcription by RNA polymerase I requires the 5′ exonuclease Rat1 in yeast. Genes Dev 22(8):1069–1081. doi:10.1101/gad.463708

Epshtein V, Cardinale CJ, Ruckenstein AE, Borukhov S, Nudler E (2007) An allosteric path to transcription termination. Mol Cell 28(6):991–1001. doi:10.1016/j.molcel.2007.10.011

Epshtein V, Dutta D, Wade J, Nudler E (2010) An allosteric mechanism of Rho-dependent transcription termination. Nature 463(7278):245–249. doi:10.1038/nature08669

Facciotti MT, Reiss DJ, Pan M, Kaur A, Vuthoori M, Bonneau R, Shannon P, Srivastava A, Donohoe SM, Hood LE, Baliga NS (2007) General transcription factor specified global gene regulation in archaea. Proc Natl Acad Sci U S A 104(11):4630–4635. doi:10.1073/pnas.0611663104

Forbes AJ, Patrie SM, Taylor GK, Kim YB, Jiang L, Kelleher NL (2004) Targeted analysis and discovery of posttranslational modifications in proteins from methanogenic archaea by top-down MS. Proc Natl Acad Sci U S A 101(9):2678–2683. doi:10.1073/pnas.0306575101

Fouqueau T, Zeller ME, Cheung AC, Cramer P, Thomm M (2013) The RNA polymerase trigger loop functions in all three phases of the transcription cycle. Nucleic Acids Res 41 (14):7048–7059. doi:10.1093/nar/gkt433

Geiduschek EP, Ouhammouch M (2005) Archaeal transcription and its regulators. Mol Microbiol 56(6):1397–1407. doi:10.1111/j.1365-2958.2005.04627.x

Gietl A, Holzmeister P, Blombach F, Schulz S, von Voithenberg LV, Lamb DC, Werner F, Tinnefeld P, Grohmann D (2014) Eukaryotic and archaeal TBP and TFB/TF(II)B follow different promoter DNA bending pathways. Nucleic Acids Res 42(10):6219–6231. doi:10.1093/nar/gku273

Grohmann D, Nagy J, Chakraborty A, Klose D, Fielden D, Ebright RH, Michaelis J, Werner F (2011) The initiation factor TFE and the elongation factor Spt4/5 compete for the RNAP clamp during transcription initiation and elongation. Mol Cell 43(2):263–274. doi:10.1016/j.molcel.2011.05.030

Han Z, Libri D, Porrua O (2016) Biochemical characterization of the helicase Sen1 provides new insights into the mechanisms of non-coding transcription termination. Nucleic Acids Res 45 (3):1355–1370. doi:10.1093/nar/gkw1230

Hein PP, Kolb KE, Windgassen T, Bellecourt MJ, Darst SA, Mooney RA, Landick R (2014) RNA polymerase pausing and nascent-RNA structure formation are linked through clamp-domain movement. Nat Struct Mol Biol 21(9):794–802. doi:10.1038/nsmb.2867

Heinicke I, Muller J, Pittelkow M, Klein A (2004) Mutational analysis of genes encoding chromatin proteins in the archaeon *Methanococcus voltae* indicates their involvement in the regulation of gene expression. Mol Genet Genomics 272(1):76–87. doi:10.1007/s00438-004-1033-5

Hirtreiter A, Damsma GE, Cheung AC, Klose D, Grohmann D, Vojnic E, Martin AC, Cramer P, Werner F (2010a) Spt4/5 stimulates transcription elongation through the RNA polymerase clamp coiled-coil motif. Nucleic Acids Res 38(12):4040–4051. doi:10.1093/nar/gkq135

Hirtreiter A, Grohmann D, Werner F (2010b) Molecular mechanisms of RNA polymerase - the F/E (RPB4/7) complex is required for high processivity in vitro. Nucleic Acids Res 38(2):585–596. doi:10.1093/nar/gkp928

Huber SM, van Delft P, Mendil L, Bachman M, Smollett K, Werner F, Miska EA, Balasubramanian S (2015) Formation and abundance of 5-hydroxymethylcytosine in RNA. Chembiochem 16(5):752–755. doi:10.1002/cbic.201500013

Jäger D, Förstner KU, Sharma CM, Santangelo TJ, Reeve JN (2014) Primary transcriptome map of the hyperthermophilic archaeon *Thermococcus kodakarensis*. BMC Genomics 15:684. doi:10.1186/1471-2164-15-684

Jäger D, Sharma CM, Thomsen J, Ehlers C, Vogel J, Schmitz RA (2009) Deep sequencing analysis of the *Methanosarcina mazei* Go1 transcriptome in response to nitrogen availability. Proc Natl Acad Sci U S A 106(51):21878–21882. doi:10.1073/pnas.0909051106

Jensen KF, Pedersen S (1990) Metabolic growth rate control in *Escherichia coli* may be a consequence of subsaturation of the macromolecular biosynthetic apparatus with substrates and catalytic components. Microbiol Rev 54(2):89–100

Kadonaga JT (2012) Perspectives on the RNA polymerase II core promoter. Wiley Interdiscip Rev Dev Biol 1(1):40–51. doi:10.1002/wdev.21

Kanai T, Akerboom J, Takedomi S, van de Werken HJ, Blombach F, van der Oost J, Murakami T, Atomi H, Imanaka T (2007) A global transcriptional regulator in Thermococcus kodakaraensis controls the expression levels of both glycolytic and gluconeogenic enzyme-encoding genes. J Biol Chem 282(46):33659–33670. doi:10.1074/jbc.M703424200

Kim J, Iyer VR (2004) Global role of TATA box-binding protein recruitment to promoters in mediating gene expression profiles. Mol Cell Biol 24(18):8104–8112. doi:10.1128/MCB.24.18.8104-8112.2004

Kim M, Krogan NJ, Vasiljeva L, Rando OJ, Nedea E, Greenblatt JF, Buratowski S (2004) The yeast Rat1 exonuclease promotes transcription termination by RNA polymerase II. Nature 432 (7016):517–522. doi:10.1038/nature03041

Knüppel R, Kuttenberger C, Ferreira-Cerca S (2017) Toward time-resolved analysis of RNA metabolism in archaea using 4-thiouracil. Front Microbiol 8:286. doi:10.3389/fmicb.2017.00286

Koide T, Reiss DJ, Bare JC, Pang WL, Facciotti MT, Schmid AK, Pan M, Marzolf B, Van PT, Lo FY, Pratap A, Deutsch EW, Peterson A, Martin D, Baliga NS (2009) Prevalence of transcription promoters within archaeal operons and coding sequences. Mol Syst Biol 5:285. doi:10.1038/msb.2009.42

Konig J, Zarnack K, Rot G, Curk T, Kayikci M, Zupan B, Turner DJ, Luscombe NM, Ule J (2011) iCLIP–transcriptome-wide mapping of protein-RNA interactions with individual nucleotide resolution. J Vis Exp 50:2638. doi:10.3791/2638

Kostrewa D, Zeller ME, Armache KJ, Seizl M, Leike K, Thomm M, Cramer P (2009) RNA polymerase II-TFIIB structure and mechanism of transcription initiation. Nature 462 (7271):323–330. doi:10.1038/nature08548

Larochelle S, Amat R, Glover-Cutter K, Sanso M, Zhang C, Allen JJ, Shokat KM, Bentley DL, Fisher RP (2012) Cyclin-dependent kinase control of the initiation-to-elongation switch of RNA polymerase II. Nat Struct Mol Biol 19(11):1108–1115. doi:10.1038/nsmb.2399

Lee SJ, Surma M, Hausner W, Thomm M, Boos W (2008) The role of TrmB and TrmB-like transcriptional regulators for sugar transport and metabolism in the hyperthermophilic archaeon Pyrococcus furiosus. Arch Microbiol 190(3):247–256. doi:10.1007/s00203-008-0378-2

Li J, Qi L, Guo Y, Yue L, Li Y, Ge W, Wu J, Shi W, Dong X (2015) Global mapping transcriptional start sites revealed both transcriptional and post-transcriptional regulation of cold adaptation in the methanogenic archaeon *Methanolobus psychrophilus*. Sci Rep 5:9209. doi:10.1038/srep09209

Lipscomb GL, Keese AM, Cowart DM, Schut GJ, Thomm M, Adams MW, Scott RA (2009) SurR: a transcriptional activator and repressor controlling hydrogen and elemental sulphur metabolism in Pyrococcus furiosus. Mol Microbiol 71(2):332–349. doi:10.1111/j.1365-2958.2008.06525.x

Liu H, Wang K, Lindas AC, Peeters E (2016) The genome-scale DNA-binding profile of BarR, a beta-alanine responsive transcription factor in the archaeon *Sulfolobus acidocaldarius*. BMC Genomics 17:569. doi:10.1186/s12864-016-2890-0

Maruyama H, Harwood JC, Moore KM, Paszkiewicz K, Durley SC, Fukushima H, Atomi H, Takeyasu K, Kent NA (2013) An alternative beads-on-a-string chromatin architecture in *Thermococcus kodakarensis*. EMBO Rep 14(8):711–717. doi:10.1038/embor.2013.94

Mayer A, Lidschreiber M, Siebert M, Leike K, Soding J, Cramer P (2010) Uniform transitions of the general RNA polymerase II transcription complex. Nat Struct Mol Biol 17(10):1272–1278. doi:10.1038/nsmb.1903

Micorescu M, Grunberg S, Franke A, Cramer P, Thomm M, Bartlett M (2008) Archaeal transcription: function of an alternative transcription factor B from *Pyrococcus furiosus*. J Bacteriol 190(1):157–167. doi:10.1128/JB.01498-07

Mooney RA, Davis SE, Peters JM, Rowland JL, Ansari AZ, Landick R (2009) Regulator trafficking on bacterial transcription units in vivo. Mol Cell 33(1):97–108. doi:10.1016/j.molcel.2008.12.021

Nagy J, Grohmann D, Cheung AC, Schulz S, Smollett K, Werner F, Michaelis J (2015) Complete architecture of the archaeal RNA polymerase open complex from single-molecule FRET and NPS. Nat Commun 6:6161. doi:10.1038/ncomms7161

Naji S, Grünberg S, Thomm M (2007) The RPB7 orthologue E' is required for transcriptional activity of a reconstituted archaeal core enzyme at low temperatures and stimulates open complex formation. J Biol Chem 282(15):11047–11057. doi:10.1074/jbc.M611674200

Nalabothula N, Xi L, Bhattacharyya S, Widom J, Wang JP, Reeve JN, Santangelo TJ, Fondufe-Mittendorf YN (2013) Archaeal nucleosome positioning in vivo and in vitro is directed by primary sequence motifs. BMC Genomics 14:391. doi:10.1186/1471-2164-14-391

Nguyen-Duc T, van Oeffelen L, Song N, Hassanzadeh-Ghassabeh G, Muyldermans S, Charlier D, Peeters E (2013) The genome-wide binding profile of the *Sulfolobus solfataricus* transcription factor Ss-LrpB shows binding events beyond direct transcription regulation. BMC Genomics 14:828. doi:10.1186/1471-2164-14-828

Ochs SM, Thumann S, Richau R, Weirauch MT, Lowe TM, Thomm M, Hausner W (2012) Activation of archaeal transcription mediated by recruitment of transcription factor B. J Biol Chem 287(22):18863–18871. doi:10.1074/jbc.M112.365742

Ouhammouch M, Dewhurst RE, Hausner W, Thomm M, Geiduschek EP (2003) Activation of archaeal transcription by recruitment of the TATA-binding protein. Proc Natl Acad Sci U S A 100(9):5097–5102. doi:10.1073/pnas.0837150100

Ouhammouch M, Geiduschek EP (2001) A thermostable platform for transcriptional regulation: the DNA-binding properties of two Lrp homologs from the hyperthermophilic archaeon *Methanococcus jannaschii*. EMBO J 20(1–2):146–156. doi:10.1093/emboj/20.1.146

Ouhammouch M, Langham GE, Hausner W, Simpson AJ, El-Sayed NM, Geiduschek EP (2005) Promoter architecture and response to a positive regulator of archaeal transcription. Mol Microbiol 56(3):625–637. doi:10.1111/j.1365-2958.2005.04563.x

Peeters E, Driessen RP, Werner F, Dame RT (2015) The interplay between nucleoid organization and transcription in archaeal genomes. Nat Rev Microbiol 13(6):333–341. doi:10.1038/nrmicro3467

Peeters E, Peixeiro N, Sezonov G (2013) Cis-regulatory logic in archaeal transcription. Biochem Soc Trans 41(1):326–331. doi:10.1042/BST20120312

Perez-Rueda E, Janga SC (2010) Identification and genomic analysis of transcription factors in archaeal genomes exemplifies their functional architecture and evolutionary origin. Mol Biol Evol 27(6):1449–1459. doi:10.1093/molbev/msq033

Phung DK, Rinaldi D, Langendijk-Genevaux PS, Quentin Y, Carpousis AJ, Clouet-d'Orval B (2013) Archaeal beta-CASP ribonucleases of the aCPSF1 family are orthologs of the eukaryal CPSF-73 factor. Nucleic Acids Res 41(2):1091–1103. doi:10.1093/nar/gks1237

Porrua O, Boudvillain M, Libri D (2016) Transcription termination: variations on common themes. Trends Genet 32(8):508–522. doi:10.1016/j.tig.2016.05.007

Porrua O, Libri D (2013) A bacterial-like mechanism for transcription termination by the Sen1p helicase in budding yeast. Nat Struct Mol Biol 20(7):884–891. doi:10.1038/nsmb.2592

Proudfoot NJ (2016) Transcriptional termination in mammals: stopping the RNA polymerase II juggernaut. Science 352(6291):aad9926. doi:10.1126/science.aad9926

Qureshi SA, Baumann P, Rowlands T, Khoo B, Jackson SP (1995) Cloning and functional analysis of the TATA-binding protein from *Sulfolobus shibatae*. Nucleic Acids Res 23(10):1775–1781. doi:10.1093/nar/23.10.1775

Qureshi SA, Bell SD, Jackson SP (1997) Factor requirements for transcription in the archaeon *Sulfolobus shibatae*. EMBO J 16(10):2927–2936. doi:10.1093/emboj/16.10.2927

Ray-Soni A, Bellecourt MJ, Landick R (2016) Mechanisms of bacterial transcription termination: all good things must end. Annu Rev Biochem 85:319–347. doi:10.1146/annurev-biochem-060815-014844

Reichelt R, Gindner A, Thomm M, Hausner W (2016) Genome-wide binding analysis of the transcriptional regulator TrmBL1 in *Pyrococcus furiosus*. BMC Genomics 17:40. doi:10.1186/s12864-015-2360-0

Rowlands T, Baumann P, Jackson SP (1994) The TATA-binding protein - a general transcription factor in eukaryotes and archaebacteria. Science 264(5163):1326–1329. doi:10.1126/science.8191287

Rudrappa D, Yao AI, White D, Pavlik BJ, Singh R, Facciotti MT, Blum P (2015) Identification of an archaeal mercury regulon by chromatin immunoprecipitation. Microbiology 161(12):2423–2433. doi:10.1099/mic.0.000189

Sandelin A, Carninci P, Lenhard B, Ponjavic J, Hayashizaki Y, Hume DA (2007) Mammalian RNA polymerase II core promoters: insights from genome-wide studies. Nat Rev Genet 8(6):424–436. doi:10.1038/nrg2026

Sandman K, Grayling RA, Dobrinski B, Lurz R, Reeve JN (1994) Growth-phase-dependent synthesis of histones in the archaeon *Methanothermus fervidus*. Proc Natl Acad Sci U S A 91(26):12624–12628

Sandman K, Krzycki JA, Dobrinski B, Lurz R, Reeve JN (1990) HMf, a DNA-binding protein isolated from the hyperthermophilic archaeon *Methanothermus fervidus*, is most closely related to histones. Proc Natl Acad Sci U S A 87(15):5788–5791

Santangelo TJ, Artsimovitch I (2011) Termination and antitermination: RNA polymerase runs a stop sign. Nat Rev Microbiol 9(5):319–329. doi:10.1038/nrmicro2560

Santangelo TJ, Cubonova L, Matsumi R, Atomi H, Imanaka T, Reeve JN (2008) Polarity in archaeal operon transcription in *Thermococcus kodakaraensis*. J Bacteriol 190(6):2244–2248. doi:10.1128/JB.01811-07

Santangelo TJ, Cubonova L, Skinner KM, Reeve JN (2009) Archaeal intrinsic transcription termination in vivo. J Bacteriol 191(22):7102–7108. doi:10.1128/JB.00982-09

Santangelo TJ, Reeve JN (2006) Archaeal RNA polymerase is sensitive to intrinsic termination directed by transcribed and remote sequences. J Mol Biol 355(2):196–210. doi:10.1016/j.jmb.2005.10.062

Sasse-Dwight S, Gralla JD (1989) KMnO4 as a probe for *lac* promoter DNA melting and mechanism in vivo. J Biol Chem 264(14):8074–8081

Schwalb B, Michel M, Zacher B, Fruhauf K, Demel C, Tresch A, Gagneur J, Cramer P (2016) TT-seq maps the human transient transcriptome. Science 352(6290):1225–1228. doi:10.1126/science.aad9841

Searcy DG (1975) Histone-like protein in the prokaryote *Thermoplasma acidophilum*. Biochim Biophys Acta 395(4):535–547

Searcy DG, Delange RJ (1980) *Thermoplasma acidophilum* histone-like protein. Partial amino acid sequence suggestive of homology to eukaryotic histones. Biochim Biophys Acta 609 (1):197–200

Seitzer P, Wilbanks EG, Larsen DJ, Facciotti MT (2012) A Monte Carlo-based framework enhances the discovery and interpretation of regulatory sequence motifs. BMC Bioinf 13:317. doi:10.1186/1471-2105-13-317

Shultzaberger RK, Chen Z, Lewis KA, Schneider TD (2007) Anatomy of *Escherichia coli* sigma70 promoters. Nucleic Acids Res 35(3):771–788. doi:10.1093/nar/gkl956

Smollett K, Blombach F, Reichelt R, Thomm M, Werner F (2017) A global analysis of transcription reveals two modes of Spt4/5 recruitment to archaeal RNA polymerase. Nat Microbiol 2:17021. doi:10.1038/nmicrobiol.2017.21

Smollett K, Blombach F, Werner F (2015) Transcription in archaea: preparation of *Methanocaldococcus jannaschii* transcription machinery. Methods Mol Biol 1276:291–303. doi:10.1007/978-1-4939-2392-2_17

Spitalny P, Thomm M (2008) A polymerase III-like reinitiation mechanism is operating in regulation of histone expression in archaea. Mol Microbiol 67(5):958–970. doi:10.1111/j.1365-2958.2007.06084.x

Straub J, Brenneis M, Jellen-Ritter A, Heyer R, Soppa J, Marchfelder A (2009) Small RNAs in haloarchaea: identification, differential expression and biological function. RNA Biol 6 (3):281–292

Tan L, Wiesler S, Trzaska D, Carney HC, Weinzierl RO (2008) Bridge helix and trigger loop perturbations generate superactive RNA polymerases. J Biol 7(10):40. doi:10.1186/jbiol98

Tang TH, Polacek N, Zywicki M, Huber H, Brugger K, Garrett R, Bachellerie JP, Huttenhofer A (2005) Identification of novel non-coding RNAs as potential antisense regulators in the archaeon *Sulfolobus solfataricus*. Mol Microbiol 55(2):469–481. doi:10.1111/j.1365-2958.2004.04428.x

Thomason MK, Bischler T, Eisenbart SK, Forstner KU, Zhang A, Herbig A, Nieselt K, Sharma CM, Storz G (2015) Global transcriptional start site mapping using differential RNA sequencing reveals novel antisense RNAs in *Escherichia coli*. J Bacteriol 197(1):18–28. doi:10.1128/JB.02096-14

Toffano-Nioche C, Ott A, Crozat E, Nguyen AN, Zytnicki M, Leclerc F, Forterre P, Bouloc P, Gautheret D (2013) RNA at 92 degrees C: the non-coding transcriptome of the hyperthermophilic archaeon *Pyrococcus abyssi*. RNA Biol 10(7):1211–1220. doi:10.4161/rna.25567

Tonner PD, Pittman AM, Gulli JG, Sharma K, Schmid AK (2015) A regulatory hierarchy controls the dynamic transcriptional response to extreme oxidative stress in archaea. PLoS Genet 11(1): e1004912. doi:10.1371/journal.pgen.1004912

Torarinsson E, Klenk HP, Garrett RA (2005) Divergent transcriptional and translational signals in archaea. Environ Microbiol 7(1):47–54. doi:10.1111/j.1462-2920.2004.00674.x

Visone V, Vettone A, Serpe M, Valenti A, Perugino G, Rossi M, Ciaramella M (2014) Chromatin structure and dynamics in hot environments: architectural proteins and DNA topoisomerases of thermophilic archaea. Int J Mol Sci 15(9):17162–17187. doi:10.3390/ijms150917162

Vvedenskaya IO, Vahedian-Movahed H, Zhang Y, Taylor DM, Ebright RH, Nickels BE (2016) Interactions between RNA polymerase and the core recognition element are a determinant of

transcription start site selection. Proc Natl Acad Sci U S A 113(21):E2899–E2905. doi:10.1073/pnas.1603271113
Wade JT (2015) Where to begin? Mapping transcription start sites genome-wide in escherichia coli. J Bacteriol 197(1):4–6. doi:10.1128/JB.02410-14
Wade JT, Grainger DC (2014) Pervasive transcription: illuminating the dark matter of bacterial transcriptomes. Nat Rev Microbiol 12(9):647–653. doi:10.1038/nrmicro3316
Werner F (2012) A nexus for gene expression-molecular mechanisms of Spt5 and NusG in the three domains of life. J Mol Biol 417(1-2):13–27. doi:10.1016/j.jmb.2012.01.031
Werner F, Grohmann D (2011) Evolution of multisubunit RNA polymerases in the three domains of life. Nat Rev Microbiol 9(2):85–98. doi:10.1038/nrmicro2507
Werner F, Weinzierl RO (2002) A recombinant RNA polymerase II-like enzyme capable of promoter-specific transcription. Mol Cell 10(3):635–646
Werner F, Weinzierl RO (2005) Direct modulation of RNA polymerase core functions by basal transcription factors. Mol Cell Biol 25(18):8344–8355. doi:10.1128/MCB.25.18.8344-8355.2005
West S, Gromak N, Proudfoot NJ (2004) Human 5′→3′ exonuclease Xrn2 promotes transcription termination at co-transcriptional cleavage sites. Nature 432(7016):522–525. doi:10.1038/nature03035
Wilbanks EG, Larsen DJ, Neches RY, Yao AI, Wu CY, Kjolby RA, Facciotti MT (2012) A workflow for genome-wide mapping of archaeal transcription factors with ChIP-seq. Nucleic Acids Res 40(10):e74. doi:10.1093/nar/gks063
Wilkinson SP, Ouhammouch M, Geiduschek EP (2010) Transcriptional activation in the context of repression mediated by archaeal histones. Proc Natl Acad Sci U S A 107(15):6777–6781. doi:10.1073/pnas.1002360107
Winkelman JT, Vvedenskaya IO, Zhang YC, Zhang Y, Bird JG, Taylor DM, Gourse RL, Ebright RH, Nickels BE (2016) Multiplexed protein-DNA cross-linking: scrunching in transcription start site selection. Science 351(6277):1090–1093. doi:10.1126/science.aad6881
Wurtzel O, Sapra R, Chen F, Zhu Y, Simmons BA, Sorek R (2010) A single-base resolution map of an archaeal transcriptome. Genome Res 20(1):133–141. doi:10.1101/gr.100396.109
Xie Y, Reeve JN (2004) Transcription by an archaeal RNA polymerase is slowed but not blocked by an archaeal nucleosome. J Bacteriol 186(11):3492–3498. doi:10.1128/JB.186.11.3492-3498.2004
Yang C, Bolotin E, Jiang T, Sladek FM, Martinez E (2007) Prevalence of the initiator over the TATA box in human and yeast genes and identification of DNA motifs enriched in human TATA-less core promoters. Gene 389(1):52–65. doi:10.1016/j.gene.2006.09.029
Zhao S, Fung-Leung WP, Bittner A, Ngo K, Liu X (2014) Comparison of RNA-Seq and microarray in transcriptome profiling of activated T cells. PLoS One 9(1):e78644. doi:10.1371/journal.pone.0078644
Zuo Y, Steitz TA (2015) Crystal structures of the *E. coli* transcription initiation complexes with a complete bubble. Mol Cell 58(3):534–540. doi:10.1016/j.molcel.2015.03.010

Chapter 2
Transcription Factor-Mediated Gene Regulation in *Archaea*

Elizabeth A. Karr, Catherine E. Isom, Vy Trinh, and Eveline Peeters

Abstract Adequate gene regulation in response to environmental and/or metabolic changes is crucial for the fitness and survival of microorganisms. Transcription factors are important elements of microbial gene regulation. Intriguingly, although the archaeal basal transcription machinery is more similar to that seen in the eukaryotic domain of life, transcription factor-mediated gene regulation largely follows the bacterial paradigm. This chapter deals with structural and functional characteristics of archaeal transcription factors. Although one-component and two-component systems are both present in the archaeal domain of life, one-component systems dominate. Different aspects of transcription factor functioning are discussed, including mechanisms of DNA binding, regulatory mechanisms and sensing and signal transduction mechanisms. Archaeal transcription factors primarily interact with DNA using a winged helix-turn-helix DNA binding motif. Transcriptional repression is achieved through a variety of promoter occlusion mechanisms like those seen in bacteria. In contrast, activation mechanisms vary from those found in bacteria and involve recruitment of the general transcription factors TATA binding protein and transcription factor B to the promoter. A variety of environmental signals are sensed through ligand binding or redox-sensing. The body of literature covering studies of archaeal transcription regulation has expanded significantly over the past 15 years. However, there is still much to be learned from future studies particularly in the area of signal transduction as well as gene regulatory networks in archaea.

E.A. Karr • V. Trinh
Price Family Foundation Institute of Structural Biology and Department of Microbiology and Plant Biology, University of Oklahoma, 770 Van Vleet Oval, Norman, OK 73019, USA
e-mail: lizkarr@ou.edu

C.E. Isom
Price Family Foundation Institute of Structural Biology and Department of Chemistry and Biochemistry, University of Oklahoma, 101 Stephenson Parkway, Norman, OK 73019, USA
e-mail: catisom@ou.edu

E. Peeters (✉)
Research Group of Microbiology, Department of Bio-engineering Sciences, Vrije Universiteit Brussel, Pleinlaan 2, 1050 Brussels, Belgium
e-mail: Eveline.Peeters@vub.be

B. Clouet-d'Orval (ed.), *RNA Metabolism and Gene Expression in Archaea*,
Nucleic Acids and Molecular Biology 32, DOI 10.1007/978-3-319-65795-0_2

2.1 Introduction

The so-called third domain of life, *Archaea*, are a diverse group of single-celled organisms (Woese et al. 1990). Originally lauded for their extreme habitats, it quickly became apparent that these microorganisms are ubiquitous in nature (DeLong and Pace 2001; Chaban et al. 2006). From Antarctic lakes to the hot springs of the Kamchatka peninsula to black smoker hydrothermal vents in the depths of the ocean and finally local landfills, archaea are vital members in their ecosystems (Franzmann et al. 1992; Karr et al. 2006; Chaban et al. 2006; Teske and Sørensen 2008; Reigstad et al. 2010). Their metabolisms are as diverse as their range of ecosystems; archaea are key players in the global sulfur, nitrogen and carbon cycles, illustrating their importance in the biosphere. While the archaeal branch of the tree of life continues to expand, our understanding of their role in many ecosystems is limited (Castelle et al. 2015). This is in part due to culturing limitations, which, in turn, limits our understanding of their physiology to the molecular detection of known functional genes (Auguet et al. 2010). On the surface, the lack of a nuclear membrane and the existence of diverse metabolic capabilities in archaea makes them resemble bacteria. However, archaeal molecular machineries harbor a "mosaic" of bacterial and eukaryotic features (Baumann et al. 1995; Bell and Jackson 1998).

This chapter focuses on the regulation of gene expression at the level of transcription in archaea. Therefore, a discussion of the molecular machinery necessary for basal transcription is of utmost importance. While the basal transcription machinery in archaea resembles a pared down version of the eukaryotic transcription machinery, the regulation of transcription is dominated by the quintessential bacterial-like transcription factors (TFs) (Kyrpides and Ouzounis 1999; Aravind and Koonin 1999; Pérez-Rueda and Collado-Vides 2001). In turn, archaeal promoter/operator architecture reflects this mosaic of features. The archaeal RNA polymerase (RNAP) is a 12–14 subunit enzyme that resembles the eukaryotic RNAP II (Baumann et al. 1995; Werner et al. 2000; Todone et al. 2001; Jun et al. 2011). RNAP is recruited to the promoter by general TFs. These general TFs are the TATA binding protein (TBP) and transcription factor B (TFB) (Gohl et al. 1995). To initiate transcription, TBP binds to the TATA box located approximately 25 nucleotides upstream of the transcription start site (TSS). TBP binding is followed by TFB binding to the B recognition element (BRE), a purine-rich segment just upstream of the TATA box. The TBP-TFB-DNA ternary complex then recruits RNAP to the promoter generating the closed initiation complex (Gohl et al. 1995; Hausner et al. 1996; Bell et al. 1999b). In contrast to eukaryotic transcription initiation, transitioning to the open complex in archaea does not require ATP. This is likely due to the absence of a TFIIH homologue whose helicase activity requires ATP in eukaryotes (Hausner and Thomm 2001). The DNA-TBP-TFB-RNAP complex represents the minimal transcription machinery in archaea and, in vitro, these components are sufficient to initiate transcription from most promoters (Darcy et al. 1999).

A homologue of the α-subunit of the eukaryal basal TF TFIIE, transcription factor E (TFE), has also been identified in archaea (Bell et al. 2001; Hanzelka et al. 2001). Although not essential for transcription, TFE increases the transcription from less robust promoters in vitro (Bell et al. 2001; Hanzelka et al. 2001; Werner and Weinzierl 2005). More recently, TFE has been shown to stabilize the initiation complex and to play a role in the regulation of RNAP activity (Grohmann et al. 2011; Walker and Santangelo 2015). Many archaea, particularly the haloarchaea, encode multiple TBP or TFB proteins (Baliga et al. 2000; Tonner et al. 2015). They utilize these alternate TBP and TFB proteins for global gene regulation in a manner reminiscent of alternative σ factors in bacteria whereby subsets of promoters are recognized optimally by a particular TBP-TFB combination (Coker and DasSarma 2007; Facciotti et al. 2007; Santangelo et al. 2007; Micorescu et al. 2007; Paytubi and White 2009; Tonner et al. 2015).

With virtually no habitat devoid of archaea, there is an abundance of environmental signals and stressors that must be integrated and transmitted into a transcriptional output. Apart from a few exceptions, archaeal promoter-specific transcription regulation follows the bacterial paradigm (Krüger et al. 1998; Kyrpides and Ouzounis 1999; Aravind and Koonin 1999; Guillière et al. 2013). Archaea share regulatory TFs with bacteria. Herein, we will focus on the role of bacterial-type TFs in promoter-specific regulation of archaeal transcription.

2.2 Transcription Factor Families in Archaea

2.2.1 One-Component Systems

The majority of prokaryotic TFs are so-called one-component systems (OCSs), in which a sensor module is combined with a DNA-binding domain (DBD) within a single protein, enabling the direct connection of the sensing of intracellular signals (e.g., metabolite concentrations) to a regulatory function. In contrast to the eukaryotic-like core components of the basal transcription machinery, archaea share regulatory TFs with bacteria (Bell and Jackson 2001; Coulson et al. 2007). This sharing can be explained in two ways (Aravind and Koonin 1999): (1) the presence of a small set of ancestral bacterial/archaeal TFs in the last universal common ancestor (LUCA) or (2) the occurrence of multiple lateral gene transfer (LGT) events from bacteria to archaea throughout evolution (Kunin et al. 2005; Nelson-Sathi et al. 2014).

The predominant class of DBDs in bacterial/archaeal OCSs harbour a helix-turn-helix (HTH) motif (Kyrpides and Ouzounis 1999; Aravind and Koonin 1999), in which the C-terminal α-helix of a three-helical bundle, called the recognition helix, establishes sequence-specific contacts with the major groove of the DNA (Aravind et al. 2005). The largest fraction of archaeal HTH-containing TFs harbour an additional secondary structure element, namely a β-hairpin unit (the so-called "wing") that flanks the HTH motif yielding a winged HTH (wHTH) motif (Kyrpides and Ouzounis 1999; Aravind and Koonin 1999).

OCSs can be further classified according to their domain architecture. As compared to bacterial OCSs, a larger fraction of the archaeal OCSs are small single-domain proteins solely composed of a DBD (Aravind and Koonin 1999; Pérez-Rueda and Janga 2010). As a consequence, the average size of archaeal TFs is smaller than that of bacterial TFs (a median size of 179 versus 236 amino acids, respectively) (Pérez-Rueda and Janga 2010). Besides single-domain OCSs, the most common organization of archaeal TFs is a two-domain architecture in which the DBD is fused to a distinct globular domain that mediates interactions with small molecule ligands (Aravind and Koonin 1999). These sensing domains have large structural variations, enabling the classification of regulatory TFs into different families (see also Sect. 2.2.3).

Only a single database of archaeal TF sequences exist (Wu et al. 2008), which is unfortunately limited to 37 archaeal genomes and has not been updated since 2008. Together with extensive phylogenomic analyses of TF genes in archaeal genomes, this resource demonstrates that the diversity of archaeal TF families is smaller than that of bacteria and that archaea share almost all their TF families with bacteria, while this is not the case vice versa (Minezaki et al. 2005; Pérez-Rueda and Janga 2010; Martínez-Núñez et al. 2013). The ancestral core of TF families, which are universally present in archaea and were most probably present in LUCA, is composed of the ArsR, AsnC, HTH-3 and TrmB families (Pérez-Rueda and Janga 2010). Together with the MarR and GntR families, which were probably acquired by LGT events early in evolution after the divergence of the bacterial and archaeal domains of life, these families are characterized by distinct pan-bacterial and pan-archaeal groups (Aravind et al. 2005; Iyer and Aravind 2012). In contrast, phylogenetic analysis suggested the occurrence of several independent LGT events (AraC-, TrpR- and TetR-family TF-encoding genes) from Bacteria to Archaea and sporadic LGT events (BirA-, ModE-, PadR- and DtxR/Fur-family TF-encoding genes) (Aravind et al. 2005). Rare examples of archaea-specific TF families that are not found in bacteria include the HTH-10 (Minezaki et al. 2005) and the Lrs14 family (Orell et al. 2013).

A linear correlation exists between the number of TF-encoding genes and genome size in archaea, similar to bacteria, which can be linked to the environmental niche in which they are found and the corresponding lifestyle (Pérez-Rueda and Janga 2010). Indeed, archaea living host-dependently in a stable environment require a smaller gene regulatory capacity for cellular survival and fitness than those living in continuously varying environmental conditions. For example, while the parasitic hyperthermophile *Nanoarchaeum equitans* has a reduced genome and is predicted to contain only eight TF-encoding genes, the metabolically versatile *Methanosarcina acetivorans* has a much larger genome and a correspondingly larger repertoire of 158 predicted TF-encoding genes (Pérez-Rueda and Janga 2010). Archaea with larger genomes and a larger number of TFs are generally not characterized by a larger diversity in TF families (Martínez-Núñez et al. 2013). Instead, gene duplications resulted in lineage-specific expansions of TF paralogues within specific families (Martínez-Núñez et al. 2013; Plaisier et al. 2014). When comparing the proportion of genes devoted to TF functions, archaea seem to have lower proportions that encode TFs as compared to bacteria (Minezaki et al. 2005; Pérez-Rueda and Janga 2010; Martínez-Núñez et al. 2013). This raises the question as to how archaea regulate similar genomes with a more limited repertoire of TFs (which are furthermore smaller and less diverse than the bacterial counterparts) and implies that these TFs function differently from a mechanistic perspective (see Sect. 2.8) or that archaea employ other gene regulatory strategies (e.g., post-transcriptional or post-translational regulation) to a greater extent (Pérez-Rueda and Janga 2010; Martínez-Núñez et al. 2013).

2.2.2 Two-Component Systems

Besides OCSs, prokaryotes harbour two-component systems (TCSs), which are regulatory systems composed of two individual proteins: a histidine kinase (HK), typically a membrane protein with an extracellular sensing domain, and a response regulator (RR) that performs the regulation in response to the detected signal. Signals are transmitted from the HK to the RR by means of phosphorylation. In bacteria, RRs generally harbour a classical HTH DBD and employ similar DNA-binding and regulatory strategies as OCSs. A major advantage of TCSs with respect to OCSs is the ability of the organism to sense the extracellular environment.

Despite the large evolutionary success and the abundant and almost ubiquitous presence of TCSs in bacterial organisms, they are less common in archaea (Koretke et al. 2000; Ashby 2006; Wuichet et al. 2010). TCS-encoding genes are predicted to be present in about 50% of archaeal species, which all belong to the *Euryarchaeota* and *Thaumarchaeota* (Ashby 2006; Wuichet et al. 2010). In contrast, TCSs are completely absent in *Crenarchaeota* and *Korarchaeota* (Ashby 2006; Coulson et al. 2007). This correlates to the organismal lifestyle in the sense that psychrophilic and mesophilic archaea use more TCSs for gene regulatory processes than (hyper-) thermophilic species (Chen et al. 2012). Furthermore, larger numbers of TCSs are found in methanogens and halophilic archaea that display great metabolic flexibility (Ashby 2006). For example, the genome of the psychrophilic methanogenic Euryarchaeote *Methanococcoides burtonii* is predicted to code for the exceptionally high number of 45 TCSs (Allen et al. 2009). Despite the importance of TCSs for the cellular regulation of these archaea, there is a great lack of information for archaeal TCSs. Currently only two systems have been experimentally characterized: LtrR/LtrK involved in cold adaptation in *M. burtonii* (Najnin et al. 2016) and FilI/FilR1/FilR2 involved in quorum sensing and the regulation of methanogenesis in *Methanosaeta harundinacea* (Li et al. 2014). Additionally, the HK component of a TCS, MsmS, from *M. acetivorans* has been initially characterized and shown to autophosphorylate in a redox-dependent manner (Molitor et al. 2013).

TCSs are postulated to have originated in *Bacteria* after the divergence of the archaeal/eukaryotic lineage and radiated into the archaeal domain of life by LGT. More specifically, phylogenetic analysis supported the occurrence of multiple LGT events in ancestral *Euryarchaeota* and more recently into methanogens (Ashby 2006). Following these LGT events, evolutionary expansions took place that explain the current existence of species-specific clusters, for example in *Methanothermobacter thermautotrophicus* (Koretke et al. 2000). As compared to bacterial TCSs, HKs and RRs are not always present proportionally in archaeal genomes (Koretke et al. 2000). Furthermore, while most bacterial RRs harbour a classical DNA-binding output domain with a wHTH motif, a large number of archaeal RRs lack an output domain and are composed solely of a receiver domain (Aravind et al. 2005; Ashby 2006; Wuichet et al. 2010). These RRs are postulated to mediate signal trafficking through protein-protein interactions with other classes of TFs.

2.2.3 Common Transcription Factor Families in Archaea

Most archaeal TFs are two-domain OCSs composed of an N-terminal wHTH DBD and a C-terminal sensing domain. As mentioned above, these belong to families that are generally present in bacteria and that are defined based on structural rather than sequence similarity (Fig. 2.1). Although the wHTH fold is shared by most of these families, the relative orientation of the fold largely differs among the different families. The structure of the sensor domain is often a unique determinant of which family a given TF belongs to. Sequence conservation of members within a specific family is typically highest in the DBD. Below, we discuss the general characteristics of four of the most widely represented TF families in archaea: Lrp/AsnC, TrmB, MarR and ArsR.

2.2.3.1 Lrp/AsnC Family

The Lrp/AsnC family is named after the prototypical regulators leucine-responsive regulatory protein (Lrp) and asparagine synthase C (AsnC) in *Escherichia coli* and is sometimes referred to as Feast Famine Regulatory Protein (FFRP). It is a well-represented family in archaea, constituting about 8% of all regulatory TF genes (Pérez-Rueda and Janga 2010). Each archaeal genome, even the parasitic *N. equitans*, harbours genes encoding Lrp-like regulators with an average of 5 (±4) copies per genome (Plaisier et al. 2014). This widespread occurrence, not only in archaea but also in bacteria, suggests the presence of an ancestral Lrp-like regulator in LUCA, which was later lost in the eukaryal lineage (Brinkman et al. 2003).

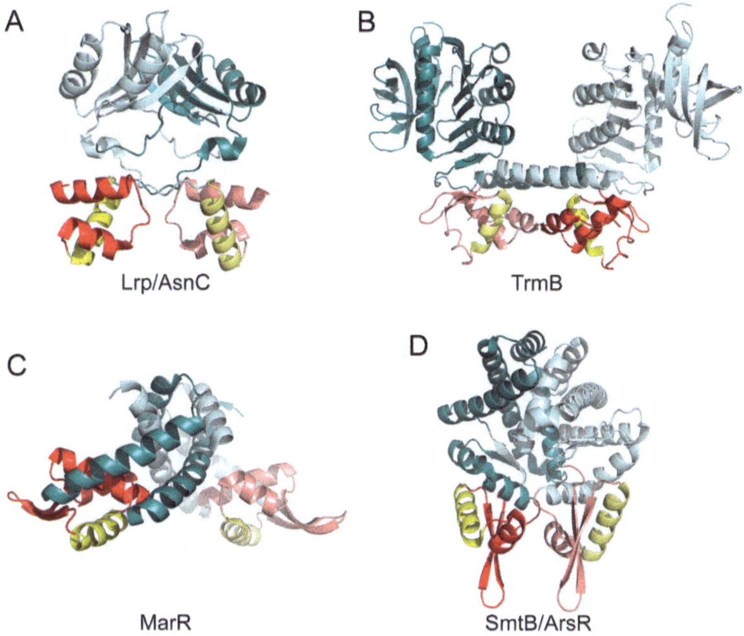

Fig. 2.1 Crystal structures of representatives of the four most common transcription factor families in archaea. (**a**) Structure of Grp, an Lrp-like protein in *Sulfolobus tokodaii* (PDB ID 2K9I) (Kumarevel et al. 2008a); (**b**) Structure of TrmB, a member of the TrmB family in *Pyrococcus furiosus* (PDB ID 3QPH) (Krug et al. 2013); (**c**) Structure of BldR from *Sulfolobus solfataricus*, a member of the MarR family (PDB ID 3F3X) (Di Fiore et al. 2009); (**d**) Structure of PH1932, a member of the SmtB/ArsR family in *Pyrococcus horikoshii* (PDB ID 1ULY) (Itou et al. 2008). DNA-binding domains are colored *red/yellow*, with the recognition helices in *yellow*, whereas ligand-binding/oligomerization domains are colored *turquoise*

The Lrp family is one of the best studied TF families in archaea (Peeters and Charlier 2010). Archaeal Lrp-like regulators act locally or globally and regulate genes involved in amino acid metabolism, central metabolism and/or transport processes (Brinkman et al. 2003; Peeters and Charlier 2010). They are generally responsive to amino acid molecules (Okamura et al. 2007; Schwaiger et al. 2010; Song et al. 2013; Vassart et al. 2013; Liu et al. 2014) and in some cases to other small molecules (Leonard et al. 2001; Yokoyama et al. 2006b; Kawashima et al. 2008; Peeters et al. 2009). These ligands interact with the C-terminal domain that is folded into an α/β sandwich structure typified by an antiparallel β-sheet surrounded by two α-helices (Fig. 2.1a) (Leonard et al. 2001; Koike et al. 2004; Okamura et al. 2007; Kumarevel et al. 2008a). This domain, called the Regulation of Amino acid Metabolism (RAM) domain, is also responsible for oligomerization. Lrp-like TFs tend to oligomerize into multimers of dimers (Peeters and Charlier 2010) and typically crystallize as octameric units (Leonard et al. 2001; Kumarevel et al. 2008a).

2.2.3.2 TrmB Family

The TrmB family is another large group of TFs. This family was originally discovered in archaea (Lee et al. 2003) and later also identified in bacteria (Kim et al. 2016). The family is named after the Transcription regulator of *mal* operon (TrmB) protein initially characterized in *Thermococcales* as a maltose-responsive TF of an operon encoding a trehalose/maltose (TM) ABC transporter (Lee et al. 2003). *Thermococcales* belong to the phylum *Euryarchaeota*, in which TrmB regulators are abundantly present. Although less widespread, TrmB members are also represented in the *Cren-*, *Thaum-*, *Kor-* and *Nanoarchaeota* (Maruyama et al. 2011; Peeters et al. 2015). TrmB-encoding genes are postulated to have been subjected to extensive LGT (Diruggiero et al. 2000).

TrmB-like TFs are typically global regulators that regulate a relatively large number of genes and operons with functions related, but not restricted, to sugar transport and metabolism (Lee et al. 2007a, 2008; Kanai et al. 2007; Schmid et al. 2009; Wagner et al. 2014; Gindner et al. 2014; Reichelt et al. 2016; Kim et al. 2016). They can function as activators or repressors or have a dual function (Lee et al. 2003; Gindner et al. 2014). Apart from the N-terminal wHTH DBD, canonical TrmB-like TFs are characterized by a long amphipathic α-helix that advances dimerization of the protein by establishing a coiled-coil structure and connects the DBD to the C-terminal ligand-binding domain (Fig. 2.1b) (Krug et al. 2013). The latter adopts a complex structure, with one subdomain consisting of an 8-stranded β-sheet flanked by multiple large α-helices and a second subdomain forming an irregular flattened 7-stranded β-barrel (Krug et al. 2006, 2013). The cleft between the two subdomains forms the ligand-binding pocket in which specific contacts are established with sugar molecules.

Besides classical TrmB homologs, proteins that lack the ligand-binding domain and thus have shorter lengths and variable ligand-binding properties have also been classified as belonging to the TrmB family (Kim et al. 2016). These include MreA from *Methanosarcina*, which is a global regulator of distinct methanogenic pathways (Reichlen et al. 2012) and TrmBL2, which is a non-specific DNA-binding protein that has a chromatin organization function (Maruyama et al. 2011) (see Sect. 2.5.2.1).

2.2.3.3 MarR Family

First identified in *E. coli* as a multiple antibiotic resistance regulator (MarR), the prokaryotic MarR family is a diverse group of TFs that regulate genes involved in a wide variety of processes ranging from stress response, metabolism and virulence to the degradation of chemicals such as phenolic compounds and antibiotics (Cohen et al. 1993). This family has vastly proliferated in archaea and gave rise to several archaeal subfamilies (Aravind et al. 2005), although thus far, only a handful of archaeal MarR-like TFs have been characterized with a focus on structural analysis.

The biological function of most of these regulators is elusive, with the exception of BldR and BldR2 in the hyperthermoacidophilic *Sulfolobus solfataricus*, which are involved in the detoxification of aromatic compounds (Fiorentino et al. 2007, 2011). Similar to bacterial MarR regulators, archaeal MarR regulators have been shown to interact with lipophilic compounds that generally have a planar structure, such as ethidium and the phenolics benzaldehyde and salicylate (Fiorentino et al. 2007; Saridakis et al. 2008; Yu et al. 2009). This ligand specificity is in accordance with a physiological role in regulating detoxification processes.

Crystal structures have been determined for four archaeal MarR members: StEmrR from *Sulfolobus tokodaii* (Miyazono et al. 2007; Kumarevel et al. 2008b), BldR from *S. solfataricus* (Di Fiore et al. 2009), MTH313 from *M. thermautotrophicus* (Saridakis et al. 2008) and PH1061 from *Pyrococcus horikoshii* (Okada et al. 2006). Despite low sequence conservation, these MarR homologs display a high structural similarity, as well as with bacterial MarR family members. They are characterized by a homodimeric structure that exhibits a triangular tweezer-like structure with two centrally located wHTH domains as tips and two closely interacting dimerization domains that are each composed of a long N-terminal α-helix and two C-terminal α-helices (Fig. 2.1c).

2.2.3.4 SmtB/ArsR Family

In bacteria, the SmtB/ArsR family consists of metalloregulatory proteins that are usually encoded in an operon together with their target genes that confer resistance to heavy metals in response to interaction with metal ions (Busenlehner et al. 2003). In archaea, it is the largest TF family, being universally present with more than 700 members identified in 52 archaeal genomes (Pérez-Rueda and Janga 2010). This finding could be related to the large number of extremophilic species in the archaeal domain of life and the importance of heavy metal detoxification for survival in harsh environments. Unlike bacteria, not all archaeal members of the SmtB/ArsR TFs are metalloregulatory TFs as is the case in bacteria. For example, MsvR in *M. thermautotrophicus* is a redox-sensitive ArsR-like TF that regulates oxidative stress response (Karr 2010). Despite the abundance of the ArsR family, very little research has been performed for these TFs: ArsR-like regulators involved in arsenite resistance have been characterized in *Halobacterium* sp. and *Ferroplasma acidarmanus* (Wang et al. 2004; Baker-Austin et al. 2007), while a mercuric ion-responsive ArsR-like protein has been studied in *S. solfataricus* (Schelert et al. 2004, 2006).

Only a single crystal structure has been reported for an archaeal ArsR-type regulator with an unknown biological function (Itou et al. 2008) (Fig. 2.1d). The C-terminal domain is composed of four α-helices, which interact to form a homodimeric structure. Furthermore, a ligand-binding pocket has been postulated to be formed on the inside of the unique hat-shaped helix-bundle formed by these C-terminal domains (Itou et al. 2008).

2.3 Experimental Strategies to Study Transcription Factor Function

Information regarding archaeal transcription regulatory systems is accumulating. Computational studies have provided accurate predictions of which genes encode TFs in archaeal genomes (Aravind and Koonin 1999; Pérez-Rueda and Janga 2010; Charoensawan et al. 2010). Structural analysis has been performed for a number of archaeal TFs and significant advances have been made in understanding the molecular mechanisms and physiological role of archaeal transcription regulation using a variety of in vitro and in vivo methodologies. Although detailed mechanistic studies of individual TFs have yielded valuable insights, the nuances lie in understanding global regulatory capabilities of TFs in addition to the interaction of these regulatory proteins with their effectors from a system-level perspective. In the current -*omics* era, system-biological studies of archaeal gene regulatory networks (GRNs) are limited as compared to similar studies in bacterial or eukaryal model organisms. Consequently, the biological role of most of the predicted TFs encoded in archaeal genomes is still unknown.

Here, we provide an overview of in vitro and in vivo methodologies that are well suited for the study of TF function and have been used for the study of archaeal regulators. Evidently, a combination of methodologies provides the most thorough approach to the study of TFs and their regulatory networks.

2.3.1 In vitro Methodologies

A first point of interest when initiating the study of an uncharacterized TF is if and how the protein interacts with DNA. The electrophoretic mobility shift assay (EMSA) was one of the first methods developed to study protein-DNA interactions in vitro, and is still commonly used to characterize archaeal TFs (Garner and Revzin 1981). EMSA analysis is a fast and straightforward assay that enables the in vitro analysis of TF-DNA interactions by the electrophoretic separation of TF-DNA complexes on a polyacrylamide or agarose gel based on size and charge, as compared to a DNA-only control. Visualization of unbound or complexed DNA molecules is accomplished using radioactive, fluorescent, biotin, or ethidium bromide labelling of the DNA.

EMSAs were used as a major experimental approach to study the function of archaeal TFs in early studies, when genetic tools and in vivo methodologies for the study of archaeal physiology were still limiting (Napoli et al. 1999; Enoru-Eta et al. 2000; Bell and Jackson 2000; Brinkman et al. 2000; Peeters et al. 2004). EMSA analysis is hypothesis-driven, and in many studies a probe is used that represents the control region of the TF-encoding gene itself since a large fraction of prokaryotic TFs perform an autoregulation. Under ideal conditions, EMSAs can provide not only qualitative but also quantitative information based on densitometric quantifications of the ratio of bound to unbound nucleic acid on the gel (van Oeffelen et al. 2014). Ideal conditions for quantification assume that the starting concentrations of both TF and DNA are known in addition to the stoichiometry of the TF-DNA complexes, which allows for the calculation of an apparent equilibrium dissociation constant K_D and even cooperativity parameters in case multiple stoichiometrically distinct complexes are formed. In this manner, a quantitative EMSA analysis of the interactions between the Lrp-like regulator Ss-LrpB in *S. solfataricus* and the three-site control region of its own gene provided detailed insights into the thermodynamics of the interaction (Peeters et al. 2013b). Other useful attributes to this classic method for assessing TF-DNA interaction are the use of competitor DNA to determine DNA-binding specificity as well as analysing the effects of putative effectors on TF-DNA interaction.

Further analysis of the DNA binding region of a TF can be accomplished using deoxyribonuclease I (DNase I) footprinting and similar high-resolution contact probing techniques. Briefly, the TF of interest is added to a pool of DNA and binding sites are identified based on the principle that DNA bound to the TF will be protected from DNase I hydrolysis (Galas and Schmitz 1978).

Although studying the binding of TFs to a collection of promoter DNAs using EMSA might eventually lead to the identification of a consensus sequence, a more appropriate in vitro method to study TF-DNA interactions without prior knowledge of specific binding sites is the method "systematic evolution of ligands by exponential enrichment" (SELEX) (Tuerk and Gold 1990). SELEX utilizes a large double-stranded DNA (dsDNA) oligonucleotide library (or genomic DNA library) with the TF of interest to progressively decipher DNA-binding sequence specificity with repeated rounds of partition and amplification of the bound sequence. It has been used for the determination of the DNA-binding specificities of a variety of archaeal regulators that belong to the Lrp/AsnC family (Ouhammouch and Geiduschek 2001; Yokoyama et al. 2006a, 2009). The combination of SELEX methodology with next-generation sequencing technology (e.g., Illumina), termed SELEX-seq or High-Throughput-SELEX (HT-SELEX) (Slattery et al. 2011; Riley et al. 2014), provides an improved resolution but has not yet been implemented for the study of DNA-binding specificities of archaeal TFs.

In vitro transcription assays answer the question as to whether a particular TF can influence template-directed synthesis of RNA molecules using an in vitro assembled transcription machinery (Hüdepohl et al. 1990; Reiter et al. 1990). In vitro transcription systems were instrumental to early studies of archaeal promoter specific transcription regulation, particularly for organisms that had not yet had genetic systems developed. One of the first demonstrations of a repressors mechanism of repression in vitro was done with the *Archaeoglobus fulgidus* TF MDR1 (Bell et al. 1999a). The *Methanocaldococcus jannaschii* in vitro transcription system was used to demonstrate the mechanism of activation for one of the first archaeal transcription activators identified, the Lrp-type Ptr2 (Ouhammouch et al. 2003). Additionally, the in vitro transcription systems for *M. thermautotrophicus* and *P. furiosus* were used to determine repression and repression/activation of the first redox-sensitive TFs MsvR and SurR, respectively (Lipscomb et al. 2009; Yang et al. 2010; Karr 2010). These are just a few select examples from expansive literature investigating archaeal transcription regulation using in vitro systems.

2.3.2 In vivo Methodologies

2.3.2.1 Chromatin Immunoprecipitation Techniques

Chromatin immunoprecipitation (ChIP) is a very powerful method to study TF function and enables to identify TF-DNA interactions, and thus the regulon, on a genome-wide scale in vivo. Antibodies specific to the TF of interest are used to enrich for DNA-chromatin extracts during ChIP. Following ChIP, the precipitated DNA is identified via hybridization to a microarray (ChIP-chip) (Aparicio et al. 2004) or by high-throughput sequencing (ChIP-seq) (Robertson et al. 2007). Genome-wide ChIP approaches have provided useful methods for identifying GRNs in archaeal genomes (see Sect. 2.8). The TF TrmB from the halophilic archaeon *Halobacterium salinarum NRC-1* was shown to coordinate the transcription of over 100 central metabolism genes further providing insight into TrmB as an "evolutionary mosaic" using ChIP-chip in conjunction with gene expression analysis (Schmid et al. 2009). ChIP-chip results can also be used to determine DNA regions involved in *cis*-regulatory binding, as was accomplished for Idr1 and Idr2, two functionally-independent TFs necessary for iron homeostasis in *H. salinarum* (Schmid et al. 2011).

The cost of ChIP-seq relative to ChIP-chip, especially considering the relatively smaller archaeal genomes, makes ChIP-seq an even more appealing approach to identify specific TF-DNA interaction and TF GRNs in archaea (Park 2009; Wilbanks et al. 2012). This cost-benefit analysis was demonstrated previously in the model archaeon *H. salinarum* whereby the GRN of natively expressed TFs was mapped at a cost of ~$15 per sample (Wilbanks et al. 2012). Indeed, complementing in vitro TF functional data with in vivo ChIP-seq analysis has recently been explored for TFs from archaea. For example, TrmBL1 from *Pyrococcus furiosus* was analysed using ChIP-seq under gluconeogenic growth conditions, which indicated TrmBL1 as more of a global regulator than initially supposed using in vitro methodologies (Reichelt et al. 2016). ChIP-seq analysis of *S. acidocaldarius* BarR revealed the involvement of this Lrp-like TF in the regulation of glutamine synthesis, in addition to the previously identified β-alanine aminotransferase regulation (Liu et al. 2016). Furthermore, ChIP-seq analysis also revealed that the BarR regulon overlaps with the regulon of other Lrp-like regulators (Liu et al. 2016). The resolution of the mapped binding sites could be improved using a relatively new extension of ChIP-seq that utilizes exonuclease trimming following immunoprecipitation (ChIP-exo) (Matteau and Rodrigue 2015) but has not yet been implemented for the study of archaeal GRNs.

2.3.2.2 Genetic Techniques

Genetic systems provide a necessary tool for gaining a deeper understanding of the effects that specific TFs have on archaeal physiology. For an in-depth review on well-established genetic systems in halophilic, methanogenic, thermophilic, and/or acidophilic archaea see (Leigh et al. 2011). Genetic systems in archaea have developed relatively slowly due to the need for anaerobic conditions and, often, slow growth that is species-dependent. Additionally, archaea often lack the antibiotic sensitivity required for the use of genetic markers. However, the first demonstration of archaeal in vivo transformation was successful over two decades ago in the halophilic archaeon *H. salinarum* (formerly *H. halobium*) (Cline and Doolittle 1987). Homologous gene replacement has since been accomplished in *H. salinarum* using the *ura3* gene as a counterselectable marker based on 5-fluoroorotic acid sensitivity (Peck et al. 2000).

The successful transformation of a methanogenic archaeon using a puromycin resistance marker (*pac* cassette) developed soon after the establishment of a genetic system in *H. salinarum* (Gernhardt et al. 1990). Methanococci provided a good starting point for developing genetic systems in methanogens because they are facultatively autotrophic and grow relatively quickly (Tumbula and Whitman 1999). The success of designing and implementing a genetic system in methanococci has led to the development of other methanogenic genetic systems. For instance, the *pac* cassette was used as a successful marker for the efficient liposome-mediated transformation of *M. acetivorans* (Metcalf et al. 1997).

The development of a genetic system for *Sulfolobales* was relatively slower because initially only two selectable markers were available, uracil auxotrophy and the *lacS* gene (Leigh et al. 2011). Currently, mainly uracil auxotrophy is used in a genetic toolbox for the model species *S. acidocaldarius* (Wagner et al. 2012).

2.4 Mechanisms of DNA Binding

2.4.1 Structural Motifs in DNA-Binding Domains

Archaeal TFs usually interact with B-DNA, a double stranded right-handed helix forming major and minor grooves with the base pairs (bps) oriented perpendicular to the helical axis. The major groove is wider than the minor groove and is often the site of sequence-specific protein-DNA interactions. As mentioned above, a common structural feature of all archaeal OCS TFs is the wHTH DBD (see also Fig. 2.1). TF function is largely determined by the affinity and sequence specificity of the protein's interaction with DNA mediated by this motif. There are three bacterial-type structural motifs that characterize the DBD of archaeal TFs: HTH (the most common motif), ribbon-helix-helix (RHH), also termed the MetJ/Arc domain, and the Zn-ribbon motif. In addition to these bacterial motifs, the typically eukaryotic leucine zipper DNA-binding motif also exists in archaeal TFs.

2.4.1.1 Winged Helix-Turn-Helix Motif

The predominant HTH motif consists of a half-open tri-helical bundle, in which the C-terminal α-helix is the recognition helix that establishes sequence-specific interactions with the major groove of dsDNA (Aravind et al. 2005). The other helices can function to stabilize the DNA-protein complex (Rohs et al. 2010). The binding geometry of the HTH domain is restricted by the essentially straight recognition helix that cannot curve around the major DNA groove allowing the recognition helix to interact with five or less consecutive bps (Suzuki et al. 1995). The wing in the wHTH motif is located at the cleft of the half-open bundle and often contributes to additional contacts with DNA, typically with the minor groove (Aravind et al. 2005; Harami et al. 2013).

LrpA from *P. furiosus* is a member of the Lrp/AsnC family and represents a typical example of a HTH-harbouring TF. The structure of LrpA was one of the first archaeal TFs to be solved (Leonard et al. 2001). The recognition helix in the HTH motif has predominately positively charged residues involved in binding the negatively charged phosphate backbone of DNA. A model of the LrpA-DNA complex shows LrpA binding to DNA as a dimer with the recognition helix making contacts with adjacent turns of the major groove of DNA (Brinkman et al. 2000). FL11, another HTH-containing Lrp-like TF from *Pyrococcus* OT3, was the first archaeal TF solved in complex with DNA (Yokoyama et al. 2007). This co-crystal structure demonstrated that not only the recognition helix (α3) but also the preceding α-helix (α2) establishes sequence-specific interactions with the DNA. More specifically, residues Ala34-Thr37, which form a loop between α-helices 2 and 3, make the majority of contacts in the major groove with five consecutive bps. Additionally, residues on α-helices 2 and 3 themselves contribute to hydrophobic interactions with the major groove. The crystal structure of the DNA-protein complex also indicates that the DNA undergoes protein-induced bending (Yokoyama et al. 2007).

Although wHTH DNA-binding motifs are often thought to have the recognition helix interacting with the major groove DNA and the wing interacting with the minor groove, not all wHTH archaeal TFs exhibit the canonical wHTH-DNA interactions. Upon solving the co-crystal structure of the MarR-like ST1710 from *S. tokodaii* complexed with DNA, a unique and atypical interaction mode became apparent in which the loop formed between the two β-strands is a major determinant of the interaction with the DNA rather than the recognition helix (Kumarevel et al. 2009). Furthermore, comparative analysis of the structures of the apo-form and the ST1710-DNA complex revealed that significant conformational changes occur upon DNA binding.

2.4.1.2 Ribbon-Helix-Helix Motif

A minor fraction of archaeal TFs are characterized by the alternative RHH DNA-binding motif, also named MetJ/Arc domain (Aravind and Koonin 1999;

Minezaki et al. 2005; Allen et al. 2009; Pérez-Rueda and Janga 2010; Chen et al. 2012; Martínez-Núñez et al. 2013). The RHH motif differs from the HTH motif in that the N-terminal α-helix is replaced by a β-strand (Aravind and Koonin 1999; Aravind et al. 2005; Ashby 2006; Pérez-Rueda and Janga 2010; Najnin et al. 2016). RHH DNA-binding proteins are often dimers formed by the N-terminal β-strands, which creates a curved β-sheet (Chothia 1984). This β-sheet makes contacts with the major groove of double stranded DNA (Gomis-Rüth et al. 1998). The curved nature of the β-sheet and the major groove allows for more contacts than the contacts established by the recognition helix in the HTH structural motif. The helices in the RHH motif are often involved in dimerization but may also contribute to additional DNA interactions.

RHH TFs have been detected in phylogenetically distinct archaea and are postulated to be so widespread because of their presence in toxin/antitoxin modules, which are subjected to extensive LGT (Aravind and Koonin 1999; Allen et al. 2009; Iyer and Aravind 2012; Li et al. 2014). *P. horikoshii* NikR is an example of an RHH TF and functions as a homotetramer that binds DNA in a nickel-dependent manner. NikR behaves as a canonical RHH motif with the positively charged β-sheet protruding into the DNA major grooves (Chivers and Tahirov 2005). The two dimeric DNA-binding domains interact with the DNA major grooves separated by two helical turns. Within the DBD there is a highly conserved arginine residue and two non-conserved serine residues, which are important for NikR DNA recognition. Although arginine is highly conserved among NikRs in other species, variations in DNA recognition sequences may be due to additional residues that interact with DNA.

2.4.1.3 Zn-Ribbon Motif

The Zn-ribbon motif is characterized by a Zn^{2+} ion coordinated by two pairs of cysteines, each associated with a pair of short β-strands (Aravind and Koonin 1999; Ashby 2006; Wang et al. 2007; Wu et al. 2008; Najnin et al. 2016). For example, PF0610 from *P. furiosus* was shown to exhibit a novel wHTH variant with a Zn-ribbon motif (Wang et al. 2007). PF0610 is unique with two CXXC motifs between the two β-strands that make part of the wing of the wHTH motif. The recognition helix and a portion of the wing/Zn-ribbon motif possesses a large number of basic residues indicating that both may be involved in the DNA-protein interaction.

2.4.1.4 Leucine Zipper Motif

As an exception to the observation that archaeal TFs are bacterial-like, rare examples of eukaryote-like archaeal TFs have been found as well. They harbour a leucine zipper DNA-binding motif, which is a dimeric parallel coiled-coil formed by the dimerization of amphipathic α-helices. The B-ZIP (basic-region leucine zipper) class of eukaryotic TFs consists of a leucine zipper and a conserved DNA-binding basic region. The transcription activator GvpE of *Halobacterium*, involved in the regulation of gas vesicle production, is an example of an archaeal TF with a motif resembling the basic leucine zipper motif (Krüger et al. 1998). Molecular modelling showed a leucine-rich C-terminal region, indicative of a dimerization function, and a basic region rich in arginine and lysine residues suggesting a DNA-binding function.

2.4.2 DNA-Binding Specificity

The DNA-binding sequence specificity can vary among archaeal TFs. Classical TFs have a high specificity, whereas DNA-binding proteins that function in both gene regulation and chromatin organization bind the DNA in a non-sequence specific manner (see Sect. 2.5.2). Due to contacts between amino acids and the base-specific hydrogen bond donor and acceptor elements, sequence specificity is usually determined by interactions in the major groove of the DNA (Fig. 2.2). The difference between A:T and T:A as well as G:C and C:G bps in the minor groove is

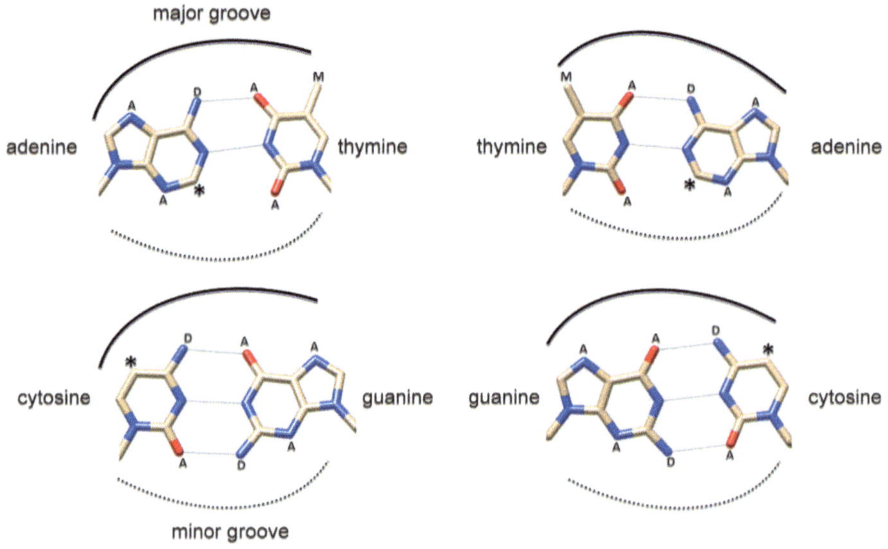

Fig. 2.2 Base pair recognition in the major and minor groove determines DNA-binding specificity. The hydrogen bond donor and acceptor (indicated with D and A, respectively) patterns enable proteins to discriminate between base pairs (A:T or T:A and G:C or C:G) in the major groove but not the minor groove. An *asterisk* indicates a base carbon hydrogen and M a methyl group. Adapted from Harteis and Schneider (2014)

indistinguishable in comparison to the major groove, which shows a distinct hydrogen bond donor and acceptor pattern that allows for a sequence-specific DNA-protein complex (Seeman et al. 1976). Hydrophobic, electrostatic and van der Waals interactions occur as well, but do not contribute as much to the specificity of DNA recognition. In addition to protein recognition of a specific DNA sequence through hydrogen bonds, sometimes mediated by water and hydrophobic interactions, the DNA structure may also contribute to protein-DNA complex formation (Rohs et al. 2010). For example, protein-DNA complexes may require DNA that is intrinsically bent rather than straight.

As most archaeal TFs have a dimeric nature with a two-fold symmetry axis, sequence-specific recognition of the DNA is typically represented by an (partial) inverted repeat in the DNA sequence (Fig. 2.3a). Recognition motifs are represented by a consensus sequence with palindromic half-sites and a non-informative but usually AT-rich center (Fig. 2.3b). The size of such a recognition motif can range from 8 bp, as observed for the ARA box motif of the *Sulfolobus* arabinose-responsive regulator (Brouns et al. 2006; Peng et al. 2009) up to 24 bp as is the case for the heat-shock regulator Phr in *P. furiosus* (Keese et al. 2010). However, most motifs have a typical size of 13–17 bps reflecting recognition in two adjacent major groove segments and the intervening minor groove segment. Indeed, base interactions in two adjacent major groove segments with two identical but symmetrically positioned wHTH recognition helices or RHH β-strands impose sequence specificity while the presence of weak bps in between specific interaction sites facilitates protein-induced DNA deformations (Fig. 2.3a). As exemplified by the FL11:DNA co-crystal structure, protein-induced DNA bending is quite common for prokaryotic TFs. The number of palindromic residues in the half sites can vary from as little as 3 bp to up to 6 bp [e.g., GTT-N_3-AAC recognition motif of SurR (Lipscomb et al. 2009) and TATCAC-N_5-GTGATA recognition motif of TrmBL1 (Reichelt et al. 2016)]. Variations in palindromic residues can be explained by the number of amino acid residues that are involved in establishing base-specific contacts, while the variations in the total size of the binding motif are related to the relative distance between the two DBDs.

Although the identification of the recognition motif of a TF is key to the understanding of its function, not only by enabling *in silico* predictions of putative target genes but also for the characterization of the molecular mechanism of regulation, detailed studies of the DNA-binding specificities of archaeal TFs are rather scarce. Initially, SELEX studies and a detailed study of the effects of saturation mutagenesis of a consensus binding site provided insights into the DNA-binding specificities of archaeal Lrp-type regulators (Yokoyama et al. 2006a, 2009; Peeters et al. 2007). These studies demonstrated that the DNA-binding motifs of Lrp-like regulators in phylogenetically distinct organisms have a similar inverted repeat consensus sequence 5′-abcdewwwedcba-3′ (with w = weak bps). This reflects a similarity in the pairwise chemical interactions established between amino acid residues in the HTH motif and major groove bases. As was observed for the FL11:DNA co-crystal structure (Yokoyama et al. 2007), amino acid residues in the loop between α2 and α3 (the recognition helix) are,

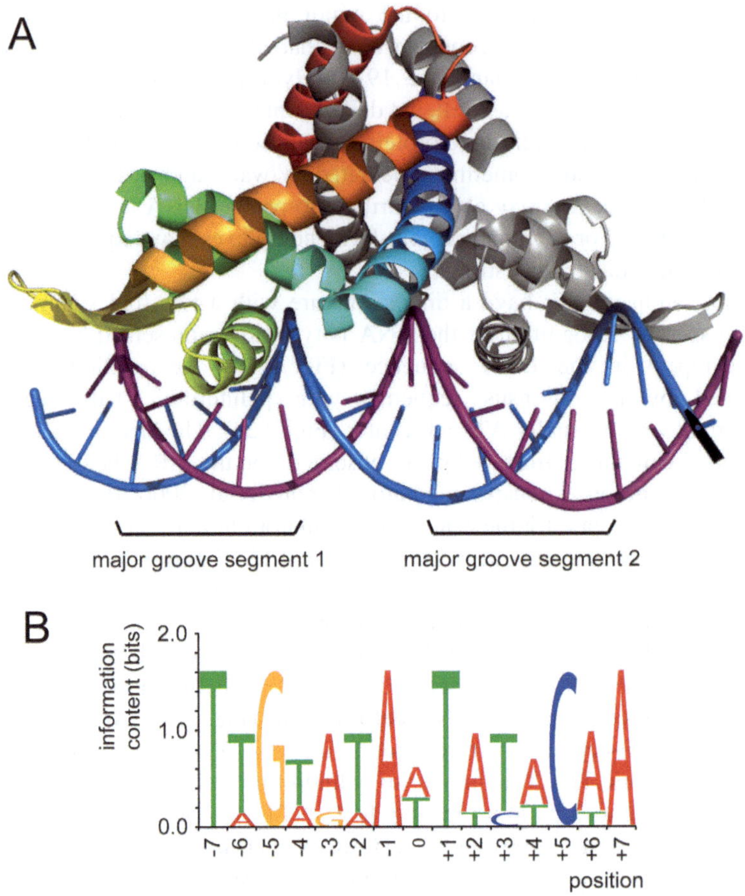

Fig. 2.3 DNA sequence specificity in transcription factor binding. (**a**) Structure of BldR of *S. sulfataricus*, a wHTH regulator belonging to the MarR family (PDB ID3F3X) (Di Fiore et al. 2009) modelled with DNA. (**b**) A sequence logo depicting the consensus sequence of a typical DNA-binding motif. This sequence logo represents the binding specificity of the Lrp-type BarR in *S. acidocaldarius* (Liu et al. 2014). Note the palindromic nature of the binding motif and the exclusive presence of weak base pairs in the center, where the protein faces the minor groove side of the DNA

together with a highly conserved arginine residue in the recognition helix, crucial determinants of the sequence specificity (Yokoyama et al. 2009). More recently, genome-wide ChIP studies enabled to systematically determine DNA-binding motifs of archaeal TFs by predicting a conserved motif in sets of TF-bound sequences (Schmid et al. 2009; Nguyen Duc et al. 2013; Song et al. 2013; Rudrappa et al. 2015; Liu et al. 2016; Reichelt et al. 2016).

2.4.3 *Operator Architecture*

Instead of interacting with a single binding site in the neighborhood of a target gene promoter, most TFs bind cooperatively to an array of multiple binding sites (e.g., Bell et al. 1999a; Peeters et al. 2004). A cooperative binding mode results in a gene regulatory response that is characterized by non-linear dynamics and a higher sensitivity. Typically, multiple binding sites are regularly spaced with center-to-center distances of two to three helical turns enabling binding of multiple interacting TF molecules on the same side of the DNA helix, although these sites are oftentimes degenerated. In these cases, binding is nucleated on a well-conserved binding site followed by cooperative binding extending in a single direction on less conserved sites (Karr et al. 2008; Peeters et al. 2009; Peixeiro et al. 2012).

Archaeal TFs have also been shown to interact with "auxiliary" binding sites in addition to the operator binding sites that evoke a regulatory response. These auxiliary sites are located at a distance from the primary operator binding site (s) and assist in an efficient binding of the operator without contributing to transcription regulation themselves. Auxiliary sites are located at a distance upstream of the main operator, as is the case for a TrmB-like TF in *S. islandicus* (Peng et al. 2009) or are located downstream of the main operator in the coding sequence of the gene under regulation. The latter has been observed for a variety of Lrp-type TFs (Ouhammouch et al. 2005; Nguyen Duc et al. 2013; Song et al. 2013; Liu et al. 2016).

2.5 Interplay with DNA Topology and Chromatin Structure

2.5.1 *Effect of Local DNA Topology on Transcription Factors and Vice Versa*

In addition to the DNA sequence, the topology of the DNA template could also be important determinants of TF binding and function. In bacteria, chromosome topology is known to be a global regulator communicating environmental and cellular cues, such as temperature, nutritional state, oxidative stress and energy levels to transcription regulatory networks (Hatfield and Benham 2002). Initial indications suggest that similar mechanisms exist in hyperthermophilic archaea, of which the global genome topology has a relaxed to positive level of superhelicity (Nadal et al. 1986; Forterre et al. 1996). As positive supercoiling stabilizes the DNA helix, in contrast to negative supercoiling, it is believed that the chromosome topology of hyperthermophiles could compensate for the denaturing effect of the high growth temperature (López-García and Forterre 1999). Furthermore, temperature has been shown to influence DNA topology in the thermophilic crenarchaeal model organism *Sulfolobus*, which in turn affects basal transcription efficiency in a temperature-dependent manner. At a temperature suboptimal for growth, negatively supercoiled templates are transcribed at higher efficiencies than positively supercoiled templates, as demonstrated with in an in vitro transcription system (Bell et al. 1998).

Vice versa, TFs sometimes change the topology of DNA upon binding, a feature that could also be related to the regulatory mechanism. For example, the tryptophan-responsive TF TrpY in *M. thermautotrophicus* introduces negative supercoiling upon assembling a large nucleoprotein complex, involving multiple TrpY molecules, with the control region of one of its target genes *trpB2* (Karr et al. 2008). It is proposed that TrpY repression is achieved by displacement of TBP and TFB. As another example, the specific TF Ss-LrpB, belonging to the Lrp/AsnC family, has been shown to wrap DNA upon binding three regularly spaced binding sites in the promoter region of its own gene (Peeters et al. 2006) and is assumed to introduce changes in the local DNA topology of the promoter region, thereby regulating transcription (Peeters et al. 2013a). Similarly, "wrapped" nucleoprotein complexes have been postulated to be formed for the archaeal Lrp-type proteins FL11 in *Pyrococcus* OT3 and LrpA in *P. furiosus* (Leonard et al. 2001; Koike et al. 2004).

2.5.2 Distinction Between Transcription Factors and Chromatin Proteins

Archaeal chromatin has a heterogeneous nature and is organized by different nucleoid-associated proteins that vary between phyla and even genera and species (e.g., histones in *Euryarchaeota*, Alba and small basic nucleoid-associated proteins in *Crenarchaeota*) (Peeters et al. 2015). In certain cases, proteins that belong to classical TF families such as Lrp/AsnC or TrmB are also involved in chromatin organization, either as their sole function or as an additional function. As with bacteria, it is sometimes difficult to know where to draw a line separating "true" TFs from "true" chromatin proteins (Fig. 2.4).

The classical TF paradigm is challenged by the observation in genome-wide ChIP studies that archaeal TFs bind to more target sites than anticipated based on their physiological role and/or regulatory target genes (see also Sect. 2.4). These additional genomic sites are located at a significant distance from transcription start sites, and are often intragenic (Schmid et al. 2009; Nguyen Duc et al. 2013; Song et al. 2013; Liu et al. 2016). For the Lrp-like TF Ss-LrpB in *S. solfataricus* it has been observed that binding to sites with a regulatory function is characterized by a higher binding affinity and cooperativity in contrast to the binding to these non-regulatory "spurious" sites (Nguyen Duc et al. 2013). Possibly, these binding events serve to sequester TF molecules as a means to improve thermodynamic regulation of the concentration of freely available regulatory molecules (Macquarrie et al. 2011). On the other hand, non-regulatory binding events might

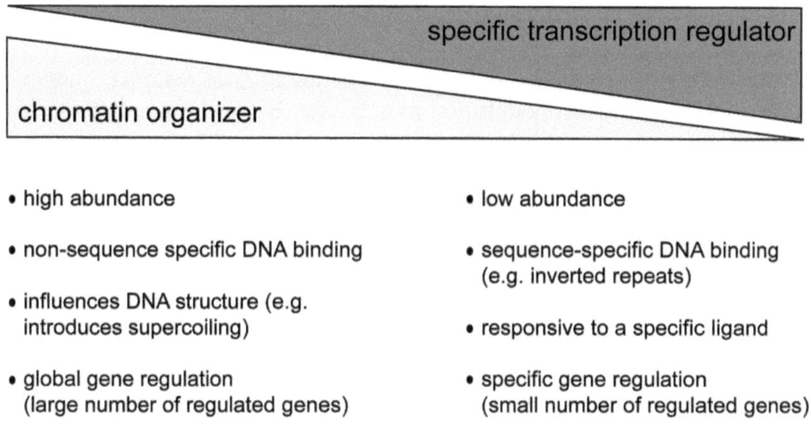

Fig. 2.4 Conceptual scheme of the continuous spectrum between a chromatin-organizer and a transcription regulator function

contribute to establishing and stabilizing the complicated three-dimensional structure of archaeal chromatin.

For bacteria, it has been postulated that early in evolution only chromatin proteins existed that were solely dedicated to performing genome compaction by bending, distorting and twisting the DNA existed (Visweswariah and Busby 2015). Indeed, fitting the entire genomic DNA into the confined space of a prokaryotic cell is essential for survival. Later in evolution, some of these chromatin proteins also took on a transcription regulatory function upon binding in the neighborhood of TSS, thereby conferring a fitness advantage to the organism. The existing TFs have thus evolved from non-specifically binding chromatin proteins and while some of these have become dedicated TFs, others still combine both functions (Fig. 2.4) (Visweswariah and Busby 2015). Given their bacterial nature (see Sect. 2.2.1), we are convinced that this evolutionary hypothesis also holds for archaeal TFs. This is nicely demonstrated by studies performed for Alba, an ancient and universally present DNA-binding protein in archaea. Whereas Alba is a crucial chromatin protein in the *Crenarchaeota* (Peeters et al. 2015), it has evolved into a sequence-specific TF of autotrophic growth in the Euryarchaeote *Methanococcus maripaludis* in which the chromatin is mainly organized by histone proteins (Heinicke et al. 2004; Liu et al. 2009). Some TF families harbour members that are on either end of the spectrum (e.g., the Lrp family (Schwaiger et al. 2010) or the TrmB family, see Sect. 2.5.2.1), but other families appear to be specialized in promoter-specific transcription regulation or in combining a chromatin-organizing function with a global gene regulatory role (e.g., the Lrs14 family, see Sect. 2.5.2.2).

2.5.2.1 TrmB-like Proteins as Chromatin Organizers

The TrmB family of TFs has been best studied in the *Euryarchaeota* (Gindner et al. 2014; Kim et al. 2016). Although prototypical TrmB family members are sugar-responsive specific transcriptional regulators (Lee et al. 2003; Gindner et al. 2014) (see Sect. 2.2.3.2), the related TrmBL2 was also classified as a member of the TrmB family while it appears to have a function in chromatin organization (Maruyama et al. 2011). In accordance with this more global chromatin-organizing function and in contrast to specific TrmB-like regulators, TrmBL2 is highly conserved in the euryarchaeal order *Thermococcales* and is characterized by a high and constitutive expression level (Lee et al. 2007a; Maruyama et al. 2011; Efremov et al. 2015).

Upon comparing TrmB and TrmBL2 from *P. furiosus*, the latter protein is somewhat smaller because it lacks the sugar-binding subdomain. Otherwise, both protein structures are characterized by the same succession of domains consisting of an N-terminal extended wHTH domain, a long amphipathic α-helix responsible for dimerization and a similar C-terminal domain exhibiting a mixed α/β nature (Krug et al. 2013; Ahmad et al. 2015). This structure justifies the classification of TrmBL2 as TrmB-like. In contrast to the homodimeric TrmB structure, TrmBL2 forms tetramers in which the C-terminal domains mediate interactions between two dimers (Ahmad et al. 2015).

TrmBL2 interacts with both dsDNA and single-stranded (ss) DNA in a salt- and topology-dependent manner (Efremov et al. 2015; Wierer et al. 2016). These protein-DNA interactions are established in a non-sequence specific manner (Maruyama et al. 2011), an observation that is supported by the TrmBL2:DNA co-crystal structure, which reveals interactions of the wHTH fold with the deoxyribose phosphate backbone but not with the bases (Ahmad et al. 2015). TrmBL2 binding results in the formation of thick filamentous nucleoprotein structures which are assumed to play a role in chromatin organization by antagonizing the packaging of DNA and competing with histones for binding (Maruyama et al. 2011; Efremov et al. 2015). TrmBL2 also acts as a global TF by repressing about 6.5% of all genes in the genome of *P. furiosus* (Maruyama et al. 2011).

2.5.2.2 Lrs14-like Proteins as Chromatin Organizers

Although originally incorrectly annotated as belonging to the Lrp/AsnC family (Napoli et al. 1999), Lrs14 DNA-binding proteins are characterized by a completely different structural topology thus necessitating their classification as a separate family (Orell et al. 2013). Lrs14-like proteins are small (about 12 kDa) with a central wHTH motif flanked on either side by an α-helix of which the C-terminal α-helix mediates dimerization (Shinkai et al. 2007). Bacterial homologs of the Lrs14 family have not yet been identified; it thus appears that Lrs14 is one of the few wHTH TF families that originated in the archaeal domain of life and has not been spread to the bacterial domain via LTG.

Lrs14-like proteins have been studied in *Sulfolobus* spp. These studies have provided strong indications that Lrs14-like proteins, which were first considered to be classical TFs (Napoli et al. 1999; Bell and Jackson 2000; Fiorentino et al. 2003), appear to act also as chromatin-organizing proteins. These indications include: (1) the observation that the structure of the Lrs14-like protein Sto12a is similar to that of the *Sulfolobus* nucleoid-associated protein Sso10a (Chen et al. 2004); (2) the non-sequence specific interaction of Lrs14-like proteins with a variety of DNA templates, resulting in a typical ladder-like pattern of multiple complexes with different electrophoretic mobilities in EMSA (Napoli et al. 1999; 2001; Orell et al. 2013) and the lack of identifying a putative recognition motif. Furthermore, Lrs14-like proteins interact with relatively long stretches of DNA (Napoli et al. 1999; Abella et al. 2007); (3) the observation of a putative DNA-structuring role for the Lrs14-like protein Smj12 in *S. solfataricus*, as this protein induces positive DNA supercoiling and protects the DNA from thermodenaturation (Napoli et al. 2001); (4) the observation that different Lrs14-like proteins have been retrieved together with (other) chromatin proteins Alba and Sso7d in pulldown assays with a variety of bait DNA fragments, indicating a high intracellular concentration and non-sequence specific DNA interaction (Napoli et al. 2001; Fiorentino et al. 2003; Kessler et al. 2006; Abella et al. 2007).

Aside from their putative role in chromatin organization and despite their small size, Lrs14-like proteins are also implicated in gene regulation in response to environmental changes (Kessler et al. 2006; Abella et al. 2007; Orell et al. 2013). In contrast to specific TFs, this gene regulatory function is highly pleiotropic and related to complex physiological adaptations. An example is the switch between planktonic and biofilm growth. Three Lrs14-like regulators have been demonstrated to play a role in regulation of biofilm morphology and cellular motility by genetic analyses (Orell et al. 2013). Also in response to heat shock and upon the transition from the exponential to stationary growth phase (two other complex physiological changes), *lrs14*-like genes have been shown to be upregulated (Napoli et al. 1999; Tachdjian and Kelly 2006; Orell et al. 2013). The combination of a nucleoid structuring role with a global gene regulatory function as seen for Lrs14 and for TrmBL2 is reminiscent of bacterial nucleoid-associated proteins such as H-NS in Gram-negative bacteria, which is also involved in the regulation of physiological processes such as motility, biofilm formation and stress response (Ayala et al. 2015; Kim and Blair 2015).

2.6 Regulatory Mechanisms

Archaeal transcription regulatory mechanisms, like bacterial transcription regulatory mechanisms, are quite simple in the sense that TFs exert their effects by interacting with DNA sites bordering or overlapping the promoter elements upstream of genes or regulons. A major research focus is to unravel how these bacterial-like TFs bind DNA and affect or interact with the multiple components of the eukaryotic-like basal transcription machinery.

The formation of the pre-initiation complex (PIC) is initiated when TBP binds to the AT-rich TATA box portion of the promoter (Rowlands et al. 1994). This is followed by a cascade of interactions: the TBP-DNA complex recruits TFB, which interacts upstream of the TATA box with a purine-rich BRE. This interaction designates the direction of transcription (Hausner et al. 1996; Bell et al. 1999a). Finally, RNAP is then recruited by the TBP-TFB-DNA complex. TFs affect the formation of the PIC at different steps, thereby causing negative or positive regulation (Geiduschek and Ouhammouch 2005) (Fig. 2.5).

Negative regulation is often realized by TFs that bind DNA at the promoter site thereby occluding TBP, TFB, or RNAP recruitment (Fig. 2.5a, b). Negative regulators can bind near or on the BRE-TATA region of a promoter (Fig. 2.5a). For example, *P. furiosus* TrmB regulates the expression of the TM and the MD

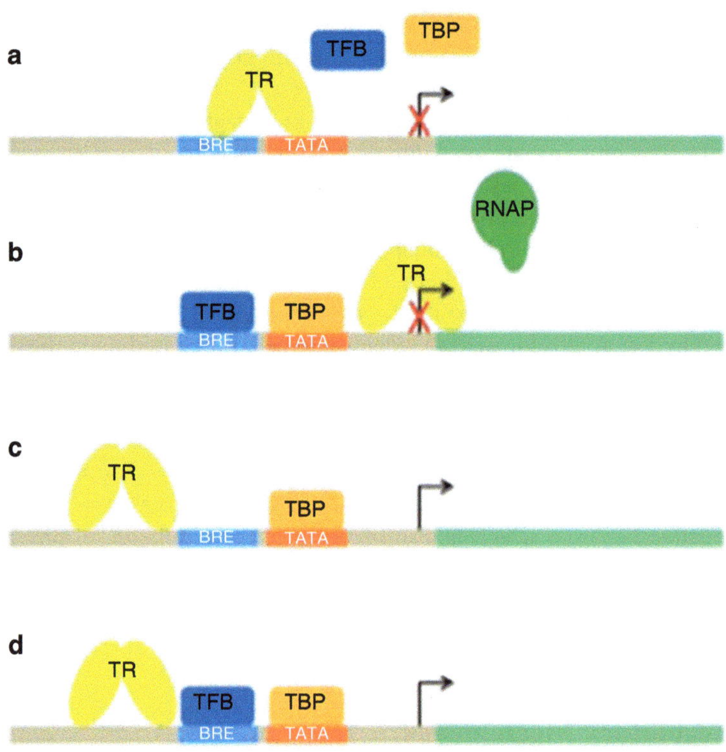

Fig. 2.5 Cartoon schematic of mechanisms of negative and positive transcription regulation exerted by archaeal transcription factors. *TR* transcription regulator. (**a**) Negative transcription regulation by inhibition of the binding of the basal transcription factors TBP and/or TFB. (**b**) Negative transcription regulation by inhibition of the recruitment of RNAP. (**c**) Positive transcription regulation by stimulation of the binding of TBP. (**d**) Positive transcription regulation by stimulation of the binding of TFB

(maltodextrin) operon by recognizing a specific semi-palindromic site within the TM promoter, which overlaps the BRE and TATA box elements (Lee et al. 2005). This suggests that the TF-DNA complex inhibits TBP-TFB-DNA formation. Interestingly, the mechanism of regulation is different for the MD promoter. Only a portion of the recognition site is located in the MD promoter. TrmB binds downstream of the BRE-TATA box, overlapping the transcriptional start site of the MD promoter, in a sugar-dependent manner (Krug et al. 2013). Other examples of transcription repressors inhibiting the recruitment of RNAP are *A. fulgidus* MDR1 and *P. furiosus* Phr. MDR1 is a metal-dependent TF and a homologue of the bacterial DxtR family. MDR1 was shown to bind multiple sites of its own promoter, overlapping its own start site. It was found to bind in a cooperative manner blocking RNAP recruitment but not the formation of the TBP-TFB-DNA complex (Bell et al. 1999a). Phr is a putative heat shock regulator, which also represses transcription by abrogating RNAP recruitment without affecting the formation of the TBP-TFB-DNA complex. Phr regulates its own expression along with two other proteins Hsp20 and AAA^{+} ATPase. Similarly to MDR1, Phr recognizes multiple sites at each promoter, which overlap the TSS (Vierke et al. 2003).

Positive regulation can occur via recruitment of TBP or TFB by the TF or when the TF-DNA complex changes the DNA conformation in order to increase the archaeal promoter binding strength. Transcription activators often bind upstream of the BRE-TATA region and facilitate TBP binding (Fig. 2.5c). An example of this is Ptr2 from *M. jannaschii*. Ptr2 is a member of the Lrp/AsnC family that recognizes two binding sites that are upstream of both *fdxA* (ferredoxin-encoding gene) and *rb2* (rubredoxin 2-encoding gene) (Ouhammouch and Geiduschek 2001). Ptr2 aids in TBP-DNA complex formation, especially for the *rb2* promoter, to which Ptr2 recruits TBP to a relatively weak TATA box (Ouhammouch et al. 2003). *S. solfataricus* BldR is a MarR family homologue and transcriptional activator, regulating expression of its own gene and *Sso2536* (alcohol dehydrogenase). The recognition sequence for BldR is located upstream of the TATA box for both the *bldR* and *sso2536* promoters. Benzaldehyde acts as a positive effector ligand promoting DNA binding (Fiorentino et al. 2007). BldR binds upstream of the TATA box, suggesting that activation occurs via recruitment of the TBP, which is mechanistically similar to transcription activation exhibited by Ptr2.

P. furiosus PF1088 (TFB-RF1) is one of the first characterized archaeal transcription activators that recruits TFB to the promoter (Fig. 2.5d). PF1088 binds directly upstream of the BRE-TATA region of the *pf1089* promoter. The BRE of the *pf1089* promoter largely deviates from the BRE consensus sequence in *Pyrococcus* making it a weak BRE (van de Werken et al. 2006). PF1088 facilitates TFB binding to the weak BRE, which forms a stable PF1088-TFB-TBP-DNA complex (Ochs et al. 2012). Within *Sulfolobus* the mechanism of arabinose-responsive activation may be similar to PF1088, with the *ara*-box-binding factor stimulating TFB binding to the weak BRE (Peng et al. 2011).

Some TFs are able to act as a repressor *and* activator, such as *P. furiosus* SurR and TrmBL1 for example. SurR activates the expression of its own gene in addition to two hydrogenase operons (*mbh1 and hydB1*) by binding either immediately upstream of the BRE-TATA region or partially overlapping the BRE. Activation may occur through recruitment of TFB or TBP to facilitate PIC formation. The repression mechanisms by SurR were addressed previously, whereby SurR blocks TBP-TFB complex formation or recruitment of RNAP (Lipscomb et al. 2009). The DNA-binding activity of SurR is dependent on redox conditions due to presence of elemental sulfur (Yang et al. 2010). Regulation from TrmBL1 is dependent on the location of TrmBL1 binding upstream or downstream of the promoter elements (Lee et al. 2008). Negative regulation occurs when TrmBL1 binds downstream of the TATA-box preventing RNAP recruitment whereas positive regulation occurs when TrmBL1 binds upstream of promoter elements (Reichelt et al. 2016). Ss-LrpB, an Lrp-type TF from *S. solfataricus* (Peeters et al. 2004), is hypothesized to be capable of acting as a repressor or activator on the same promoter depending on the protein concentration and binding site occupancy. Depending on the concentration of Ss-LrpB, TF-DNA interactions can cause DNA wrapping, thus changing DNA conformation and forming nucleoprotein complexes. This may determine whether negative or positive autoregulation occurs (Peeters et al. 2006, 2013a).

2.7 Sensing and Signal Transduction Mechanisms

Given that the largest fraction of archaeal TFs are OCSs, these proteins combine their DNA-binding and transcription regulation function with a sensing function. As a consequence, archaeal TF-mediated transcription regulation mostly occurs in response to intracellular conditions such as metabolite concentrations and redox and energy status.

2.7.1 Interaction with Small-Molecule Ligands

Two-domain OCSs typically have a C-terminal domain in which ligand-binding pockets are formed that specifically bind small molecules, called ligands or effectors (see also Sect. 2.2). Although for many archaeal TFs the ligand identity remains enigmatic, small-molecule ligands identified thus far can be classified in four groups:

1. Amino acids. Most of the archaeal members of the widespread Lrp/AsnC family bind α-amino acids (Okamura et al. 2007; Schwaiger et al. 2010; Song et al. 2013; Vassart et al. 2013) or the β-amino acid β-alanine in the case of BarR from *S. acidocaldarius*, which is a precursor of coenzyme A (Liu et al. 2014). Amino-acid responsive transcription regulation of transport and biosynthesis of these important metabolites enables the maintenance of amino acid homeostasis in the cell.

2. Sugars. Canonical members of the TrmB family are characterized by a sugar-binding pocket in the ligand-binding domain to which different types of sugars can bind (Gindner et al. 2014). In chemoorganotrophic archaea, this regulatory response enables the optimization of sugar transport, catabolism and biosynthesis similarly as catabolite repression in bacteria.
3. Aromatic compounds. Archaeal MarR regulators have been shown to interact with aromatic compounds, such as ethidium and the phenolics benzaldehyde and salicylate (Fiorentino et al. 2007; Saridakis et al. 2008; Yu et al. 2009). This ligand response is linked to detoxification processes.
4. Metal ions. Nickel has been shown to be the ligand of the RHH TF NikR in *P. horikoshii* (Chivers and Tahirov 2005). Although direct ligand-TF interactions have not yet been demonstrated, ArsR-like TFs in *S. solfataricus* and *F. acidarmus* have been postulated to bind and respond to specific metal ligands (mercuric ion and arsenite, respectively) (Schelert et al. 2006; Baker-Austin et al. 2007).

Whereas some TFs bind a single ligand, other TFs have a broader ligand-binding specifity. This has been well-studied for amino-acid interacting Lrp-like TFs (Peeters and Charlier 2010). Differences in specificity are contrasted by LrpA1 in *H. salinarum*, which only binds aspartate (Schwaiger et al. 2010), and FL11 in *Pyrococcus* OT3 and Sa-Lrp in *S. acidocaldarius*, which bind multiple amino acids (Yokoyama et al. 2007; Song et al. 2013). A detailed ligand-TF contact analysis of a variety of co-crystal structures enabled the deduction of a "structural code" for the archaeal Lrp family, i.e., a code that correlates ligand-binding amino acid residues to specific ligands (Okamura et al. 2007). This code has been leveraged and validated for the prediction of ligands for novel Lrp-like TFs (Peeters and Charlier 2010; Plaisier et al. 2014).

Different ligand response patterns are recognized, in which DNA binding and thus regulation is either inhibited or stimulated upon ligand interaction. For example, DNA binding of the crenarchaeal MarR-type TFs BldR2 and ST1710 is abrogated upon binding of salicylate and similar ligands causing a derepression (Yu et al. 2009; Fiorentino et al. 2011), while in contrast benzaldehyde stimulates DNA binding of the related BldR in *S. solfataricus* (Fiorentino et al. 2007). For certain TFs, ligand binding does not result in observable changes in the DNA-binding behavior but instead causes subtle conformational changes in the TF-DNA complex leading to regulatory effects on transcription initiation. As an example, the β-alanine responsive TF BarR in *S. acidocaldarius* performs a β-alanine-dependent activation of its target gene while this small-molecule ligand does not affect the formation of BarR-DNA complexes in any way (Liu et al. 2014, 2016). Similarly, the genome-wide binding profile of LysM is similar upon growing the cells in the presence or absence of the major effector lysine (Song et al. 2013).

Ligand binding affects the DNA-binding properties and/or transcription regulatory function typically through inducing allosteric conformational changes in the protein structure. For the MarR-like TF ST1710 and the Lrp-like Grp, both from *S. tokodaii*, the overall structures of the apo- and holo-forms are very similar and only minor conformational changes are observed (Kumarevel et al. 2008a, 2009). In the case of ST1710, the ligand salicylate binds in between the DBD and dimerization helix, thereby causing small conformational changes in the wHTH DNA-binding motifs that result in an inhibition of DNA binding (Kumarevel et al. 2009). In contrast, ligand binding could induce larger conformational changes (Chivers and Tahirov 2005; Yamada et al. 2009). In these cases, a closed octameric conformation could adopt an open conformation upon binding the ligand or vice versa. Finally, ligand binding can cause more dramatic effects, as shown for a variety of Lrp-type TFs. Here, amino acid binding induces changes in the oligomeric state of the protein, often octameric association from dimers (Yokoyama et al. 2006b, 2007; Okamura et al. 2007).

For the TF TrmB in *Thermococcales*, a complex ligand response has been unraveled (Lee et al. 2003, 2005, 2007b; Krug et al. 2013). TrmB regulates the TM transport system in response to trehalose and maltose and the MD transport system in response to maltotriose and sucrose. In these cases, TrmB exerts a transcriptional repression while the specific sugar ligands act as inducers. However, while maltose is an inducer for TM regulation, it acts in contrast as a co-repressor for MD regulation. Furthermore, glucose acts as a co-repressor for both targets. A maltose-bound co-crystal structure of TrmB (Krug et al. 2006) leads to the following explanation of this complex differential ligand response: while the α-glucosyl moiety shared by all sugar molecules interacts with six amino acid residues in the ligand-binding pocket, the variable moiety of the sugars interact with the so-called "sugar binding helix" that exerts differential allosteric effects on DNA binding. Given the different nature of the TM and MD operator sites (a pseudo-palindromic binding site *versus* a non-palindromic binding site), this differential response also varies between the different targets (Lee et al. 2007b; Krug et al. 2013).

2.7.2 Redox-Sensing Transcription Factors

Like members of the bacterial domain, both aerobic and anaerobic archaea must respond and adapt to changing redox conditions in their environment as well as combat the impacts of reactive oxygen species on macromolecules. By the early 2000s there was a plethora of information on redox-sensitive TFs in the bacterial domain (Zheng and Storz 2000; Paget and Buttner 2003). Thiol-based regulatory switches dominate amongst bacteria where oxidized cysteines form disulfide bonds in a reversible manner (Wouters et al. 2010; Hillion and Antelmann 2015). The primary difference is that some thiol-based regulators incorporate a metal within the switch. In such a case, the cysteines in the reduced form of the protein are coordinating a metal and, upon oxidation, the metal is released and a disulfide is formed (Paget and Buttner 2003). Despite the abundance of information on redox-sensing regulation in bacteria, there is a paucity of information available in the archaeal domain of life. To date, a thiol-based regulatory switch has been proposed for the archaeal TFs MsvR and SurR (Yang et al. 2010; Karr 2010;

Isom et al. 2013; Sheehan et al. 2015). Additionally, a heme-based HK, MsmS, that autophosphorylates in a redox-dependent manner was recently identified (Molitor et al. 2013). Lastly, RosR is a haloarchaeal TF that controls expression of the oxidative stress response in *H. salinarum* although the mechanism or whether it senses redox directly has not been demonstrated (Sharma et al. 2012; Tonner et al. 2015).

2.7.2.1 SurR

The sulfur response regulator, SurR, is a dual regulator of sulfur and hydrogen metabolism in *P. furiosus*. SurR is a TrmB family regulator that contains both an N- and C-terminal wHTH DBD. These DBDs are separated by a third domain that appears to be involved in dimerization. In the presence of colloidal S^0, SurR represses expression of key genes for hydrogen metabolism while activating expression of key genes in sulfur metabolism based on in vitro transcription assays (Lipscomb et al. 2009; Yang et al. 2010). When the 3-D structure of SurR was determined by X-ray crystallography, a disulfide bond was discovered in the DBD between two cysteine residues in a CXXC motif. Yang and colleagues went on to show that the disulfide between those cysteines impacted DNA binding in a redox-dependent fashion (Yang et al. 2010). This would mark the first of two descriptions of redox-sensitive TFs in the archaeal domain that year (Yang et al. 2010; Karr 2010).

2.7.2.2 MsvR

The second description of a redox-sensitive TF in the archaeal domain was that of the methanogen specific V4R domain containing regulator, MsvR, in *M. thermautotrophicus* (Karr 2010). Full-length homologs of MsvR are confined to methanogenic archaea (Karr 2010; Isom et al. 2013). *M. thermautotrophicus* MsvR (MthMsvR) was shown to regulate its own transcription as well as the divergently transcribed operon encoding an $F_{420}H_2$ oxidase, rubredoxin and rubrerythrin that are postulated to play a role in the oxidative stress response in select methanogens (Seedorf et al. 2004; 2007; Kato et al. 2008; Karr 2010). MthMsvR was shown to tightly regulate its own expression under reducing conditions as well as repress the adjacent operon in an in vitro transcription system. Limitations of the *M. thermautotrophicus* in vitro transcription system prevented the determination of the role of oxidized MthMsvR in regulation of these promoters. Nevertheless, MthMsvR does bind the overlapping promoter regions under non-reducing and reducing conditions based on EMSA assays. However, the DNA shifting pattern and protected regions based on DNaseI footprinting differ between the two conditions. It was hypothesized that cysteine residues in the V4R domain were responsible for the redox-dependent behavior of MthMsvR (Karr 2010).

Karr and colleagues went on to characterize an MsvR homologue from the metabolically versatile *M. acetivorans* (MaMsvR) (Isom et al. 2013; Sheehan et al. 2015). MaMsvR only binds DNA under reducing conditions. Alanine substitutions of cysteine residues in the V4R domain implicated these cysteines in the redox-dependent behavior of MaMsvR. However, the MsvR family members amongst the *Methanosarcinales* have several additional cysteine residues compared to MthMsvR (Isom et al. 2013). Follow up work demonstrated that a thioredoxin system from *M. acetivorans* could switch MaMsvR from the non-DNA binding oxidized state to the reduced DNA-binding site. Free thiol quantitation performed in this study indicated that additional cysteine residues outside the V4R domain may also be involved in the redox-sensing behavior of MaMsvR (Sheehan et al. 2015). However, thus far the biological role of MaMsvR and its role in regulation of other promoters has not been described.

2.7.2.3 MsmS

MsmS is the HK component of an archaeal TCS in *M. acetivorans*. However, its cognate RR has not been identified. MsmS contains a heme cofactor that is involved in sensing changes in redox conditions. In fact, MsmS autophosphorylates in a redox-dependent manner. An *M. acetivorans* Δ*msmS* strain constitutively expressed the methyltransferase, MtsF suggesting that MsmS is likely involved in redox-dependent regulation of select methanogenesis pathways (Molitor et al. 2013). With a paucity of information on TCSs and redox-sensing in archaea, further study of this system is likely to yield valuable insights into the roles of TCSs in archaeal metabolism.

2.7.2.4 RosR

RosR-like proteins are unique to the halophilic branch of the archaeal domain. In *H. salinarum*, RosR is responsible for regulation of the oxidative stress response and functions as both a transcription activator and repressor (Sharma et al. 2012). However, RosR does not contain any cysteine residues and no mechanism for redox-sensing has been postulated (Sharma et al. 2012; Tonner et al. 2015). Despite this, RosR bears mention amongst this category of TFs because it's in vivo role and promoter occupancy in response to oxidant treatment has been extensively characterized (Tonner et al. 2015). Given its intricate role in the redox-dependent response to oxidative stress it is exciting to hypothesize that RosR represents a novel mechanism for redox sensing or that post-translational modifications or regulatory cascades in response to redox changes alter its regulatory behaviour.

While our understanding of transcriptional responses to redox fluctuations in the environment is limited in archaea, the limited knowledge available suggests that redox-sensing may be lineage and metabolism specific in archaea. SurR homologs are unique to the *Thermococcales*, a group of organisms that can reduce elemental sulfur to hydrogen sulfide (Lipscomb et al. 2009; Yang et al. 2010). MsvR is unique to the methanogenic archaea and even within this group its role and behavior appears to differ between the hydrogenotrophic and acetoclastic methanogens (Karr 2010; Isom et al. 2013). And finally, RosR is unique to the metabolically versatile halophilic archaea (Sharma et al. 2012; Tonner et al. 2015). Future work will be instrumental to understanding the metabolically and environmentally diverse approach archaea take to dealing with redox changes and reactive oxygen species.

2.7.3 Temperature Sensing

Since many archaeal species live in habitats that are characterized by extreme temperatures and frequent temperature variations, an efficient gene regulation in response to sub- or supra-optimal temperature shifts is crucial for the survival and fitness. Two heat-shock TFs have been characterized, HSR1 in *A. fulgidus* and Phr in *P. furiosus* although it is unclear how these proteins sense a temperature increase (Vierke et al. 2003; Rohlin et al. 2005). The LtrR/LtrK TCS in *M. burtonii*, a psychrophile living in Antarctic lakes, performs regulation in response to a decrease in temperature below the optimal growth temperature (Najnin et al. 2016). The LtrK HK has inherent thermosensing properties as its cytoplasmic domain, containing the kinase and phosphatase active sites, displays higher enzymatic activities upon a temperature decrease. Besides this, the membrane-bound LtrK might also respond to changes in membrane structure and lipid composition (Najnin et al. 2016).

2.7.4 Post-Translational Phosphorylation

Reversible protein phosphorylation is an important signal transduction mechanism, also in the context of TF-mediated gene regulation. Whereas *Euryarchaeota* harbour TCSs that employ phosphorylation to transduce signals from HKs to the RRs, *Crenarchaeota* lack TCSs. Instead, phosphoproteome studies in the model crenarchaeotes *S. solfataricus* and *S. acidocaldarius* have demonstrated that a high fraction of the proteome is phosphorylated, including a variety of OCS TFs (in the case of *S. acidocaldarius*, 18 TFs were found to be phosphorylated) (Esser et al. 2012, 2016; Reimann et al. 2013). This observation suggests that direct phosphorylation of OCSs by eukaryotic-type kinases is a major signal transduction strategy in *Crenarchaeota* which could compensate for the lack of TCSs (Esser et al. 2012). For most TFs that were found to be phosphorylated in phosphoproteome studies, the effects of phosphorylation on their function are unknown.

The archaellum regulatory network in *S. acidocaldarius* is a nice example of the importance of post-translational regulation in combination with transcription regulation. The archaellum, as the motility structure is named, is under tight regulation in response to a variety of conditions, including starvation (Reimann et al. 2012; Lassak et al. 2013; Haurat et al. 2016). Multiple TFs are involved in regulating archaellum gene expression, including the Zn-finger TF ArnA and a von Willebrand type A domain TF ArnB, which perform a repression, and two paralogous activators ArnR and ArnR1 (Reimann et al. 2012; Lassak et al. 2013). The TF network is under control of the eukaryotic-type kinases ArnC, ArnD and ArnS: while ArnC phosphorylates both ArnA and ArnB, ArnD phosphorylates only ArnB (Reimann et al. 2012).

2.8 Networks of Transcription Factor Regulation

From a system-level perspective, transcription regulation is a complex process in which signal-responsive regulatory effects from multiple TFs are integrated in order to obtain an appropriate spatiotemporal gene expression output. As described above, we are gaining a better understanding of the function of individual TFs in archaea. Nevertheless, we are lacking integrative system-level insights into archaeal GRNs and many questions remain regarding the architecture of these networks and their evolutionary rewiring. The largest research efforts to describe GRNs have been undertaken for the model Euryarchaeon *H. salinarum*. An impressive transcriptomic study yielded a computational model for this archaeon that links more than 70 TFs to specific environmental changes and putative regulons (Bonneau et al. 2007).

Although the currently available information exposes only the tip of the iceberg of archaeal GRN structure, preliminary indications suggest that these GRNs consist of dense overlapping regulons, in which multiple TFs that typically belong to the same family regulate the same target genes or operons (Schmid et al. 2011; Nguyen Duc et al. 2013; Plaisier et al. 2014; Liu et al. 2016). For example, two iron-responsive TFs in *H. salinarum* cooperate to maintain iron homeostasis by binding to the same genomic loci in each others neighborhood (Schmid et al. 2011). Furthermore, a systematic analysis of eight Lrp-like TFs in this same organism revealed that the control regions of nearly half of all targeted genes/operons are bound by two or more TFs (Plaisier et al. 2014). A similar observation has been made for Lrp-like TFs in *Sulfolobales*. Here, the *gltB* gene encoding a glutamate synthase subunit is a binding and regulatory "hotspot" for multiple Lrp-like TFs (Nguyen Duc et al. 2013; Song et al. 2013; Vassart et al. 2013; Liu et al. 2016). Several of these TFs have different DNA-binding specificities, operator structures and thus differently positioned binding sites in the *gltB* control region and it is postulated that regulatory effects are interdependent (Liu et al. 2016). Furthermore, paralogous TFs could share the same DNA-binding specificity and compete for the same binding site (Plaisier et al. 2014). Interestingly, it has also been observed that

Lrp-like TFs co-associate on the same genomic location through protein-protein interactions in which only one of the TFs in the complex binds a recognition site (Nguyen Duc et al. 2013). This can be explained by hetero-oligomerization of the TFs, which is supported by in vitro analysis of *Pyrococcus* Lrp-like TFs (Yokoyama et al. 2006b; Okamura et al. 2007). Hetero-oligomerization of TFs leads to a combinatorial regulatory response, in which different hetero-oligomeric assemblies are characterized by different ligand and DNA binding characteristics and could explain how archaeal cells make effective use of a relative limited repertoire of TFs (Peeters and Charlier 2010). This observation suggests that while basic regulatory mechanisms of TF functioning in archaea resemble those found in bacteria, archaeal TFs have also evolved the typical eukaryotic feature of hetero-oligomerization (*cfr*. heterodimeric leucine zipper TFs in eukaryotes).

2.9 Conclusions and Perspectives

In conclusion, our understanding of archaeal transcription regulation has expanded in recent years. This is due in part to the development of genetic systems in model archaea as well as the implementation of traditional techniques for studying DNA-protein interactions. The past 20 years have taken us from the first description of a repression mechanism of an archaeal TF to defining archaeal regulons. A wealth of—mainly reductionistic- studies have provided valuable insights into how bacterial-like TFs interact with a eukaryotic-like basal transcription machinery to enable archaeal cells to respond to continuously changing environmental conditions and to be competitive in their habitats.

Thus far only a limited number of studies have begun to unravel GRNs in archaea and future research will thus be instrumental to further our understanding of the interplay of multiple TFs and hierarchical regulatory networks. Given the existing genetic tools for several archaeal model organisms and well-established *omics* technologies in the current post-genomic era, it would be very informative to perform system-level systematic mapping of entire TF-mediated GRNs for archaeal model organisms other than *H. salinarum*. Comparative analysis to well-known GRNs in bacterial and eukaryotic model organisms will enable us to evaluate similarities and/or differences in the topology and architecture of archaeal networks. Holistic analyses should nevertheless be combined with reductionistic approaches given the relative lack of information on archaeal biology. For example, it would be interesting to further unravel the role and signal transduction pathways of direct phosphorylation of OCSs in *Crenarchaeota*, which lack TCSs. The execution of systems biology studies of archaeal TF-mediated gene regulation will not only allow to catch up with similar knowledge in bacteria and eukaryotes, but will also result in future opportunities to engineer GRNs in biotechnologically interesting archaea using synthetic biology approaches.

References

Abella M, Rodríguez S, Paytubi S et al (2007) The *Sulfolobus solfataricus radA* paralogue sso0777 is DNA damage inducible and positively regulated by the Sta1 protein. Nucleic Acids Res 35:6788–6797

Ahmad MUD, Waege I, Hausner W et al (2015) Structural insights into nonspecific binding of DNA by TrmBL2, an archaeal chromatin protein. J Mol Biol 427:3216–3229

Allen MA, Lauro FM, Williams TJ et al (2009) The genome sequence of the psychrophilic archaeon, *Methanococcoides burtonii*: the role of genome evolution in cold adaptation. ISME J 3:1012–1035

Aparicio O, Geisberg JV, Struhl K (2004) Chromatin immunoprecipitation for determining the association of proteins with specific genomic sequences in vivo. Curr Protoc Cell Biol Chap 17:Unit 17.7

Aravind L, Koonin EV (1999) DNA-binding proteins and evolution of transcription regulation in the archaea. Nucleic Acids Res 27:4658–4670

Aravind L, Anantharaman V, Balaji S et al (2005) The many faces of the helix-turn-helix domain: transcription regulation and beyond. FEMS Microbiol Rev 29:231–262

Ashby MK (2006) Distribution, structure and diversity of "bacterial" genes encoding two-component proteins in the *Euryarchaeota*. Archaea 2:11–30

Auguet J-C, Barberan A, Casamayor EO (2010) Global ecological patterns in uncultured archaea. ISME J 4:182–190

Ayala JC, Wang H, Silva AJ, Benitez JA (2015) Repression by H-NS of genes required for the biosynthesis of the *Vibrio cholerae* biofilm matrix is modulated by the second messenger cyclic diguanylic acid. Mol Microbiol 97:630–645

Baker-Austin C, Dopson M, Wexler M et al (2007) Extreme arsenic resistance by the acidophilic archaeon "*Ferroplasma acidarmanus*" Fer1. Extremophiles 11:425–434

Baliga NS, Goo YA, Ng WV et al (2000) Is gene expression in *Halobacterium* NRC-1 regulated by multiple TBP and TFB transcription factors? Mol Microbiol 36:1184–1185

Baumann P, Qureshi SA, Jackson SP (1995) Transcription: new insights from studies on archaea. Trends Genet 11:279–283

Bell SD, Jackson SP (1998) Transcription and translation in archaea: a mosaic of eukaryal and bacterial features. Trends Microbiol 6:222–228

Bell SD, Jackson SP (2000) Mechanism of autoregulation by an archaeal transcriptional repressor. J Biol Chem 275:31624–31629

Bell SD, Jackson SP (2001) Mechanism and regulation of transcription in archaea. Curr Opin Microbiol 4:208–213

Bell SD, Jaxel C, Nadal M et al (1998) Temperature, template topology, and factor requirements of archaeal transcription. Proc Natl Acad Sci USA 95:15218–15222

Bell SD, Cairns SS, Robson RL, Jackson SP (1999a) Transcriptional regulation of an archaeal operon in vivo and in vitro. Mol Cell 4:971–982

Bell SD, Kosa PL, Sigler PB, Jackson SP (1999b) Orientation of the transcription preinitiation complex in archaea. Proc Natl Acad Sci USA 96:13662–13667

Bell SD, Brinkman AB, van der Oost J, Jackson SP (2001) The archaeal TFIIEalpha homologue facilitates transcription initiation by enhancing TATA-box recognition. EMBO Rep 2:133–138

Bonneau R, Facciotti MT, Reiss DJ et al (2007) A predictive model for transcriptional control of physiology in a free living cell. Cell 131:1354–1365

Brinkman AB, Dahlke I, Tuininga JE et al (2000) An Lrp-like transcriptional regulator from the archaeon *Pyrococcus furiosus* is negatively autoregulated. J Biol Chem 275:38160–38169

Brinkman AB, Ettema TJG, de Vos WM, van der Oost J (2003) The Lrp family of transcriptional regulators. Mol Microbiol 48:287–294

Brouns SJJ, Walther J, Snijders APL et al (2006) Identification of the missing links in prokaryotic pentose oxidation pathways: evidence for enzyme recruitment. J Biol Chem 281:27378–27388

Busenlehner LS, Pennella MA, Giedroc DP (2003) The SmtB/ArsR family of metalloregulatory transcriptional repressors: structural insights into prokaryotic metal resistance. FEMS Microbiol Rev 27:131–143

Castelle CJ, Wrighton KC, Thomas BC et al (2015) Genomic expansion of domain archaea highlights roles for organisms from new phyla in anaerobic carbon cycling. Curr Biol 25:690–701

Chaban B, Ng SYM, Jarrell KF (2006) Archaeal habitats--from the extreme to the ordinary. Can J Microbiol 52:73–116

Charoensawan V, Wilson D, Teichmann SA (2010) Genomic repertoires of DNA-binding transcription factors across the tree of life. Nucleic Acids Res 38:7364–7377

Chen L, Chen L-R, Zhou XE et al (2004) The hyperthermophile protein Sso10a is a dimer of winged helix DNA-binding domains linked by an antiparallel coiled coil rod. J Mol Biol 341:73–91

Chen Z, Yu H, Li L et al (2012) The genome and transcriptome of a newly described psychrophilic archaeon, *Methanolobus psychrophilus* R15, reveal its cold adaptive characteristics. Environ Microbiol Rep 4:633–641

Chivers PT, Tahirov TH (2005) Structure of *Pyrococcus horikoshii* NikR: nickel sensing and implications for the regulation of DNA recognition. J Mol Biol 348:597–607

Chothia C (1984) Principles that determine the structure of proteins. Annu Rev Biochem 53:537–572

Cline SW, Doolittle WF (1987) Efficient transfection of the archaebacterium *Halobacterium halobium*. J Bacteriol 169:1341–1344

Cohen SP, Hächler H, Levy SB (1993) Genetic and functional analysis of the multiple antibiotic resistance (mar) locus in *Escherichia coli*. J Bacteriol 175:1484–1492

Coker JA, DasSarma S (2007) Genetic and transcriptomic analysis of transcription factor genes in the model halophilic archaeon: coordinate action of TbpD and TfbA. BMC Genet 8:61

Coulson RMR, Touboul N, Ouzounis CA (2007) Lineage-specific partitions in archaeal transcription. Archaea 2:117–125

Darcy TJ, Hausner W, Awery DE et al (1999) *Methanobacterium thermoautotrophicum* RNA polymerase and transcription in vitro. J Bacteriol 181:4424–4429

DeLong EF, Pace NR (2001) Environmental diversity of bacteria and archaea. Syst Biol 50:470–478

Di Fiore A, Fiorentino G, Vitale RM et al (2009) Structural analysis of BldR from *Sulfolobus solfataricus* provides insights into the molecular basis of transcriptional activation in archaea by MarR family proteins. J Mol Biol 388:559–569

Diruggiero J, Dunn D, Maeder DL et al (2000) Evidence of recent lateral gene transfer among hyperthermophilic archaea. Mol Microbiol 38:684–693

Efremov AK, Qu Y, Maruyama H et al (2015) Transcriptional repressor TrmBL2 from *Thermococcus kodakarensis* forms filamentous nucleoprotein structures and competes with histones for DNA binding in a salt- and DNA supercoiling-dependent manner. J Biol Chem 290:15770–15784

Enoru-Eta J, Gigot D, Thia-Toong TL et al (2000) Purification and characterization of Sa-lrp, a DNA-binding protein from the extreme thermoacidophilic archaeon *Sulfolobus acidocaldarius* homologous to the bacterial global transcriptional regulator Lrp. J Bacteriol 182:3661–3672

Esser D, Pham KT, Reimann J et al (2012) Change of carbon source causes dramatic effects in the phospho-proteome of the archaeon *Sulfolobus solfataricus*. J Proteome Res 11:4823–4833

Esser D, Hoffmann L, Pham TK et al (2016) Protein phosphorylation and its role in archaeal signal transduction. FEMS Microbiol Rev 40:625–647

Facciotti MT, Reiss DJ, Pan M et al (2007) General transcription factor specified global gene regulation in archaea. Proc Natl Acad Sci USA 104:4630–4635

Fiorentino G, Cannio R, Rossi M, Bartolucci S (2003) Transcriptional regulation of the gene encoding an alcohol dehydrogenase in the archaeon *Sulfolobus solfataricus* involves multiple factors and control elements. J Bacteriol 185:3926–3934

Fiorentino G, Ronca R, Cannio R et al (2007) MarR-like transcriptional regulator involved in detoxification of aromatic compounds in *Sulfolobus solfataricus*. J Bacteriol 189:7351–7360
Fiorentino G, Del Giudice I, Bartolucci S et al (2011) Identification and physicochemical characterization of BldR2 from *Sulfolobus solfataricus*, a novel archaeal member of the MarR transcription factor family. Biochemistry 50:6607–6621
Forterre P, Bergerat A, López-García P (1996) The unique DNA topology and DNA topoisomerases of hyperthermophilic archaea. FEMS Microbiol Rev 18:237–248
Franzmann PD, Springer N, Ludwig W, Rohde M (1992) A methanogenic archaeon from ace lake, antarctica: *Methanococcoides burtonii* sp. nov. Syst Appl Microbiol 15:573–581
Galas DJ, Schmitz A (1978) DNAse footprinting: a simple method for the detection of protein-DNA binding specificity. Nucleic Acids Res 5:3157–3170
Garner MM, Revzin A (1981) A gel electrophoresis method for quantifying the binding of proteins to specific DNA regions: application to components of the *Escherichia coli* lactose operon regulatory system. Nucleic Acids Res 9:3047–3060
Geiduschek EP, Ouhammouch M (2005) Archaeal transcription and its regulators. Mol Microbiol 56:1397–1407
Gernhardt P, Possot O, Foglino M et al (1990) Construction of an integration vector for use in the archaebacterium *Methanococcus voltae* and expression of a eubacterial resistance gene. Mol Gen Genet 221:273–279
Gindner A, Hausner W, Thomm M (2014) The TrmB family: a versatile group of transcriptional regulators in archaea. Extremophiles 18:925–936
Gohl HP, Gröndahl B, Thomm M (1995) Promoter recognition in archaea is mediated by transcription factors: identification of transcription factor aTFB from *Methanococcus thermolithotrophicus* as archaeal TATA-binding protein. Nucleic Acids Res 23:3837–3841
Gomis-Rüth FX, Solá M, Acebo P et al (1998) The structure of plasmid-encoded transcriptional repressor CopG unliganded and bound to its operator. EMBO J 17:7404–7415
Grohmann D, Nagy J, Chakraborty A et al (2011) The initiation factor TFE and the elongation factor Spt4/5 compete for the RNAP clamp during transcription initiation and elongation. Mol Cell 43:263–274
Guillière F, Danioux C, Jaubert C et al (2013) Solution structure of an archaeal DNA binding protein with an eukaryotic zinc finger fold. PLoS One 8:e52908
Hanzelka BL, Darcy TJ, Reeve JN (2001) TFE, an archaeal transcription factor in *Methanobacterium thermoautotrophicum* related to eucaryal transcription factor TFIIE. J Bacteriol 183:1813–1818
Harami GM, Gyimesi M, Kovács M (2013) From keys to bulldozers: expanding roles for winged helix domains in nucleic-acid-binding proteins. Trends Biochem Sci 38:364–371
Harteis S, Schneider S (2014) Making the bend: DNA tertiary structure and protein-DNA interactions. Int J Mol Sci 15:12335–12363
Hatfield GW, Benham CJ (2002) DNA topology-mediated control of global gene expression in *Escherichia coli*. Annu Rev Genet 36:175–203
Haurat MF, Figueiredo AS, Hoffmann L et al (2016) ArnS, a kinase involved in starvation-induced archaellum expression. Mol Microbiol 103:181–194
Hausner W, Thomm M (2001) Events during initiation of archaeal transcription: open complex formation and DNA-protein interactions. J Bacteriol 183:3025–3031
Hausner W, Wettach J, Hethke C, Thomm M (1996) Two transcription factors related with the eucaryal transcription factors TATA-binding protein and transcription factor IIB direct promoter recognition by an archaeal RNA polymerase. J Biol Chem 271:30144–30148
Heinicke I, Muller J, Pittelkow M, Klein A (2004) Mutational analysis of genes encoding chromatin proteins in the archaeon *Methanococcus voltae* indicates their involvement in the regulation of gene expression. Mol Genet Genomics 272:76–87
Hillion M, Antelmann H (2015) Thiol-based redox switches in prokaryotes. Biol Chem 396:415–444

Hüdepohl U, Reiter WD, Zillig W (1990) In vitro transcription of two rRNA genes of the archaebacterium *Sulfolobus* sp. B12 indicates a factor requirement for specific initiation. Proc Natl Acad Sci USA 87:5851–5855

Isom CE, Turner JL, Lessner DJ, Karr EA (2013) Redox-sensitive DNA binding by homodimeric *Methanosarcina acetivorans* MsvR is modulated by cysteine residues. BMC Microbiol 13:163

Itou H, Yao M, Watanabe N, Tanaka I (2008) Crystal structure of the PH1932 protein, a unique archaeal ArsR type winged-HTH transcription factor from *Pyrococcus horikoshii* OT3. Proteins 70:1631–1634

Iyer LM, Aravind L (2012) Insights from the architecture of the bacterial transcription apparatus. J Struct Biol 179:299–319

Jun S-H, Reichlen MJ, Tajiri M, Murakami KS (2011) Archaeal RNA polymerase and transcription regulation. Crit Rev Biochem Mol Biol 46:27–40

Kanai T, Akerboom J, Takedomi S et al (2007) A global transcriptional regulator in *Thermococcus kodakaraensis* controls the expression levels of both glycolytic and gluconeogenic enzyme-encoding genes. J Biol Chem 282:33659–33670

Karr EA (2010) The methanogen-specific transcription factor MsvR regulates the *fpaA-rlp-rub* oxidative stress operon adjacent to *msvR* in *Methanothermobacter thermautotrophicus*. J Bacteriol 192:5914–5922

Karr EA, Ng JM, Belchik SM et al (2006) Biodiversity of methanogenic and other archaea in the permanently frozen Lake Fryxell, Antarctica. Appl Environ Microbiol 72:1663–1666

Karr EA, Sandman K, Lurz R, Reeve JN (2008) TrpY regulation of trpB2 transcription in *Methanothermobacter thermautotrophicus*. J Bacteriol 190:2637–2641

Kato S, Kosaka T, Watanabe K (2008) Comparative transcriptome analysis of responses of *Methanothermobacter thermautotrophicus* to different environmental stimuli. Environ Microbiol 10:893–905

Kawashima T, Aramaki H, Oyamada T et al (2008) Transcription regulation by feast/famine regulatory proteins, FFRPs, in archaea and eubacteria. Biol Pharm Bull 31:173–186

Keese AM, Schut GJ, Ouhammouch M et al (2010) Genome-wide identification of targets for the archaeal heat shock regulator phr by cell-free transcription of genomic DNA. J Bacteriol 192:1292–1298

Kessler A, Sezonov G, Guijarro JI et al (2006) A novel archaeal regulatory protein, Sta1, activates transcription from viral promoters. Nucleic Acids Res 34:4837–4845

Kim EA, Blair DF (2015) Function of the histone-like protein H-NS in motility of *Escherichia coli*: multiple regulatory roles rather than direct action at the flagellar motor. J Bacteriol 197:3110–3120

Kim M, Park S, Lee S-J (2016) Global transcriptional regulator TrmB family members in prokaryotes. J Microbiol 54:639–645

Koike H, Ishijima SA, Clowney L, Suzuki M (2004) The archaeal feast/famine regulatory protein: potential roles of its assembly forms for regulating transcription. Proc Natl Acad Sci USA 101:2840–2845

Koretke KK, Lupas AN, Warren PV et al (2000) Evolution of two-component signal transduction. Mol Biol Evol 17:1956–1970

Krug M, Lee S-J, Diederichs K et al (2006) Crystal structure of the sugar binding domain of the archaeal transcriptional regulator TrmB. J Biol Chem 281:10976–10982

Krug M, Lee S-J, Boos W et al (2013) The three-dimensional structure of TrmB, a transcriptional regulator of dual function in the hyperthermophilic archaeon *Pyrococcus furiosus* in complex with sucrose. Protein Sci 22:800–808

Krüger K, Hermann T, Armbruster V, Pfeifer F (1998) The transcriptional activator GvpE for the halobacterial gas vesicle genes resembles a basic region leucine-zipper regulatory protein. J Mol Biol 279:761–771

Kumarevel T, Nakano N, Ponnuraj K et al (2008a) Crystal structure of glutamine receptor protein from *Sulfolobus tokodaii* strain 7 in complex with its effector L-glutamine: implications of effector binding in molecular association and DNA binding. Nucleic Acids Res 36:4808–4820

Kumarevel T, Sakamoto K, Gopinath SCB et al (2008b) Crystal structure of an archaeal specific DNA-binding protein (Ape10b2) from *Aeropyrum pernix* K1. Proteins 71:1156–1162
Kumarevel T, Tanaka T, Umehara T, Yokoyama S (2009) ST1710-DNA complex crystal structure reveals the DNA binding mechanism of the MarR family of regulators. Nucleic Acids Res 37:4723–4735
Kunin V, Goldovsky L, Darzentas N, Ouzounis CA (2005) The net of life: reconstructing the microbial phylogenetic network. Genome Res 15:954–959
Kyrpides NC, Ouzounis CA (1999) Transcription in archaea. Proc Natl Acad Sci USA 96:8545–8550
Lassak K, Peeters E, Wróbel S, Albers S-V (2013) The one-component system ArnR: a membrane-bound activator of the crenarchaeal archaellum. Mol Microbiol 88:125–139
Lee S-J, Engelmann A, Horlacher R et al (2003) TrmB, a sugar-specific transcriptional regulator of the trehalose/maltose ABC transporter from the hyperthermophilic archaeon *Thermococcus litoralis*. J Biol Chem 278:983–990
Lee S-J, Moulakakis C, Koning SM et al (2005) TrmB, a sugar sensing regulator of ABC transporter genes in *Pyrococcus furiosus* exhibits dual promoter specificity and is controlled by different inducers. Mol Microbiol 57:1797–1807
Lee S-J, Surma M, Seitz S et al (2007a) Characterization of the TrmB-like protein, PF0124, a TGM-recognizing global transcriptional regulator of the hyperthermophilic archaeon *Pyrococcus furiosus*. Mol Microbiol 65:305–318
Lee S-J, Surma M, Seitz S et al (2007b) Differential signal transduction via TrmB, a sugar sensing transcriptional repressor of *Pyrococcus furiosus*. Mol Microbiol 64:1499–1505
Lee S-J, Surma M, Hausner W et al (2008) The role of TrmB and TrmB-like transcriptional regulators for sugar transport and metabolism in the hyperthermophilic archaeon *Pyrococcus furiosus*. Arch Microbiol 190:247–256
Leigh JA, Albers S-V, Atomi H, Allers T (2011) Model organisms for genetics in the domain archaea: methanogens, halophiles, *Thermococcales* and *Sulfolobales*. FEMS Microbiol Rev 35:577–608
Leonard PM, Smits SH, Sedelnikova SE et al (2001) Crystal structure of the Lrp-like transcriptional regulator from the archaeon *Pyrococcus furiosus*. EMBO J 20:990–997
Li J, Zheng X, Guo X et al (2014) Characterization of an archaeal two-component system that regulates methanogenesis in *Methanosaeta harundinacea*. PLoS One 9:e95502
Lipscomb GL, Keese AM, Cowart DM et al (2009) SurR: a transcriptional activator and repressor controlling hydrogen and elemental sulphur metabolism in *Pyrococcus furiosus*. Mol Microbiol 71:332–349
Liu Y, Guo L, Guo R et al (2009) The Sac10b homolog in *Methanococcus maripaludis* binds DNA at specific sites. J Bacteriol 191:2315–2329
Liu H, Orell A, Maes D et al (2014) BarR, an Lrp-type transcription factor in *Sulfolobus acidocaldarius*, regulates an aminotransferase gene in a β-alanine responsive manner. Mol Microbiol 92:625–639
Liu H, Wang K, Lindås A-C, Peeters E (2016) The genome-scale DNA-binding profile of BarR, a β-alanine responsive transcription factor in the archaeon *Sulfolobus acidocaldarius*. BMC Genomics 17:569
López-García P, Forterre P (1999) Control of DNA topology during thermal stress in hyperthermophilic archaea: DNA topoisomerase levels, activities and induced thermotolerance during heat and cold shock in *Sulfolobus*. Mol Microbiol 33:766–777
Macquarrie KL, Fong AP, Morse RH, Tapscott SJ (2011) Genome-wide transcription factor binding: beyond direct target regulation. Trends Genet 27:141–148
Martínez-Núñez MA, Poot-Hernandez AC, Rodríguez-Vázquez K, Pérez-Rueda E (2013) Increments and duplication events of enzymes and transcription factors influence metabolic and regulatory diversity in prokaryotes. PLoS One 8:e69707

Maruyama H, Shin M, Oda T et al (2011) Histone and TK0471/TrmBL2 form a novel heterogeneous genome architecture in the hyperthermophilic archaeon *Thermococcus kodakarensis*. Mol Biol Cell 22:386–398

Matteau D, Rodrigue S (2015) Precise identification of DNA-binding proteins genomic location by exonuclease coupled chromatin immunoprecipitation (ChIP-exo). Methods Mol Biol 1334:173–193

Metcalf WW, Zhang JK, Apolinario E et al (1997) A genetic system for Archaea of the genus *Methanosarcina*: liposome-mediated transformation and construction of shuttle vectors. Proc Natl Acad Sci USA 94:2626–2631

Micorescu M, Grunberg S, Franke A et al (2007) Archaeal transcription: function of an alternative transcription factor B from *Pyrococcus furiosus*. J Bacteriol 190:157–167

Minezaki Y, Homma K, Nishikawa K (2005) Genome-wide survey of transcription factors in prokaryotes reveals many bacteria-specific families not found in archaea. DNA Res 12:269–280

Miyazono K-I, Tsujimura M, Kawarabayasi Y, Tanokura M (2007) Crystal structure of an archaeal homologue of multidrug resistance repressor protein, EmrR, from hyperthermophilic archaea *Sulfolobus tokodaii* strain 7. Proteins 67:1138–1146

Molitor B, Stassen M, Modi A et al (2013) A heme-based redox sensor in the methanogenic archaeon *Methanosarcina acetivorans*. J Biol Chem 288:18458–18472

Nadal M, Mirambeau G, Forterre P et al (1986) Positively supercoiled DNA in a virus-like particle of an archaebacterium. Nature 321:256–258

Najnin T, Siddiqui KS, Taha et al (2016) Characterization of a temperature-responsive two component regulatory system from the Antarctic archaeon, *Methanococcoides burtonii*. Sci Rep 6:24278

Napoli A, van der Oost J, Sensen CW et al (1999) An Lrp-like protein of the hyperthermophilic archaeon *Sulfolobus solfataricus* which binds to its own promoter. J Bacteriol 181:1474–1480

Napoli A, Kvaratskelia M, White MF et al (2001) A novel member of the bacterial-archaeal regulator family is a nonspecific DNA-binding protein and induces positive supercoiling. J Biol Chem 276:10745–10752

Nelson-Sathi S, Sousa FL, Roettger M et al (2014) Origins of major archaeal clades correspond to gene acquisitions from bacteria. Nature 517:77–80

Nguyen Duc T, van Oeffelen L, Song N et al (2013) The genome-wide binding profile of the *Sulfolobus solfataricus* transcription factor Ss-LrpB shows binding events beyond direct transcription regulation. BMC Genomics 14:828

Ochs SM, Thumann S, Richau R et al (2012) Activation of archaeal transcription mediated by recruitment of transcription factor B. J Biol Chem 287:18863–18871

Okada U, Sakai N, Yao M et al (2006) Structural analysis of the transcriptional regulator homolog protein from *Pyrococcus horikoshii* OT3. Proteins 63:1084–1086

Okamura H, Yokoyama K, Koike H et al (2007) A structural code for discriminating between transcription signals revealed by the feast/famine regulatory protein DM1 in complex with ligands. Structure 15:1325–1338

Orell A, Peeters E, Vassen V et al (2013) Lrs14 transcriptional regulators influence biofilm formation and cell motility of Crenarchaea. ISME J 7:1886–1898

Ouhammouch M, Geiduschek EP (2001) A thermostable platform for transcriptional regulation: the DNA-binding properties of two Lrp homologs from the hyperthermophilic archaeon *Methanococcus jannaschii*. EMBO J 20:146–156

Ouhammouch M, Dewhurst RE, Hausner W et al (2003) Activation of archaeal transcription by recruitment of the TATA-binding protein. Proc Natl Acad Sci USA 100:5097–5102

Ouhammouch M, Langham GE, Hausner W et al (2005) Promoter architecture and response to a positive regulator of archaeal transcription. Mol Microbiol 56:625–637

Paget MSB, Buttner MJ (2003) Thiol-based regulatory switches. Annu Rev Genet 37:91–121

Park PJ (2009) ChIP-seq: advantages and challenges of a maturing technology. Nat Rev Genet 10:669–680

Paytubi S, White MF (2009) The crenarchaeal DNA damage-inducible transcription factor B paralogue TFB3 is a general activator of transcription. Mol Microbiol 72:1487–1499
Peck RF, DasSarma S, Krebs MP (2000) Homologous gene knockout in the archaeon *Halobacterium salinarum* with ura3 as a counterselectable marker. Mol Microbiol 35:667–676
Peeters E, Charlier D (2010) The Lrp family of transcription regulators in archaea. Archaea 2010:750457
Peeters E, Thia-Toong T-L, Gigot D et al (2004) Ss-LrpB, a novel Lrp-like regulator of *Sulfolobus solfataricus* P2, binds cooperatively to three conserved targets in its own control region. Mol Microbiol 54:321–336
Peeters E, Willaert R, Maes D, Charlier D (2006) Ss-LrpB from *Sulfolobus solfataricus* condenses about 100 base pairs of its own operator DNA into globular nucleoprotein complexes. J Biol Chem 281:11721–11728
Peeters E, Wartel C, Maes D, Charlier D (2007) Analysis of the DNA-binding sequence specificity of the archaeal transcriptional regulator Ss-LrpB from *Sulfolobus solfataricus* by systematic mutagenesis and high resolution contact probing. Nucleic Acids Res 35:623–633
Peeters E, Albers S-V, Vassart A et al (2009) Ss-LrpB, a transcriptional regulator from *Sulfolobus solfataricus*, regulates a gene cluster with a pyruvate ferredoxin oxidoreductase-encoding operon and permease genes. Mol Microbiol 71:972–988
Peeters E, Peixeiro N, Sezonov G (2013a) Cis-regulatory logic in archaeal transcription. Biochem Soc Trans 41:326–331
Peeters E, van Oeffelen L, Nadal M et al (2013b) A thermodynamic model of the cooperative interaction between the archaeal transcription factor Ss-LrpB and its tripartite operator DNA. Gene 524:330–340
Peeters E, Driessen RPC, Werner F, Dame RT (2015) The interplay between nucleoid organization and transcription in archaeal genomes. Nat Rev Microbiol 13:333–341
Peixeiro N, Keller J, Collinet B et al (2012) Structure and function of AvtR, a novel transcriptional regulator from a hyperthermophilic archaeal lipothrixvirus. J Virol 87:124–136
Peng N, Xia Q, Chen Z et al (2009) An upstream activation element exerting differential transcriptional activation on an archaeal promoter. Mol Microbiol 74:928–939
Peng N, Ao X, Liang YX, She Q (2011) Archaeal promoter architecture and mechanism of gene activation. Biochem Soc Trans 39:99–103
Pérez-Rueda E, Collado-Vides J (2001) Common history at the origin of the position-function correlation in transcriptional regulators in archaea and bacteria. J Mol Evol 53:172–179
Pérez-Rueda E, Janga SC (2010) Identification and genomic analysis of transcription factors in archaeal genomes exemplifies their functional architecture and evolutionary origin. Mol Biol Evol 27:1449–1459
Plaisier CL, Lo F-Y, Ashworth J et al (2014) Evolution of context dependent regulation by expansion of feast/famine regulatory proteins. BMC Syst Biol 8:1–14
Reichelt R, Gindner A, Thomm M, Hausner W (2016) Genome-wide binding analysis of the transcriptional regulator TrmBL1 in *Pyrococcus furiosus*. BMC Genomics 17:40
Reichlen MJ, Vepachedu VR, Murakami KS, Ferry JG (2012) MreA functions in the global regulation of methanogenic pathways in *Methanosarcina acetivorans*. MBio 3:e00189
Reigstad LJ, Jorgensen SL, Schleper C (2010) Diversity and abundance of Korarchaeota in terrestrial hot springs of Iceland and Kamchatka. ISME J 4:346–356
Reimann J, Lassak K, Khadouma S et al (2012) Regulation of archaella expression by the FHA and von Willebrand domain-containing proteins ArnA and ArnB in *Sulfolobus acidocaldarius*. Mol Microbiol 86:24–36
Reimann J, Esser D, Orell A et al (2013) Archaeal signal transduction: impact of protein phosphatase deletions on cell size, motility and energy metabolism in *Sulfolobus acidocaldarius*. Mol Cell Proteomics 12:3908–3923
Reiter WD, Hüdepohl U, Zillig W (1990) Mutational analysis of an archaebacterial promoter: essential role of a TATA box for transcription efficiency and start-site selection in vitro. Proc Natl Acad Sci USA 87:9509–9513

Riley TR, Slattery M, Abe N et al (2014) SELEX-seq: a method for characterizing the complete repertoire of binding site preferences for transcription factor complexes. Methods Mol Biol 1196:255–278

Robertson G, Hirst M, Bainbridge M et al (2007) Genome-wide profiles of STAT1 DNA association using chromatin immunoprecipitation and massively parallel sequencing. Nat Methods 4:651–657

Rohlin L, Trent JD, Salmon K et al (2005) Heat shock response of *Archaeoglobus fulgidus*. J Bacteriol 187:6046–6057

Rohs R, Jin X, West SM et al (2010) Origins of specificity in protein-DNA recognition. Annu Rev Biochem 79:233–269

Rowlands T, Baumann P, Jackson SP (1994) The TATA-binding protein: a general transcription factor in eukaryotes and archaebacteria. Science 264:1326–1329

Rudrappa D, Yao AI, White D et al (2015) Identification of an archaeal mercury regulon by chromatin immunoprecipitation. Microbiology 161:2423–2433

Santangelo TJ, Cubonová L, James CL, Reeve JN (2007) TFB1 or TFB2 is sufficient for *Thermococcus kodakaraensis* viability and for basal transcription in vitro. J Mol Biol 367:344–357

Saridakis V, Shahinas D, Xu X, Christendat D (2008) Structural insight on the mechanism of regulation of the MarR family of proteins: high-resolution crystal structure of a transcriptional repressor from *Methanobacterium thermoautotrophicum*. J Mol Biol 377:655–667

Schelert J, Dixit V, Hoang V et al (2004) Occurrence and characterization of mercury resistance in the hyperthermophilic archaeon *Sulfolobus solfataricus* by use of gene disruption. J Bacteriol 186:427–437

Schelert J, Drozda M, Dixit V et al (2006) Regulation of mercury resistance in the crenarchaeote *Sulfolobus solfataricus*. J Bacteriol 188:7141–7150

Schmid AK, Reiss DJ, Pan M et al (2009) A single transcription factor regulates evolutionarily diverse but functionally linked metabolic pathways in response to nutrient availability. Mol Syst Biol 5:282

Schmid AK, Pan M, Sharma K, Baliga NS (2011) Two transcription factors are necessary for iron homeostasis in a salt-dwelling archaeon. Nucleic Acids Res 39:2519–2533

Schwaiger R, Schwarz C, Furtwangler K et al (2010) Transcriptional control by two leucine-responsive regulatory proteins in *Halobacterium salinarum* R1. BMC Mol Biol 11:40

Seedorf H, Dreisbach A, Hedderich R et al (2004) F420H2 oxidase (FprA) from *Methanobrevibacter arboriphilus*, a coenzyme F420-dependent enzyme involved in O2 detoxification. Arch Microbiol 182:126–137

Seedorf H, Hagemeier CH, Shima S et al (2007) Structure of coenzyme F420H2 oxidase (FprA), a di-iron flavoprotein from methanogenic Archaea catalyzing the reduction of O2 to H2O. FEBS J 274:1588–1599

Seeman NC, Rosenberg JM, Rich A (1976) Sequence-specific recognition of double helical nucleic acids by proteins. Proc Natl Acad Sci USA 73:804–808

Sharma K, Gillum N, Boyd JL, Schmid AK (2012) The RosR transcription factor is required for gene expression dynamics in response to extreme oxidative stress in a hypersaline-adapted archaeon. BMC Genomics 13:351

Sheehan R, McCarver AC, Isom CE et al (2015) The *Methanosarcina acetivorans* thioredoxin system activates DNA binding of the redox-sensitive transcriptional regulator MsvR. J Ind Microbiol Biotechnol 42:965–969

Shinkai A, Sekine S-I, Urushibata A et al (2007) The putative DNA-binding protein Sto12a from the thermoacidophilic archaeon *Sulfolobus tokodaii* contains intrachain and interchain disulfide bonds. J Mol Biol 372:1293–1304

Slattery M, Riley T, Liu P et al (2011) Cofactor binding evokes latent differences in DNA binding specificity between hox proteins. Cell 147:1270–1282

Song N, Nguyen Duc T, van Oeffelen L et al (2013) Expanded target and cofactor repertoire for the transcriptional activator LysM from *Sulfolobus*. Nucleic Acids Res 41:2932–2949

Suzuki M, Brenner SE, Gerstein M, Yagi N (1995) DNA recognition code of transcription factors. Protein Eng 8:319–328
Tachdjian S, Kelly RM (2006) Dynamic metabolic adjustments and genome plasticity are implicated in the heat shock response of the extremely thermoacidophilic archaeon *Sulfolobus solfataricus*. J Bacteriol 188:4553–4559
Teske A, Sørensen KB (2008) Uncultured archaea in deep marine subsurface sediments: have we caught them all? ISME J 2:3–18
Todone F, Brick P, Werner F et al (2001) Structure of an archaeal homolog of the eukaryotic RNA polymerase II RPB4/RPB7 complex. Mol Cell 8:1137–1143
Tonner PD, Pittman AMC, Gulli JG et al (2015) A regulatory hierarchy controls the dynamic transcriptional response to extreme oxidative stress in archaea. PLoS Genet 11:e1004912
Tuerk C, Gold L (1990) Systematic evolution of ligands by exponential enrichment: RNA ligands to bacteriophage T4 DNA polymerase. Science 249:505–510
Tumbula DL, Whitman WB (1999) Genetics of *Methanococcus*: possibilities for functional genomics in archaea. Mol Microbiol 33:1–7
van de Werken HJG, Verhees CH, Akerboom J et al (2006) Identification of a glycolytic regulon in the archaea *Pyrococcus* and *Thermococcus*. FEMS Microbiol Lett 260:69–76
van Oeffelen L, Peeters E, Nguyen Le Minh P, Charlier D (2014) The "densitometric image analysis software" and its application to determine stepwise equilibrium constants from electrophoretic mobility shift assays. PLoS One 9:e85146
Vassart A, van Wolferen M, Orell A et al (2013) Sa-Lrp from *Sulfolobus acidocaldarius* is a versatile, glutamine-responsive, and architectural transcriptional regulator. MicrobiologyOpen 2:75–93
Vierke G, Engelmann A, Hebbeln C, Thomm M (2003) A novel archaeal transcriptional regulator of heat shock response. J Biol Chem 278:18–26
Visweswariah SS, Busby SJW (2015) Evolution of bacterial transcription factors: how proteins take on new tasks, but do not always stop doing the old ones. Trends Microbiol 23:463–467
Wagner M, van Wolferen W et al (2012) Versatile genetic tool box for the crenarchaeote *Sulfolobus acidocaldarius*. Front Microbiol 3:214
Wagner M, Wagner A, Ma X et al (2014) Investigation of the malE promoter and MalR, a positive regulator of the maltose regulon, for an improved expression system in *Sulfolobus acidocaldarius*. Appl Environ Microbiol 80:1072–1081
Walker JE, Santangelo TJ (2015) Analyses of in vivo interactions between transcription factors and the archaeal RNA polymerase. Methods 86:73–79
Wang G, Kennedy SP, Fasiludeen S et al (2004) Arsenic resistance in *Halobacterium* sp. strain NRC-1 examined by using an improved gene knockout system. J Bacteriol 186:3187–3194
Wang X, Lee H-S, Sugar FJ et al (2007) PF0610, a novel winged helix-turn-helix variant possessing a rubredoxin-like Zn ribbon motif from the hyperthermophilic archaeon, *Pyrococcus furiosus*. Biochemistry 46:752–761
Werner F, Weinzierl ROJ (2005) Direct modulation of RNA polymerase core functions by basal transcription factors. Mol Cell Biol 25:8344–8355
Werner F, Eloranta JJ, Weinzierl RO (2000) Archaeal RNA polymerase subunits F and P are bona fide homologs of eukaryotic RPB4 and RPB12. Nucleic Acids Res 28:4299–4305
Wierer S, Daldrop P, Ud Din Ahmad M et al (2016) TrmBL2 from *Pyrococcus furiosus* interacts both with double-stranded and single-stranded DNA. PLoS One 11:e0156098
Wilbanks EG, Larsen DJ, Neches RY et al (2012) A workflow for genome-wide mapping of archaeal transcription factors with ChIP-seq. Nucleic Acids Res 40:e74
Woese CR, Kandler O, Wheelis ML (1990) Towards a natural system of organisms: proposal for the domains archaea, bacteria, and eucarya. Proc Natl Acad Sci USA 87:4576–4579
Wouters MA, Fan SW, Haworth NL (2010) Disulfides as redox switches: from molecular mechanisms to functional significance. Antioxid Redox Signal 12:53–91
Wu J, Wang S, Bai J et al (2008) ArchaeaTF: an integrated database of putative transcription factors in archaea. Genomics 91:102–107

Wuichet K, Cantwell BJ, Zhulin IB (2010) Evolution and phyletic distribution of two-component signal transduction systems. Curr Opin Microbiol 13:219–225

Yamada M, Ishijima SA, Suzuki M (2009) Interactions between the archaeal transcription repressor FL11 and its coregulators lysine and arginine. Proteins 74:520–525

Yang H, Lipscomb GL, Keese AM et al (2010) SurR regulates hydrogen production in *Pyrococcus furiosus* by a sulfur-dependent redox switch. Mol Microbiol 77:1111–1122

Yokoyama K, Ihara M, Ebihara S, Suzuki M (2006a) A SELEX study of the DNA-binding specificity of archaeal FFRPs: 2. FL4 (pot1613368). Proc Jpn Acad Ser B 82:1–12

Yokoyama K, Ishijima SA, Clowney L et al (2006b) Feast/famine regulatory proteins (FFRPs): *Escherichia coli* Lrp, AsnC and related archaeal transcription factors. FEMS Microbiol Rev 30:89–108

Yokoyama K, Ishijima SA, Koike H et al (2007) Feast/famine regulation by transcription factor FL11 for the survival of the hyperthermophilic archaeon *Pyrococcus* OT3. Structure 15:1542–1554

Yokoyama K, Nogami H, Kabasawa M et al (2009) The DNA-recognition mode shared by archaeal feast/famine-regulatory proteins revealed by the DNA-binding specificities of TvFL3, FL10, FL11 and Ss-LrpB. Nucleic Acids Res 37:4407–4419

Yu L, Fang J, Wei Y (2009) Characterization of the ligand and DNA binding properties of a putative archaeal regulator ST1710. Biochemistry 48:2099–2108

Zheng M, Storz G (2000) Redox sensing by prokaryotic transcription factors. Biochem Pharmacol 59:1–6

Chapter 3
Translation Regulation: The Archaea-Eukaryal Connection

Dario Benelli, Anna La Teana, and Paola Londei

Abstract Translation, as an essential cellular process, is very well conserved through evolution. Nonetheless, the translational apparatus, namely the ribosomes and the accessory protein factors that assist all the steps of translation, have incurred a certain divergence in the three domains of cellular descent, the Bacteria, the Archaea and the Eukarya. The strongest evolutionary divergence is seen at the level of the initiation step, during which the ribosomes identify the start codon on the mRNA and set the correct reading frame for decoding. Initiation is a crucial event that sets the general rate of translation and is the target of most mechanisms of translational regulation.

While the Bacteria have a very streamlined translational apparatus, especially as regards the translation initiation factors, the other prokaryotic domain, the Archaea, displays an unexpected degree of complexity. Moreover, the components of the archaeal translational apparatus are evolutionarily closer to those of the Eukarya, and the Archaea share with the Eukarya certain translation factors that are not found in Bacteria. This chapter reviews the similarities and the differences of the several steps of translation in the three domains of life, with special emphasis on the still poorly understood connection between Archaea and Eukarya.

3.1 The Translational Apparatus in the Three Domains of Life

Regulation of gene expression at the translational level has received relatively little attention for a long time. Recently, however, especially after the discovery of the small regulator RNAs, miRNA and siRNA, the scientific community has begun to

D. Benelli • P. Londei (✉)
Dipartimento di Biotecnologie Cellulari ed Ematologia, Policlinico Umberto I, Istituto Pasteur-Fondazione Cenci Bolognetti, Università di Roma Sapienza, Viale Regina Elena 324, 00161 Rome, Italy
e-mail: paola.londei@uniroma1.it

A. La Teana
Dipartimento di Scienze della Vita e dell'Ambiente, Università Politecnica delle Marche, Via Brecce Bianche, 60131 Ancona, Italy

B. Clouet-d'Orval (ed.), *RNA Metabolism and Gene Expression in Archaea*, Nucleic Acids and Molecular Biology 32, DOI 10.1007/978-3-319-65795-0_3

realize that translational regulation is more widespread and important than previously thought, and that it impacts importantly on many essential cell functions.

Translation is known to consist of several distinct steps, initiation, elongation, termination and ribosome recycling. During initiation, the ribosomes, with the aid of a set of proteins termed translation initiation factors, identify the start codon on the mRNA thereby defining the correct reading frame for decoding. This process is rather complex and is fundamental in determining the general rate of translation and the relative abundance of the final protein product.

Translational elongation is itself divided in three steps. The first is adaptation, during which an amino-acylated tRNA enters the ribosomal A site and recognizes the correct codon on the mRNA with the aid of elongation factor 1 (EF1, termed EFTu in Bacteria). The second is trans-peptidation, during which the amino acid carried by the tRNA in the A site is added to the growing peptide chain carried by the tRNA in the P site. The catalytic activity for this reaction is provided by the ribosome itself, specifically by the peptidyl-transferase center of the large ribosomal subunit. The third and final step of elongation, translocation, entails a reciprocal movement of the ribosome and the mRNA, whose final result is a three-nucleotide shift of the mRNA that places the next codon in the A site. Elongation is assisted by elongation factor 2 (EF2, termed EFG in Bacteria).

Termination and ribosome recycling are the final steps of translation, ensuring that the completed polypeptide chain is released from the ribosomes and that the monomeric ribosome is again split into subunits dissociating from the tRNA and the mRNA. As the former ones, this step is also assisted by accessory factors, the termination (or release) factors and the recycling factors.

All of the stages of translation include factors that are G proteins and require therefore the hydrolysis of GTP.

A general scheme of the translation steps and of the factors assisting them in the three domains of cell descent is depicted in Fig. 3.1.

Translation may be regulated at any of the above described steps. However, the majority of the regulatory mechanisms act at the level of initiation, influencing the ease with which the ribosomes access the mRNA and/or identify the initiation codon, and thus determining the general rate of decoding. The elongation step may also be subjected to regulation, especially in the case of certain proteins with an idiosyncratic amino acid composition. In the three domains of life, however, translational regulation has attained different levels of accuracy and complexity, and the translational apparatus has diverged accordingly.

To ensure a sophisticated and accurate regulation of protein synthesis, eukaryotic cells have a correspondingly complex translational apparatus. Compared with the bacterial one, the eukaryotic translational apparatus (Fig. 3.1) has a plethora of components, especially as regards the accessory protein factors that assist and modulate the initiation step (Hinnebusch and Lorsch 2012). Recently, it has also emerged that certain eukaryotic cells may synthesize specialized ribosomes, having a slightly altered protein complement, that preferentially translate specific classes of mRNA, incrementing the production of proteins poorly translated by the normal particles (Xue and Barna 2012; Preiss 2016).

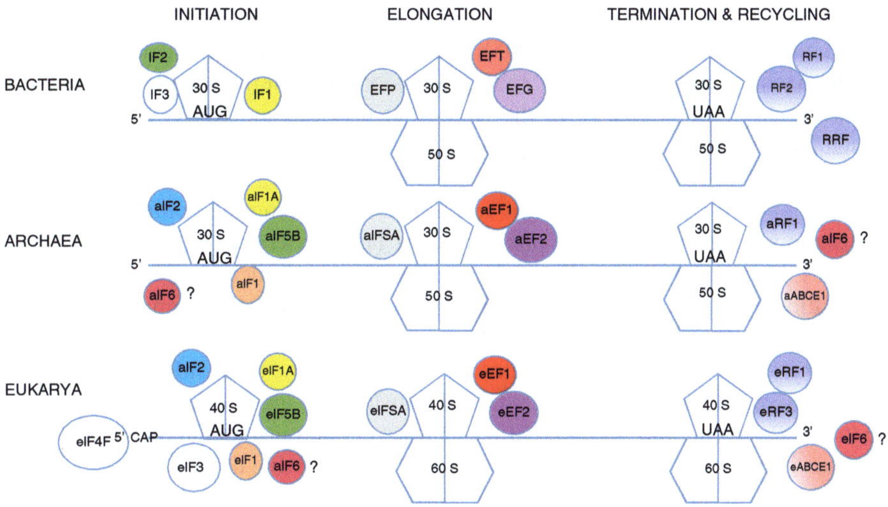

Fig. 3.1 Overview of the translation steps and of the factors participating in each of them in the three domains of life. *Top*: Bacterial translation; *middle*: Archaeal translation; *bottom*: Eukaryal translation. The *straight line* holding the ribosomes represents the mRNA, oriented as illustrated in a 5′-3′ direction. The AUG start codon and one of the possible stop codons (UAA) are shown. The three sets of ribosomes on each mRNA are, from left to right, those engaged in initiation, elongation and termination, respectively. The ribosomal subunits are schematized as divided in two sectors, which represent the P site (*on the left*) and the A site (*on the right*). The E site is not shown for simplicity. Only the small ribosomal subunit is shown for the initiation step, since it carries out by itself most of this process. The protein factors participating in each of the steps of translation are shown as spheres close to the ribosomes. The names of the various factors are indicated; their positions relative to the ribosome indicates approximately the main site of interaction. The homologous factors in the different domains are evidenced with the same color; colorless factors are those unique to the domain considered. The *question mark* for a/eIF6 means that the role of this protein in a specific translation step is still uncertain; therefore, a/eIF6 is shown as a participant in the initiation step or in the termination/recycling step in both Archaea and Eukarya

Thus, as regards translational regulation, a wide gap appears to exist separating eukaryotic and bacterial cells. The latter have a much simpler translational apparatus, and make use of a minimum of accessory factors for assisting the main target of translational regulation, namely the initiation step. Moreover, transcription and translation are simultaneous events, which restricts the possibility of much sophistication in regulating decoding.

What is true for the Bacteria, however, is not true for the entire prokaryotic world. It has long been known that the Archaea have a translational apparatus that is more complex than the bacterial one and that includes components found in the eukaryotes but not in the bacteria.

The Archaea have ribosomes that are both bigger and richer in proteins than the bacterial ones, even if there is a pronounced variability depending on the archaeal species (Lecompte et al. 2002). Moreover, the Archaea have a set of translation factors decidedly more complex that the bacterial one, especially as regards translation initiation factors (Londei 2005). Some of these factors are specifically shared by the Archaea and the Eukarya, to the exclusion of Bacteria. Finally, the archaeal translational components, from the ribosomal proteins and RNAs to the translational factors, are closer in primary sequence to their eukaryal than to their bacterial counterparts.

The similarity between Archaea and Eukarya regarding the components of the translational apparatus is still puzzling to a large extent, even after over three decades of archaeal studies. Since the Archaea have no nucleus, transcription and translation happen simultaneously as in Bacteria; moreover, the Archaea are known to have polycistronic mRNAs as the Bacteria, implying the ability for the archaeal ribosomes to perform repeated cycles of initiation on the same mRNA.

All this would point to a mechanism of translational regulation generally similar to the bacterial one. Indeed, what little is known about translational regulation in the archaea is in line with this prediction, even if the available data are very scarce.

Yet, the presence of distinct "eukaryotic" features in archaeal translation is undeniable. Some of them have been studied and understood to some extent, while others are still mysterious. In the following, we will try to highlight the similarities between archaeal and eukaryotic translation, taking into account the individual steps of the protein synthesis process.

3.2 Evolutionary Divergence in Translational Initiation

During the initiation step of translation, the ribosomes must identify the correct starting point for decoding on the mRNA, and convey the initiator tRNA on the initiation codon. This apparently simple feat is in actuality tremendously complex, and this is why most of the mechanisms that control speed and efficiency of translation operate at the initiation step.

A staggering amount of research has been performed on eukaryotic as well as on prokaryotic translational initiation. Summarizing in the extreme, the generally accepted model in Eukaryotes is that termed "ribosome scanning". The small ribosomal subunit (40S), in a complex with several protein factors and with initiator tRNA (met-tRNAi) lands in the vicinity of the capped 5′ end of the mRNA and moves along it until the initiation codon (generally AUG) is found. Then the scanning complex stops, the 60S subunits joins, the initiation factors leave the ribosome and elongation begins (Hinnebusch 2014).

While this model applies to the majority of mRNAs, there are also alternative initiation pathways that take place on uncapped mRNAs. The best known is the internal initiation model, relying upon the presence of special regions on the mRNA to which the ribosome can bind directly (the IRES or ribosome landing pads). Ribosome binding to an IRES may or may not be followed by scanning, but does not require the cap-binding initiation factors (Johnson et al. 2017).

In stark contrast with the complexity of eukaryotic initiation, Bacteria employ an extremely streamlined mechanism, in which the small ribosomal subunit (30S) interacts directly with mRNAs, often polycistronic, through the so-called TIR (Translation Initiation Region). This includes the initiation codon preceded, in many but not in all cases, by the Shine-Dalgarno sequence which specifically pairs with the 3′end of the 16S rRNA. In this process just three initiation factors are sufficient for the identification of the start codon and for correctly positioning the ribosome on it.

Interestingly, of the three bacterial initiation factors, two are universally conserved proteins, found in all three domains of life. However, one of these, the factor called IF2, functions as the tRNAi binding factor, a role that is not conserved in either Archaea or eukaryotes (Gualerzi and Pon 2015).

The Archaea have many apparent similarities with the Bacteria, such as being endowed with prokaryotic-sized ribosomes (70S). Their mRNAs also share common characteristics with those of Bacteria, in fact, they are often polycistronic and may contain Shine-Dalgarno sequences, albeit these are infrequent in certain archaeal species (Benelli et al. 2016). Archaeal mRNAs also hold unique features such as, in many cases, the lack of a 5′UTR (leaderless mRNAs). Leaderless mRNAs are unevenly distributed among Archaea: they are the majority of mRNAs in certain species of the phylum Crenarcheota, while being much less frequent in Euryarcheota such as methanogens.

However, with respect to the Bacteria, the Archaea have an enlarged set of translation initiation factors, although it is unclear why it should be so. To date, the recognized translation initiation factors in Archaea are the proteins termed aIF2, aIF1, aIF1A, aIF5B. Another factor, aIF6, is certainly involved in translation but its function is still uncertain. Two of these five proteins, aIF1A and aIF5B, are also found in all Bacteria. They are, respectively, homologous to the bacterial factors IF1 and IF2. The factor termed IF1 (or SUI1) in Archaea and Eukarya is also present in some, but not all, bacterial phyla, being sometimes termed YCiH. The remaining two proteins, a/eIF2 and a/eIF6, are shared exclusively by the Archaea and the Eukarya, and presumably the latter have inherited them from their archaeal ancestor (Benelli et al. 2016).

The Eukarya, of course, have many more initiation factors that are not found in either of the prokaryotic domains. Among these, the factors that interact with the cap at the mRNA 5′ and that guide the 40S subunits during scanning. An overview of the translation initiation factors in the three domains of life is presented in Table 3.1.

Clearly, the most interesting question is why the Archaea should have a translation initiation apparatus more complex than the bacterial one, and particularly why they should share with the eukaryotes a specific set of factors. While the answer to this question still remains elusive, the progress of our knowledge on archaeal initiation has begun to elucidate the similarities and differences existing in the features of translation initiation that employ factors specifically shared by the bacteria and the archaea.

Table 3.1 Translation initiation factors in the three domains of life

	Homologue in other domains	Function
Bacteria		
IF1	aIF1A, eIF1A	Occupies A site during initiation
IF2	aIF5B, eIF5B	Promotes binding of tRNAi in P site
IF3	None	Helps choice of correct initiation codon
Eukarya		
eIF2 (trimer)	aIF2 (trimer)	Promotes tRNAi binding in P site
eIF1	aIF1, SUI1	Helps choice of correct initiation codon
eIF1A	IF1, aIF1A	Occupies A site during initiation
eIF5B	IF2, aIF5B	Stabilizes tRNAi in P site and promotes subunit joining
eIF5	None	GTPase of eIF2
eIF4F (trimer)	None	Binds cap, unwinds mRNA, aids scanning
eIF3 (octamer)	None	Subunit anti-association factor
eIF6	aIF6	Ribosome biogenesis factor subunit anti-association factor ribosome recycling factor?
Archaea		
aIF1	eIF1	Helps choice of correct initiation codon
aIF1A	IF1, eIF1A	Occupies A site during initiation
aIF2 (trimer)	eIF2	Promotes tRNAi binding in P site
aIF5B	IF2, eIF5B	Stabilizes tRNAi in P site promotes subunit joining?
aIF6	eIF6	Subunit anti-association factor ribosome recycling factor?

3.2.1 The a/eIF1/SUI1 Factors

The protein termed a/eIF1 or SUI1 is one of the translation initiation factors, universally shared by the Archaea and the Eukarya, but lacking in most Bacteria. To be sure, genes encoding homologues of a/eIF1 have been found in certain bacterial phyla, such as the proteobacteria and the cyanobacteria, but are apparently missing in all other species (Kyrpides and Woese 1998). Studies in *E. coli* have established that the SUI1 homologue is not essential (Baba et al. 2006) and that it probably does not participate in translational initiation, although it may be involved in the expression of certain stress-related genes (Osterman et al. 2015).

The peculiar evolutionary distribution of a/eIF1/SUI1 is compatible with the idea that this gene was originally present in the common ancestor of all cell domains, but was subsequently lost by the Bacteria, probably because it was replaced by another, bacterial specific factor, better adapted to perform its appointed function.

But what is the function of a/eIF1/SUI1? In Eukaryotes, where it has been studied extensively, eIF1 is known to have important roles in translational initiation. It binds to the 40S subunits and prevents the premature joining of the 60S particle. Also, and more importantly, it discriminates against non-canonical initiation codons, helping to ensure the fidelity of translational initiation. Moreover, eIF1 is essential for the process of ribosome scanning, whereby the 43S initiation complex, consisting of the 40S ribosomal subunit and of several initiation factors (including IF1 and eIF1A, another universally conserved protein) moves along the mRNA to locate the translation start codon. Cryoelectron microscopy studies suggested that eIF1 and eIF1A maintain the initiating ribosome in an "open", scanning-competent, conformation until the start codon is located, and the first codon/anti-codon base-pairing has been established (Passmore et al. 2007). Then the complex undergoes a conformational change and eIF1 is released (Maag et al. 2005; Cheung et al. 2007).

The release of eIF1 is believed to free the C terminus of eIF1A for interactions with eIF5, which stabilizes the closed state of the complex (Maag et al. 2006).

In Archaea, the function of aIF1 has been studied to some extent in the extreme thermophile *Sulfolobus solfataricus*. It has been established that, as in the Eukaryotes, the factor binds specifically to the 30S ribosomal subunits and is not found on elongating 70S ribosomes, arguing for a specific role in translation initiation (Hasenöhrl et al. 2006).

The binding site of aIF1 on the 30S subunits has also been defined, and found to coincide with that occupied by the corresponding eukaryotic factor on the 40S subunit. Experiments of hydroxy-radical probing have identified helices 23 and 24 of the 16S RNA as the region protected by aIF1 binding, a region that corresponds with that protected by eIF1 on the 40S subunits (Hasenöhrl et al. 2009).

The function of the archaeal factor also apparently corresponds to that of its eukaryal counterpart, specifically regarding the role in determining the fidelity of initiation codon choice. Indeed, aIF1 discriminates against ribosome binding to a mRNA having the non-canonical initiation codon AUU (Hasenöhrl et al. 2009). It is interesting to note that this important "fidelity function" also exists in Bacteria, but it is performed by the bacterial-specific factor IF3, that has no evident homology with a/eIF1. As said above, IF3 probably has replaced IF1/SUI1 in the course of bacterial evolution. The reason for discarding a universal factor for a new one is not evident, but it is probably due to the progressive streamlining of the translation initiation mechanism (and in general, of the gene expression process) that took place once the bacterial lineage separated from the common stem of the tree of life. The lack of a comparable streamlining in Archaea is conceivably due to the fact that the Archaea mostly occupy "extreme" ecological niches where competition for fast growing is not so hard as in the bacterial world.

3.2.2 The a/eIF2 and IF2/IF5B Factors

a/eIF2 is a trimeric protein specifically shared by the Archaea and the Eukarya but lacking in Bacteria. Although the Bacteria do possess a translation initiation factor termed IF2, this is not homologous to the same-named archaeal/bacterial protein but to the factor termed IF5B in the other two domains. Therefore, IF2/5B is a universal factor, while a/eIF2 is specific of the archaeal and eukaryal domains only (Kyrpides and Woese 1998) (Table 3.1).

The terminology of these proteins is already confusing, but the confusion is even greater when it comes to their function. Regarding IF2/IF5B, since this protein is one of the two universally conserved initiation factor, one would expect a correspondingly universal and presumably essential function. But it is not so. In Bacteria, IF2 is a truly central player in translational initiation: it interacts with the initiator tRNA (fmet-tRNAi) and promotes its accommodation in the ribosomal P site, at the same time favoring subunit joining (Gualerzi and Pon 2015). By contrast, in Archaea and Eukarya, the initiator tRNA (met-tRNAi) binding factor is the trimeric IF2 (Pedullà et al. 2005; Schmitt et al. 2010), that, as said above, has no counterpart in Bacteria. The archaeal/eukaryal homologue of bacterial IF2, a/eIF5B, does not bind met-tRNAi but apparently still promotes subunit joining, also stabilizing the interaction of met-tRNAi in the P site (Maone et al. 2007).

Therefore, when it comes to the fundamental task of recognizing the specific initiator tRNA and promoting its interaction with the ribosomal P site, there seems to be a clear-cut evolutionary divergence separating the bacterial domain from the archaeal and eukaryal ones.

In the Archaea, moreover, the trimeric IF2 has a peculiar and unexpected function. It interacts specifically with the tri-phosphate 5′ end of the mRNA protecting it against 5′-end degradation Such interaction takes place both with the trimeric form of aIF2 and with its individual subunit γ and is favored when the factor is in a cytoplasmic, free state (Hasenöhrl et al. 2008). Instead, when aIF2 is in a ribosome-bound state, it has a much stronger affinity for met-tRNAi. This dual function of aIF2 is thought to prevent mRNA degradation under unfavorable nutritional conditions, when ribosome synthesis temporarily stops and ribosomes become fewer (Hasenöhrl et al. 2008).

These observations have led to speculate that, in Archaea, IF2 evolved originally to protect the mRNAs against 5′-end degradation, thus prefiguring a sort of cap-binding protein system reminiscent of that seen in modern eukaryotic cells. Archaeal mRNAs have no real “caps”, but their free 5′ tri-phosphate end, interacting specifically with aIF2, would perform the same protective function as the eukaryotic cap. However, the “capping” system seen in modern eukaryotic cells must have evolved de novo during the separate evolution of eukaryal translation, since it is based on components that are specific of the Eukarya and have no counterparts in the other cell domains.

A possible evolutionary history of the tRNAi binding proteins in the three domains of life has been recently described in detail elsewhere (Benelli et al. 2016).

3.2.3 The a/eIF6 Factors

In both Archaea and Eukarya, the translation factor IF6 is a small (27 kDa), monomeric protein that binds specifically to the large ribosomal subunit. The role initially proposed for this factor in eukaryotes was that of preventing the association of the 40S and 60S subunits until the pre-initiation complex was correctly positioned on the start codon (Valenzuela et al. 1982). However, it was later observed that eIF6 is located also in the nucleolus and that its loss affects the biogenesis of 60S particles, suggesting that the protein has an important role in ribosome biosynthesis (Si and Maitra 1999).

aIF6, the archaeal homologue, is a few amino acids shorter than its eukaryal counterpart, but shares otherwise a high degree of homology with it. Its three-dimensional structure has been solved (Groft et al. 2000). It shows a peculiar fold, termed "pentein" because it is composed by a repetition of five very similar domains. The structure of the eukaryal counterpart, modelled on the basis of the archaeal one, is essentially the same.

a/e IF6 binds with high affinity to the large ribosomal subunit, either 50S or 60S. The binding site, first determined for the archaeal factor (Benelli et al. 2009; Greber et al. 2012) and later also for the eukaryal one (Klinge et al. 2011), lies on the surface of the large subunit that interacts with the small subunits, thus justifying its role as an anti-association factor. This region of the ribosome is rather protein-poor; however, IF6 is located in the vicinity of L14p and L24e, and, in Archaea at least, interacts with the former (Benelli et al. 2009).

To date, the role in translation of IF6 remains puzzling. In both the Archaea and the Eukarya, only about 1 in 10 large ribosomal subunits carry a/eIF6 in the cytoplasm. Moreover, the interaction of a/eIF6 with the ribosome is quite strong, and specific factors are required for its release. As regards eukaryotic ribosomes, two different mechanisms have been proposed for eIF6 release. One posits that eIF6 detachment from the 60S subunits is promoted by the GTPase, Efl1, which acts in concert with the ribosome-binding factor Sdo1 (also called SBDS) to couple GTP hydrolysis with IF6 release (Weis et al. 2015). Another proposed mechanism suggests that eIF6 release is triggered by the phosphorylation of the factor, in turn promoted by translation-stimulating signalling transduced by the ribosome-bound kinase RACK1 (Ceci et al. 2003). It is unclear whether these mechanisms co-exist or operate in different circumstances or in different cells.

In eukaryotes, the current consensus model for eIF6 function has it that the factor intervenes in the final maturation steps of the large ribosomal subunit. Immature 60S ribosomes would be shipped to the cytoplasm carrying bound eIF6. The release of the factor, by whichever mechanism, would allow the particles to participate in translation. In this model, the main role of eIF6 would be that of fine-tuning translation by regulating the amount of available 60S subunits.

Compared to Eukarya, much fewer data are available on the function of archaeal IF6. It is known that aIF6 binds tightly and specifically to the 50S ribosomal subunit and thereby inhibits subunit association (Benelli et al. 2009). 50S subunits carrying aIF6 are unable to participate in translation, since they are not found in either 80S ribosomes or in polysomes. However, the mechanism for aIF6 release from the 50S subunit is still unknown. Phosphorylation is in all probability not involved, since efforts to determine whether aIF6 undergoes this type of modification have been unsuccessful (Benelli and Londei, unpublished work). However, the Archaea do harbour a homologue of eukaryal Sdo1/SBDS, which closely corresponds to its eukaryal counterpart in sequence and structure.

The function of aSdo1/SBDS is currently under scrutiny in our laboratories. Preliminary experiments performed with the thermophilic archaeon *S. solfataricus* seem to indicate that addition of recombinant aSdo1 to ribosomes or cell lysates promotes the release of aIF6 in a GTP-dependent manner. Moreover, aSdo1 appears to bind stoichiometrically to the 50S subunit (Benelli, La Teana and Londei, unpublished work). However, the Archaea do not have any evident homologue of the Efl1 protein, suggesting that another, archaeal-specific, GTPase must be involved in the process. Experiments currently under way in our laboratory are aimed at identifying such a GTPase, and at elucidating the mechanism promoting aIF6 release from archaeal large subunits.

As regards the function of archaeal aIF6, there is very little solid evidence so far. Undoubtedly, the protein prevents subunit association and inhibits protein synthesis when added in excess to cell lysates (Benelli et al. 2009), but the physiological significance of this remains elusive. A later study suggests that the main role of aIF6 might be that of promoting ribosome recycling, stimulating the dissociation of 70S ribosomes at the end of each translation cycle (Barthelme et al. 2011). Further data are, however, needed to confirm this surmise. Finally, as suggested for the eukaryotic homologue, aIF6 might participate in ribosome biosynthesis, but the issue remains entirely to be explored experimentally.

The confusion about the role in translation of a/eIF6 is all the more frustrating since this factor, in eukaryotes at least, has clearly a very important role in regulating certain crucial cellular processes. Remarkably, eIF6 over-expression is observed in many natural cancers, while, conversely, eIF6 haplo-insufficiency protects against certain types of tumours (Gandin et al. 2008). Moreover, the over-expression of eIF6 has been described to produce developmental defects in *Xenopus* (De Marco et al. 2010, 2011).

Unfortunately, there are no data in Archaea to show whether aIF6 imbalances have any kind of physiological effects. It is only known that aIF6 is over-expressed under stress conditions, a fact that may suggest a role in controlling cellular behaviour roughly similar to that observed for its eukaryotic counterpart. However, we are still a long way from understanding all the functional facets of this fascinating factor, let alone the motive for its evolutionary conservation in the archaeal/eukaryal line. It is to be hoped that a better understanding of the function of aIF6, the evolutionary forerunner of eIF6, will also help in elucidating the function of the latter.

3.3 Elongation

Elongation is the most conserved among the steps of protein synthesis. In all organisms, the elongation cycle entails the participation of two accessory factors. The first of these, called EF1 in Eukarya and Archaea and EFTu in bacteria, accompanies aminoacyl-tRNA in the ribosomal A site and controls the correctness of codon-anticodon interaction. The second, called EF2 in Archaea and in Eukarya, and EFG in Bacteria, assists the process of translocation, i.e. the movement of the ribosome one codon further down the mRNA. Both elongation factors are G proteins that hydrolyze GTP as an essential part of their function. The prokaryotic proteins are somewhat smaller than the eukaryotic ones, but they are clearly homologous and their mechanism of action is strictly conserved throughout evolution.

Recent research, however, has unveiled certain specialized aspects of elongation that have received scanty attention until now. They regard the function of another evolutionarily conserved translation factor, the protein called EFP in bacteria and IF5A in Archaea and Eukarya. The existence of this protein has been known for a long time, but the details of its function have only recently been analyzed. It is interesting to review the relevant data, since this factor may have many functional facets which we are only just beginning to understand.

The eukaryotic and the bacterial proteins were discovered in the '70s. Their different names are due to the fact that initially the eukaryal protein was classified as an initiation factor (then named eIF4D) and the bacterial one as an elongation factor. In spite of this, both proteins were characterized as having the same activity: the ability to stimulate the formation of (f)Met-Puromycin *in vitro* (Benne et al. 1978; Glick and Ganoza 1975).

The eukaryal protein was found to contain a unique post- translational modification: hypusination. This modification is carried out in two successive enzymatic reactions. In the first deoxyhypusine synthase (DHS) transfers the aminobutyl moiety of spermidine to the ε-amino group of a specific lysine located in the N-terminal domain of the protein, in the second reaction the intermediate, deoxyhypusine, is transformed into hypusine by deoxyhypusine hydroxylase (DOHH) (Cooper et al. 1983; Park et al. 2010).

More recently, a post-translational modification, lysinylation, has been identified also in the bacterial protein. This modification occurs in three steps catalyzed by the following enzymes: YjeK, which converts a free S-α-Lys to R-β-Lys, YjeA, a paralog of lysyl-tRNA synthetases that transfers the R- β-Lys to the ε-amino group of a specific lysine and, finally, YfcM which hydroxylates the lysysl-lysine residue (Reviewed in Rossi et al. 2014).

3D structures are available from Bacteria, Archaea, Protozoa, yeast and human, and they show overall a similar organization: the bacterial protein folds into three domains while in all other cases the proteins are organized in two domains whose structure is superimposable with the first two bacterial domains (Fig. 3.2). The basic N-terminal domain contains the site of post-translational modification in an exposed loop while the acidic C-terminal domain is characterized by an OB-fold (Reviewed in Dever et al. 2014).

A complete functional characterization has been obtained, for both proteins, only in more recent years. Studies have established a role for both eIF5A and EF-P in translation elongation more than in initiation: they are able to promote the synthesis of proteins containing successive residues of proline (PPP or PPG)

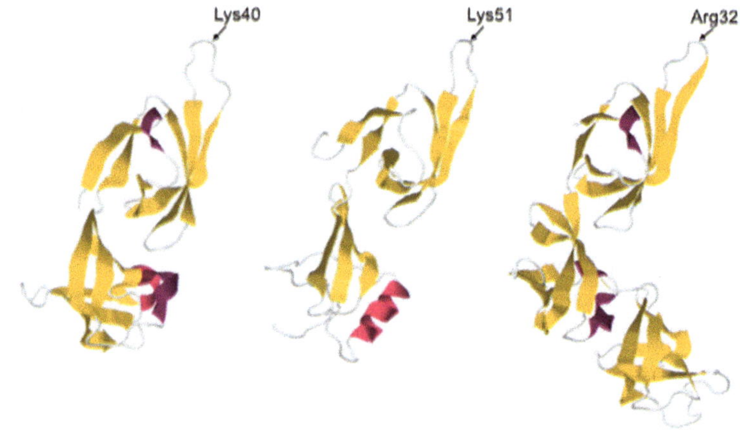

Fig. 3.2 The three-dimensional structures of archaeal (*Pyrococcus horikoshii*, PDB: 1IZ6) and eukaryal (*Saccharomyces cerevisiae*, PDB: 3ER0) IF5A compared with their bacterial homologue (*Thermus thermophilus*, PDB: 1UEB) EF-P. The protein regions with a β-sheet conformation are depicted in *yellow*, the α-helices are *purple*. The *arrows* point at the sites of modification: hypusination for aIF5A and eIF5A and lysinylation for EF-P

(Gutierrez et al. 2013; Ude et al. 2013; Doerfel et al. 2013). Synthesis of these type of sequences, in fact, causes the ribosome to stall, and only the intervention of eIF5A/EF-P guarantees the recovery of the elongation process.

According to the model suggested by structural studies (Schmidt et al. 2016; Melnikov et al. 2016), the factor would bind to stalled ribosomes, trapped in a pre-translocational state and with a free E-site, and interact with A76 residue of a P-site tRNA, via its hypusine moiety. This interaction stabilizes the tRNA facilitating in this way peptide bond formation.

In addition to its direct role as a translation factor, eIF5A has been related to a variety of cellular processes including: mRNA decay (Zuk and Jacobson 1998), cell cycle progression (Hanauske-Abel et al. 1994), apoptosis (Caraglia et al. 2003), cell polarity (Chatterjee et al. 2006; Zanelli and Valentini 2005), retroviral infection (Hoque et al. 2009) and stress responses (Gosslau et al. 2009). Whether eIF5A is endowed with different functions or if this pleiotropic behavior results from secondary effects of its role as a translation factor, remains to be clarified.

The eukaryal and the bacterial proteins, as described above, have been extensively characterized while knowledge on the archaeal one is still very limited.

The presence of a hypusinated protein in Archaea was discovered several years ago: the protein was purified from *Sulfolobus acidocaldarius* DSM 639, it is a protein of 135 AA with a mass of about 15 KDa and pI of 7.8, mainly present in the post-ribosomal fraction (Bartig et al. 1992).

All archaea analyzed so far contain aIF5A, but some have a hypusinated factor, while some others contain the deoxyhypusinated form and very few, both versions of the protein (Bartig et al. 1990). Despite the presence of the different forms of the modified protein, the mechanism of hypusination remains a mystery since so far only the first enzyme involved in aIF5A modification, the DHS enzyme, has been identified in archaeal genomes while no homologs of the second enzyme, DOHH, seem to be present.

A recent paper has shed some light on the *Haloferax volcanii* modification pathway (Prunetti et al. 2016); this organism contains exclusively the deoxyhypusinylated version of aIF5A, spermidine is absent while agmatine and cadaverine represent the main polyamines present. The authors, therefore, propose a model of deoxyhypusine synthesis in *H. volcanii* that differs substantially from the canonical eukaryotic pathway: in the first reaction, the DHS enzyme transfers agmatine to the aIF5A lysine while in the second reaction the agmatinase enzyme leads to production of deoxyhypusine.

The situation in other Archaea might be similar with the involvement of enzymes completely unrelated to the eukaryal DOHH; in alternative the two modification reactions could be catalyzed by a single enzyme, DHS, endowed with a bifunctional activity. The latter possibility is supported by the recent discovery and characterization of such an enzyme in *T. vaginalis* (Quintas-Granados et al. 2016).

In any case modification of the lysine seems to be important since at least some archaea (*S. acidocaldarius*) are sensitive to the DHS inhibitor GC7, which causes a rapid and reversible arrest of growth (Jansson et al. 2000).

The *aif5a* gene appears to be essential at least in *H. volcanii* (Gäbel et al. 2013) and in *S. acidocaldarius* (La Teana, Londei and Albers, unpublished results). Its participation in the translation process has been inferred on the basis of its homology to the other factors but has not yet been demonstrated.

Experiments carried out in our laboratories and aimed at clarifying its role have confirmed that in *S. solfataricus* cell lysates aIF5A is mainly present in the post-ribosomal supernatant. However, when the lysates are programmed for translation by addition of an exogenous mRNA and fractionated on sucrose density gradients, hypusinated aIF5A becomes localized on 70 ribosomes, suggesting a conserved role in translation.

The hypothesis of a participation in the translation process strictly linked to the rescue of proteins containing polyproline motives, however, needs to be verified.

A genome-wide analysis has shown that these proteins are not so common in both Bacteria and Archaea but their abundance increases with the complexity of the organism going from Prokaryotes to Eukaryotes (Mandal et al. 2014). The frequencies of proteins containing PPP and PPG motifs range from 2.0 to 2.5% in Bacteria and Archaea to more than 20% in *H. sapiens*.

On the other hand, polyproline might not be the only motif whose translation is dependent on this factor, as it has been reported for EF-P in some bacterial species (Hersch et al. 2013).

As mentioned above, eIF5A could be involved in processes other than translation; in particular, several reports have characterized it as an RNA binding protein. The homology between the eukaryal and the archaeal protein suggests that this activity might be shared also by aIF5A and the finding that in some species of

Halobacterium the protein shows an RNA binding and degrading activity is in agreement with this hypothesis (Wagner and Klug 2007).

The most interesting aspect about this elongation factor is the conservation of its modification. Both proteins, EF-P and a/eIF5A, are targets of unique modifications, and the most conserved regions with the highest sequence homology are located around the site of modification. The modifications nevertheless are different, β-lysinylation and hypusination, and are catalyzed by completely unrelated enzymes, which are themselves highly conserved within each domain.

As said above, one hypothesis is that these factors and their enzymes have co-evolved to guarantee the synthesis of some essential protein containing proline-rich sequences. Starosta et al. (2014) have analyzed the number and conservation of polyproline-containing proteins across 1273 bacterial, 205 archaeal and 98 eukaryotic genomes finding one proline triplet which is invariant in the Valyl-tRNA synthetase (ValS) genes from all organisms. It may be that this essential protein sufficed by itself to induce the evolution of a factor specifically devoted to stimulate its synthesis. However, it is also possible that in prokaryotes EFP/aIF5A plays some other important role in addition to promoting the translation of poly-pro containing proteins. A fuller investigation of the function of archaeal aIF5A should help to answer this question.

3.4 Conclusions

Despite the considerable advances in our knowledge in the last decade or two, the evolutionary history of the translation process remains to be written in many essential aspects. The unexpected complexity detected in the Archaea regarding some features of translation, and the general closeness between Archaea and Eukarya in the sequences of many translational components, further confirm the now generally accepted idea that Archaea and Eukarya are closely related in evolution.

The prevalent view of the general evolutionary tree of life, that envisages Bacteria as the most antique branch thereof, with Archaea and Eukarya sharing a common evolutionary path before separating in their turn, is in theory open to two different interpretations. The common ancestor of all cells might have had a translational apparatus with a minimal set of components (small, protein poor ribosomes, the two universally conserved initiation factors IF2/IF5B and IF1/IF1A, two elongation factors). After the separation of the bacterial domain, other components could have been added during the common evolution of Archaea and Eukarya, and still others during the separate evolution of the Eukarya.

Alternatively, the last common ancestor of the three domains of life might have had a translation apparatus very similar to that of thermophilic Crenarchaea, deemed to be the oldest branch of the archaeal tree: relatively larger and protein-richer ribosomes and an enlarged set of translation initiation factors, including a/eIF2 (or at least its gamma-subunit), IF1/SUI1 and perhaps a/eIF6. The possible presence of a/eIF2, in either a trimeric or monomeric form, in the common ancestor, is also suggested by the fact that the majority of crenarcheal mRNAs are leaderless, i.e. lack entirely or almost so a 5′ untranslated region. It has long been known (Grill et al. 2000) that leaderless mRNAs are universally translatable by the ribosomes of all extant cells, and that therefore are the likely ancestral form of genetic message. If most ancestral mRNAs had no 5′ leader, it might have been important to protect their 5′ termini until they could be translated.

Under the above scenario, the Bacteria lost some translational components during their separate evolution, remodeling others to perform new functions. This would be, for instance, the case of IF2/5B, which would have acquired the ability of interacting with the initiator tRNA (formerly performed by the lost a/eIF2 or by its gamma subunit, see Benelli et al. 2016), retaining at the same time the ability of promoting subunit joining. IF1/SUI1 is another likely case of a factor lost in the bacterial line, replaced by the new entry IF3. The presence of the IF1/SUI1 gene, most probably in an inactive form, in certain bacterial phyla, might be considered a relic of this loss-replacement process.

References

Baba T, Ara T, Hasegawa M, Takai Y, Okumura Y, Baba M, Datsenko KA, Tomita M, Wanner BL, Mori H (2006) Construction of *Escherichia coli* K-12 in-frame, single-gene knockout mutants: the Keio collection. Mol Syst Biol 2:2006.0008

Barthelme D, Dinkelaker S, Albers SV, Londei P, Ermler U, Tampé R (2011) Ribosome recycling depends on a mechanistic link between the FeS cluster domain and a conformational switch of the twin-ATPase ABCE1. Proc Natl Acad Sci U S A 108:3228–3233

Bartig D, Schumann H, Klink F (1990) The unique posttranslational modification leading to deoxyhypusine or hypusine is a general feature of the archaebacterial kingdom. Syst Appl Microbiol 13:112–116

Bartig D, Lemkemeier K, Frank J, Lottspeich F, Klink F (1992) The archaebacterial hypusine-containing protein. Eur J Biochem 204:751–758

Benelli D, Marzi S, Mancone C, Alonzi T, la Teana A, Londei P (2009) Function and ribosomal localization of aIF6, a translational regulator shared by archaea and eukarya. Nucleic Acids Res 37:256–267

Benelli D, La Teana A, Londei P (2016) Evolution of translational initiation: from Archaea to Eukarya. In: Hernández G, Jagus R (eds) Evolution of the protein synthesis machinery and its regulation. Springer International Publishing, Switzerland, pp 61–80

Benne R, Brown-Luedi ML, Hershey JW (1978) Purification and characterization of protein synthesis initiation factors eIF-1, eIF-4C, eIF-4D, and eIF-5 from rabbit reticulocytes. J Biol Chem 253:3070–3077

Caraglia M, Marra M, Giuberti G, D'Alessandro AM, Baldi A, Tassone P, Venuta S, Tagliaferri P, Abbruzzese A (2003) The eukaryotic initiation factor 5A is involved in the regulation of proliferation and apoptosis induced by interferon-alpha and EGF in human cancer cells. J Biochem 133:757–765

Ceci M, Gaviraghi C, Gorrini C, Sala LA, Offenhauser N, Marchisio PC, Biffo S (2003) Release of eIF6 (p27BBP) from the 60S subunit allows 80S ribosome assembly. Nature 426:579–584

Chatterjee I, Gross SR, Kinzy TG, Chen KY (2006) Rapid depletion of mutant eukaryotic initiation factor 5A at restrictive temperature reveals connections to actin cytoskeleton and cell cycle progression. Mol Gen Genomics 275:264–276

Cheung YN, Maag D, Mitchell SF, Fekete CA, Algire MA, Takacs JE, Shirokikh N, Pestova T, Lorsch JR, Hinnebusch AG (2007) Dissociation of eIF1 from the 40S ribosomal subunit is a key step in start codon selection *in vivo*. Genes Dev 21:1217–1230

Cooper HL, Park MH, Folk JE, Safer B, Braverman R (1983) Identification of the hypusine-containing protein hy+ as translation initiation factor eIF-4D. Proc Natl Acad Sci U S A 80:1854–1857

De Marco N, Iannone L, Carotenuto R, Biffo S, Vitale A, Campanella C (2010) p27(BBP)/eIF6 acts as an anti-apoptotic factor upstream of Bcl-2 during *Xenopus laevis* development. Cell Death Differ 17:360–372

De Marco N, Tussellino M, Vitale A, Campanella C (2011) Eukaryotic initiation factor 6 (eif6) overexpression affects eye development in *Xenopus laevis*. Differentiation 82(2):108–115

Dever TE, Gutierrez E, Shin BS (2014) The hypusine-containing translation factor eIF5A. Crit Rev Biochem Mol Biol 49:413–425

Doerfel LK, Wohlgemuth I, Kothe C, Peske F, Urlaub H, Rodnina MV (2013) EF-P is essential for rapid synthesis of proteins containing consecutive proline residues. Science 339:85–88

Gäbel K, Schmitt J, Schulz S, Näther DJ, Soppa J (2013) A comprehensive analysis of the importance of translation initiation factors for Haloferax volcanii applying deletion and conditional depletion mutants. PLoS One 8:e77188

Gandin V, Miluzio A, Barbieri AM, Beugnet A, Kiyokawa H, Marchisio PC, Biffo S (2008) Eukaryotic initiation factor 6 is rate-limiting in translation, growth and transformation. Nature 455:684–688

Glick BR, Ganoza MC (1975) Identification of a soluble protein that stimulates peptide bond synthesis. Proc Natl Acad Sci U S A 72:4257–4260

Gosslau A, Jao DL, Butler R, Liu AY, Chen KY (2009) Thermal killing of human colon cancer cells is associated with the loss of eukaryotic initiation factor 5A. J Cell Physiol 219:485–493

Greber BJ, Boehringer D, Godinic-Mikulcic V, Crnkovic A, Ibba M, Weygand-Durasevic I, Ban N (2012) Cryo-EM structure of the archaeal 50S ribosomal subunit in complex with initiation factor 6 and implications for ribosome evolution. J Mol Biol 418:145–160

Grill S, Gualerzi CO, Londei P, Bläsi U (2000) Selective stimulation of translation of leaderless mRNA by initiation factor 2: evolutionary implications for translation. EMBO J 19:4101–4110

Groft CM, Beckmann R, Sali A, Burley SK (2000) Crystal structures of ribosome anti-association factor IF6. Nat Struct Biol 7:1156–1164

Gualerzi CO, Pon CL (2015) Initiation of mRNA translation in bacteria: structural and dynamic aspects. Cell Mol Life Sci 72:4341–4367

Gutierrez E, Shin BS, Woolstenhulme CJ, Kim JR, Saini P, Buskirk AR, Dever TE (2013) eIF5A promotes translation of polyproline motifs. Mol Cell 51:35–45

Hanauske-Abel HM, Park MH, Hanauske AR, Popowicz AM, Lalande M, Folk JE (1994) Inhibition of the G1-S transition of the cell cycle by inhibitors of deoxyhypusine hydroxylation. Biochim Biophys Acta 1221:115–124

Hasenöhrl D, Benelli D, Barbazza A, Londei P, Bläsi U (2006) *Sulfolobus solfataricus* translation initiation factor 1 stimulates translation initiation complex formation. RNA 12:674–682

Hasenöhrl D, Lombo T, Kaberdin V, Londei P, Bläsi U (2008) Translation initiation factor a/eIF2 (-gamma) counteracts 5' to 3' mRNA decay in the archaeon *Sulfolobus solfataricus*. Proc Natl Acad Sci U S A 105:2146–2150

Hasenöhrl D, Fabbretti A, Londei P, Gualerzi CO, Bläsi U (2009) Translation initiation complex formation in the crenarchaeon *Sulfolobus solfataricus*. RNA 15:2288–2298

Hersch SJ, Wang M, Zou SB, Moon K, Foster LJ, Ibba M, Navarre WW (2013) Divergent protein motifs direct elongation factor P-mediated translational regulation in *Salmonella enterica* and *Escherichia coli*. mBio 4:e00180–e00113

Hinnebusch AG (2014) The scanning mechanism of eukaryotic translation initiation. Annu Rev Biochem 83:779–812

Hinnebusch AG, Lorsch JR (2012) The mechanism of eukaryotic translation initiation: new insights and challenges. Cold Spring Harb Perspect Biol 4(10)

Hoque M, Hanauske-Abel HM, Palumbo P, Saxena D, D'Alliessi Gandolfi D, Park MH, Pe'ery T, Mathews MB (2009) Inhibition of HIV-1 gene expression by Ciclopirox and Deferiprone, drugs that prevent hypusination of eukaryotic initiation factor 5A. Retrovirology 6:90–106

Jansson BPM, Malandrin L, Johansson HE (2000) Cell cycle arrest in archaea by the hypusination inhibitor N^1-guanyl-1,7-diaminoheptane. J Bacteriol 182:1158–1161

Johnson AG, Grosely R, Petrov AN, Puglisi JD (2017) Dynamics of IRES-mediated translation. Philos Trans R Soc B 372. pii: 20160177

Klinge S, Voigts-Hoffmann F, Leibundgut M, Arpagaus S, Ban N (2011) Crystal structure of the eukaryotic 60S ribosomal subunit in complex with initiation factor 6. Science 334:941–948

Kyrpides NC, Woese CR (1998) Universally conserved translation initiation factors. Proc Natl Acad Sci U S A 95:224–228

Lecompte O, Ripp R, Thierry JC, Moras D, Poch O (2002) Comparative analysis of ribosomal proteins in complete genomes: an example of reductive evolution at the domain scale. Nucleic Acids Res 30:5382–5390

Londei P (2005) Evolution of translational initiation: new insights from the archaea. FEMS Microbiol Rev 29:185–200

Maag D, Fekete CA, Gryczynski Z, Lorsch JR (2005) A conformational change in the eukaryotic translation preinitiation complex and release of eIF1 signal recognition of the start codon. Mol Cell 17:265–275

Maag D, Algire MA, Lorsch JR (2006) Communication between eukaryotic translation initiation factors 5 and 1A within the ribosomal pre-initiation complex plays a role in start site selection. J Mol Biol 356:724–737

Mandal A, Mandal S, Park MH (2014) Genome-wide analyses and functional classification of proline repeat-rich proteins: potential role of eIF5A in eukaryotic evolution. PLoS One 9: e111800

Maone E, Di Stefano M, Berardi A, Benelli D, Marzi S, La Teana A, Londei P (2007) Functional analysis of the translation factor aIF2/5B in the thermophilic archaeon *Sulfolobus solfataricus*. Mol Microbiol 65:700–713

Melnikov S, Mailliot J, Shin BS, Rigger L, Yusupova G, Micura R, Dever TE, Yusupov M (2016) Crystal structure of hypusine-containing translation factor eIF5A bound to a rotated eukaryotic ribosome. J Mol Biol 428:3570–3576

Osterman IA, Evfratov SA, Dzama MM, Pletnev PI, Kovalchuk SI, Butenko IO, Pobeguts OV, Golovina AY, Govorun VM, Bogdanov AA, Sergiev PV, Dontsova OA (2015) A bacterial homolog YciH of eukaryotic translation initiation factor eIF1 regulates stress-related gene expression and is unlikely to be involved in translation initiation fidelity. RNA Biol 12:966–971

Park MH, Nishimura K, Zanelli CF, Valentini SR (2010) Functional significance of eIF5A and its hypusine modification in eukaryotes. Amino Acids 38:491–500

Passmore LA, Schmeing TM, Maag D, Applefield DJ, Acker MG, Algire MA, Lorsch JR, Ramakrishnan V (2007) The eukaryotic translation initiation factors eIF1 and eIF1A induce an open conformation of the 40S ribosome. Mol Cell 26:41–50

Pedullà N, Palermo R, Hasenöhrl D, Bläsi U, Cammarano P, Londei P (2005) The archaeal eIF2 homologue: functional properties of an ancient translation initiation factor. Nucleic Acids Res 33:1804–1812

Preiss T (2016) All Ribosomes are created equal. Really? Trends Biochem Sci 41:121–123

Prunetti L, Graf M, Blaby IK, Peil L, Makkay AM, Starosta AL, Papke RT, Oshima T, Wilson DN, de Crécy-Lagard V (2016) Deciphering the translation initiation factor 5A modification pathway in halophilic Archaea. Archaea 2016:7316725

Quintas-Granados LI, Carvajal Gamez BI, Villalpando JL, Ortega-Lopez J, Arroyo R, Azuara-Liceaga E, Álvarez-Sánchez ME (2016) Bifunctional activity of deoxyhypusine synthase/hydroxylase from Trichomonas vaginalis. Biochimie 123:37–51

Rossi D, Kuroshu R, Zanelli CF, Valentini SR (2014) eIF5A and EF-P: two unique translation factors are now traveling the same road. Wiley Interdiscip Rev RNA 5:209–222

Schmidt C, Becker T, Heuer A, Braunger K, Shanmuganathan V, Pech M, Berninghausen O, Wilson DN, Beckmann R (2016) Structure of the hypusinylated eukaryotic translation factor eIF-5A bound to the ribosome. Nucleic Acids Res 44:1944–1951

Schmitt E, Naveau M, Mechulam Y (2010) Eukaryotic and archaeal translation initiation factor 2: a heterotrimeric tRNA carrier. FEBS Lett 584:405–412

Si K, Maitra U (1999) The *Saccharomyces cerevisiae* homologue of mammalian translation initiation factor 6 does not function as a translation initiation factor. Mol Cell Biol 19:1416–1426

Starosta AL, Lassak J, Peil L, Atkinson GC, Woolstenhulme CJ, Virumäe K, Buskirk A, Tenson T, Remme J, Jung K, Wilson DN (2014) A conserved proline triplet in Val-tRNA synthetase and the origin of elongation factor P. Cell Rep 9:476–483

Ude S, Lassak J, Starosta AL, Kraxenberger T, Wilson DN, Jung K (2013) Translation elongation factor EF-P alleviates ribosome stalling at polyproline stretches. Science 339:82–85

Valenzuela DM, Chaudhuri A, Maitra U (1982) Eukaryotic ribosomal subunit anti-association activity of calf liver is contained in a single polypeptide chain protein of $M\mathrm{r} = 25{,}500$ (eukaryotic initiation factor 6). J Biol Chem 257:7712–7719

Wagner S, Klug G (2007) An archaeal protein with homology to the eukaryotic translation initiation factor shows ribonucleolytic activity. J Biol Chem 282:13966–13976

Weis F, Giudice E, Churcher M, Jin L, Hilcenko C, Wong CC, Traynor D, Kay RR, Warren AJ (2015) Mechanism of eIF6 release from the nascent 60S ribosomal subunit. Nat Struct Mol Biol 22:914–919

Xue S, Barna M (2012) Specialized ribosomes: a new frontier in gene regulation and organismal biology. Nat Rev Mol Cell Biol 13:355–369

Zanelli CF, Valentini SR (2005) Pkc1 acts through Zds1 and Gic1 to suppress growth and cell polarity defects of a yeast eIF5A mutant. Genetics 171:1571–1581

Zuk D, Jacobson A (1998) A single amino acid substitution in yeast eIF-5A results in mRNA stabilization. EMBO J 17:2914–2925

Chapter 4
An Overview of Ribonuclease Repertoire and RNA Processing Pathways in Archaea

Duy Khanh Phung, Marie Bouvier, and Béatrice Clouet-d'Orval

Abstract RNA processing plays a crucial role in post-transcriptional regulation of gene expression. Work conducted in Bacteria and Eukarya has defined the predominant mRNA maturation and decay pathways, as well as enzymes and cofactors responsible for these processes. In contrast, our knowledge of the mechanisms controlling RNA quality and processing in Archaea is more fragmentary. In essence, the major actors of RNA processing are ribonucleases acting in cleaving or trimming RNA molecules according to their nature and fate, making these enzymes fascinating and important players to study. More than 30 families of ribonucleases have been described in Bacteria and Eukarya, while only few have been identified in Archaea. This chapter is focused on the major ribonucleases in Archaea. After an overview of archaeal cellular RNA biotypes, we present synthetic up-to-date repertoire of the archaeal ribonuclease families as well as our state of knowledge on their roles in dedicated RNA processing pathways. In addition to this general description of archaeal RNA processing actors, specific pathways involved in processing of rRNAs, tRNAs, crRNAs and C/D sRNAs are detailed in other chapters.

4.1 Introduction

The survival of a cell depends on its ability to rapidly adapt to changing environmental conditions in which RNA biology plays a critical role as it directs multiple mechanistically distinct processes. Indeed, expression of genetic information relies on three major types of RNAs: messenger RNAs (mRNAs) that are intermediates in defining the proteome; ribosomal RNAs (rRNAs) and transfer RNAs (tRNAs) that are effector molecules acting together to decode mRNA sequence information; and

D.K. Phung • M. Bouvier
Laboratoire de Microbiologie et de Génétique Moléculaires (LMGM), Centre de Biologie Intégrative (CBI), Université de Toulouse, CNRS, Université Paul Sabatier, 31062 Toulouse, France

B. Clouet-d'Orval (✉)
Laboratoire de Microbiologie et de Génétique Moléculaires (LMGM), Centre de Biologie Intégrative (CBI), Université de Toulouse, CNRS, 31062 Toulouse, France
e-mail: Beatrice.Clouet-Dorval@ibcg.biotoul.fr

B. Clouet-d'Orval (ed.), *RNA Metabolism and Gene Expression in Archaea*,
Nucleic Acids and Molecular Biology 32, DOI 10.1007/978-3-319-65795-0_4

finally a plethora of noncoding regulatory RNAs (ncRNAs) that can regulate mRNA (sRNA/microRNA/siRNA), have key roles in mRNA splicing (snRNAs) or guide RNA modifications (box C/D and H/ACA guide RNAs).

Regulation of gene expression occurs at multiple levels including post-transcriptional processes, referred to as RNA processing, which are fundamental in defining the phenotypic characteristic of a cell. In Archaea, this level of gene regulation remains to be largely explored. Identification and characterization of actors involved in archaeal RNA maturation and decay pathways are still in their early stages. While it is now well established that archaeal and eukaryotic cells share many key components of their replication, transcription, and translation machineries (Brochier-Armanet et al. 2011; Spang et al. 2015), it remains to assess in which extend this feature apply to RNA processing and RNA regulatory factors. Providing answers to these points will bring perspectives in understanding the evolutionary route of the players acting in the biology of functional archaeal RNAs.

The level of which any RNA is expressed is determined to the same extent by its rates of transcription and by its decay. In addition, primary transcripts undergo multiple covalent modifications to reach their functional state. These processes involve the orderly action of a battery of cellular enzymes from RNA synthesis and maturation to decay to precisely monitor their quality and level. Therefore, cells depend on RNA processing systems to produce mature RNA from immature precursors as well as to discard any transcriptional byproducts and malformed transcripts. The enzymes directly responsible of these processes are ribonucleases (RNases) that catalyze the exo- or endoribonucleolytic cleavage of a phosphodiester bond from an RNA molecule. They are essential both for nonspecific RNA degradation and for numerous forms of RNA processing. Indeed, they control the fate of cellular mRNAs and make mature rRNAs, tRNAs and regulatory ncRNAs from precursors which undergo complex processes that require a high number of modifications in addition to 5′- and 3′-end processing and, often, intron splicing. Most of the known RNases are protein enzymes, but, in several cases, the catalytic moiety is an RNA molecule.

While the action of endoribonucleases (endoRNases) initiates RNA maturation and decay through internal cleavage within the transcript sequence, the exoribonucleases (exoRNases) digest RNA from either end. Consequently, cleavage by endoRNases generates substrate for 3′-5′ and 5′-3′ exoribonucleolytic decay, resulting in a tight cooperation between different RNA degrading systems. In this framework, the nature of the 5′ and 3′ ends of an RNA molecule defining its accessibility to ribonucleases is critical in its outcome. Further characteristics of ribonucleases included their ability to cleave single and/or double stranded RNA or RNA/DNA hybrid, the nature of their byproduct, their specificity for substrates of defined shape and sequence, their ability to digest DNA besides RNA, and their processive or distributive action. Systematically, exoRNases function by one of two mechanisms, either hydrolytically yielding nucleoside monophosphate products or phosphorolytically using orthophosphate as a nucleophile and producing nucleoside diphosphates in a reversible reaction.

The stability of RNA is among other things determined by a cell-specific set of RNases that can be specific or shared by several RNA processing pathways. In addition, other RNA-related proteins or auxiliary enzymes including RNA helicases, poly(A) polymerases and pyrophosphohydrolases can form RNA degrading complexes or machines to enhance and specify their activity. Moreover, the action of RNases, are tightly controlled through several means to prevent incorrect processing. For instance, the recruitment of RNA-modifying enzymes to their respective targets is specified by number of cis-acting RNA sequence elements, as well as a large repertoire of trans-acting proteins and ncRNAs.

In this context, this is not surprising that RNases, acknowledged as precision tools for RNA, are often essential in cell viability [for review see Arraiano et al. (2013)]. Understanding the unique and shared roles of these enzymes in the cell is a foremost challenge to decipher post-transcriptional regulation pathways of gene expression. Excellent in-depth reviews covering enzymes involved in critical aspects of RNA processing in Bacteria and Eukarya were highlighted in a special issue of "Biochimica and Biophysica Acta" edited by Stoecklin and Muhlemann (2013) dedicated to RNA Decay mechanisms. In light of the crucial role of RNases in bacterial and eukaryal cellular metabolism, it is of major importance to identify their archaeal counterparts to deeply understand post-transcriptional gene expression regulation in Archaea.

As mentioned above, archaeal and eukaryotic cells share many key components of their informational machineries (Brochier-Armanet et al. 2011). It is interesting to note that this affiliation extends to several of RNA processing enzymes and RNA regulatory factors which include the exosome and other translation-related systems, as snoRNP guide-dependent modifications and tRNA intron processing. However, homologues to bacterial-specific enzymes have also being identified in archaeal genomes. To this respect, archaeal cells possess original RNA processing, turnover and quality control settings with mosaic features which remain to be explored. This highlights the advantage of developing archaeal models to gain further mechanistic and evolutionary information of fundamental processes across the three domains of life.

4.2 Archaeal RNA Biotypes

The common archaeal RNA biotype that is represented in Fig. 4.1 includes mRNAs, rRNAs, tRNAs and diverse group of ncRNAs such as box C/D and H/ACA guide RNAs, the RNA moiety of RNase P, CRISPR RNAs and diverse specific regulatory ncRNAs. Archaea undoubtedly use a variety of enzymatic pathways to mature and degrade all the cellular types of RNA (Fig. 4.1).

In contrast to Eukarya which uses several classes of RNA polymerases (RNAPs) that specifically transcribe distinct and non-overlapping subsets of genes, unique transcription machinery is in charge of the transcription of all types of RNAs in Archaea (Fig. 4.1). Nevertheless, the archaeal RNAP is closely related to the eukaryotic RNA polymerase II (RNAPII) in terms of subunit composition, structure, use of general factors and mechanism of action (Werner 2007; Werner and

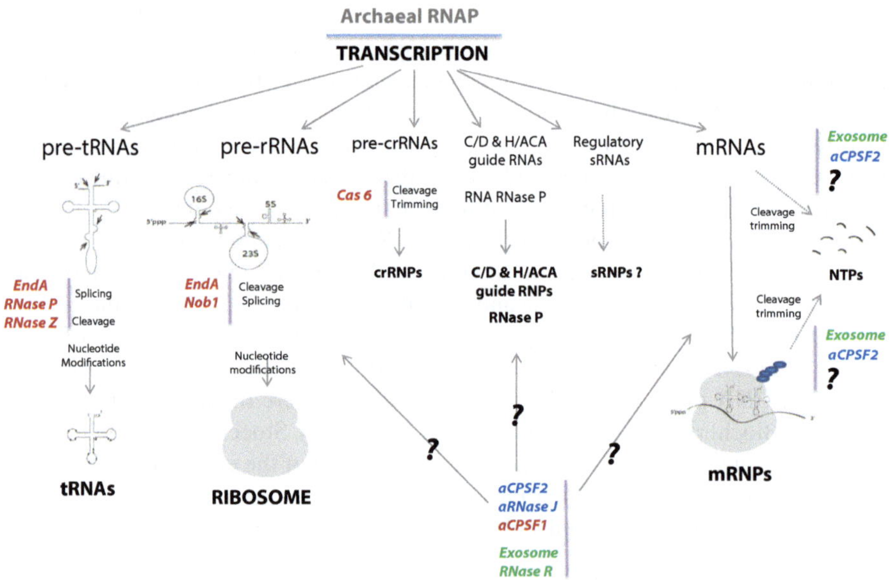

Fig. 4.1 Archaeal RNA biotypes. The unique archaeal RNAP transcribes all the different biotypes of RNAs. Only few tRNA, rRNA and CRISPR RNA maturation steps have been identified of up-to-date in Archaea. The few known RNA-modifying reactions (Cleavage/Splicing/Nucleotide modifications) are in *blue* and corresponding enzymes in *red*. *Question marks* indicate that some RNases remained to be identified. Note that many nucleotide modifications such as 2′-O-methylations and pseudourydilations are guided by C/D and H/ACA sRNPs, respectively. The *stars* (*) indicate that the catalytic moiety of the RNP is an RNA molecule. The exosome, aCPSF1, aCPSF2 and aRNase J have not being yet attributed to any specific RNA processing pathways

Grohmann 2011). The core of the archaeal and eukaryotic transcription machineries is conserved and the minimal promoter (TATA box & B-recognition element) and site-specific transcription start sites are identical. Briefly, binding of TATA-binding protein (TBP) to the promoter induces a cascade that leads of protein recruitment to transcription initiation; transcription factor B (TFB) binds to the TBP-DNA complex, and the DNA–TBP–TFB complex subsequently recruits RNAP and transcription factor E (TFE) (discussed in Chap. 1). In contrast to the basal transcription machinery, gene-specific signal-responsive transcription regulation is mediated by small bacterial-like transcription factors (Aravind 1999; Peeters et al. 2013). Recently, principles underlying how these factors cooperate in nucleoid structuring and gene regulation emerged. Accumulating evidence suggests that nucleosomal organization in Archaea is interlinked with transcription processes and is unique in many aspects (Peeters et al. 2015) (discussed in Chap. 2).

Less is understood on transcription termination and its regulation remains a mystery in Archaea. Whether the formation of RNA 3′ end depends on transcription termination factors as in eukaryotic cells (Dominski et al. 2013; Proudfoot 2011) needs to be determined. The first experimental evidence for intrinsic termination of archaeal transcription show that archaeal transcription termination is stimulated by oligo(T) sequences (Santangelo et al. 2009). More recently a broader sequence diversity of termination motifs was revealed by the term-seq sequencing method to comprehensively map RNA 3′ termini of hundreds of genes suggesting an unexpected complexity of archaeal mRNA 3′ ends (Dar et al. 2016; Garrett 2016).

Despite the similarities between archaeal and eukaryal transcription, archaeal mRNAs share bacterial mRNA characteristics with no introns, no 5′ methyl caps and no long 3′ poly A tails but with 5′ triphosphorylated ends and often polycistronic structure (Fig. 4.1). However, early data suggest that some archaeal mRNAs have significantly longer half-life than most mRNAs in bacteria (Evguenieva-Hackenberg and Klug 2009). It was reported that transcripts half-lives vary from 7 to 57 min in *Methanococcus vannielii* (Hennigan and Reeve 1994), from 6 to 120 min in *Sulfolobus solfataricus* (Bini et al. 2002) and from 4 to 80 min *in Haloferax mediterranei* (Jager et al. 2002). These initial studies proposed that the transcript 3′ ends are generated by an endonucleolytic cleavage followed by a directional 3′ to 5′ degradation. However, no specific endonucleolytic cleavage signal has been characterized so far that could support this early model of mRNA decay. In addition, more recently, *in vitro* and *in vivo* evidence suggested that the translation initiation factor a/eIF2-γ subunit binds to RNA 5′-triphosphorylated ends and protects transcripts from a 5′ end-dependent degradation pathway in the crenarchaeota *S. solfataricus* (Arkhipova et al. 2015; Hasenohrl et al. 2008). In this case, a/eIF2-γ subunit exhibits an additional function with resemblance to the eukaryotic cap-complex. This strongly suggests that a directional 5′ to 3′ degradation pathway may exist in archaeal cells. Nevertheless, pathways and factors controlling mRNA stability in Archaea are still hardly documented.

More generally, in Archaea, all tRNAs are issued from primary transcripts which are extensively processed to yield the mature and functional forms (Fig. 4.1). This process requires the universal ribonucleoprotein RNase P (Frank and Pace 1998) (discussed in Chap. 7) and RNase Z (Holzle et al. 2008; Redko et al. 2007; Schiffer et al. 2002). In addition an RNA-splicing endoRNase (EndA) removes introns from archaeal pre-tRNA and also participates in rRNA processing (Li et al. 1998; Tang et al. 2002; Thompson and Daniels 1988) (Fig. 4.1). All mature tRNAs harbored at their 3′ end a CCA triplet that is essential for both aminoacylation and recognition of the tRNA by the ribosome. For many bacterial and some archaeal tRNAs, the CCA is encoded by their respective genes and therefore is already present in the precursor (Marck and Grosjean 2002). In other cases, including all eukaryotic and most archaeal tRNAs, the CCA is added post-transcriptionally as a part of the maturation process (Cho et al. 2005; Weiner 2004).

As in Eukarya and Bacteria, rRNAs are produced from polycistronic RNA precursor (Fig. 4.1) (discussed in Chap. 6). The number of rRNA operons per genome varies from one to four depending on the archaeal specie. rRNA operons contain the 16S and 23S genes in crenarchaea with the addition of, and 5S genes in euryarchaea with a tRNA Ala gene located in the internal transcribed spacer (ITS) and a tRNA Cys gene in the distal portion of the precursor (Klug et al. 2007).

Modifying guide RNAs, abundant in Archaea and named C/D box and H/ACA box sRNAs (Fig. 4.1), guide 2′O-methylation at the ribose moiety and pseudouridylation at specific rRNA and tRNA nucleotides, respectively (Dennis and Omer 2005; Dennis et al. 2001). They are central in the landscape of RNA modifications in the eukaryal and archaeal cells. Both archaeal C/D box and H/ACA box sRNAs associate with a set of proteins to form active ribonucleproteins (RNPs) in which an intact Kink-turn or Kink-loop RNA motif is critical (Charpentier et al. 2005; Lapinaite et al. 2013; Li and Ye 2006; Nolivos et al. 2005; Yip et al. 2016; Zago et al. 2005) (discussed in Chap. 9).

Finally, crRNAs called CRISPR RNAs are, so far, exclusively involved in cellular defense against infectious genetic elements, such as plasmids and viruses (Fig. 4.1). Several types of CRISPR-cas immunity systems have been described and are found in the majority of archaeal genomes (Makarova et al. 2015). A hallmark of defense mechanisms based on clustered regularly interspaced short palindromic repeats (CRISPR) and associated proteins (Cas) are the crRNAs that guide these complexes in the destruction of invading DNA or RNA [for review, Barrangou et al. (2013)]. The proteins encoded by the *cas* genes include predicted RNA-binding proteins, nucleases, helicases and polymerases. The transcription of the CRISPR arrays produces long pre-crRNA, which are further cleaved at each repeat into individual crRNA (Fig. 4.1). Finally, during the interference phase, invading nucleic acids that match a crRNA are cleaved by effector RNP complexes consisting of dedicated Cas proteins associated to individual crRNAs (discussed in Chap. 11).

Altogether only a subset of archaeal ribonucleolytic enzymes has been assigned to specific pathways dictating the biogenesis and the fate of archaeal RNA biotypes. Many pathways and associated actors, especially those controlling mRNA processing and decay wait to be characterized (Fig. 4.1).

4.3 Ribonuclease Families

4.3.1 RNA Processing Actors in Bacteria and Eukarya

One of the main approaches used to decipher the main actors of RNA processing in Archaea is based on experience and knowledge accumulated in Eukarya and Bacteria. In Eukarya, it is now well established that multiple parallel, and partially redundant, mRNA turnover and surveillance pathways take place in the nucleus and cytoplasm of the cells. These processes are mainly 3′-to-5′ and 5′-to-3′ pathways that are respectively mediated by the 3′-5′ Exosome complex and the 5′-3′ XRN exonuclease family [for review see Brooks (2010), Henras et al. (2015), Nagarajan et al. (2013), Stoecklin and Muhlemann (2013)]. Ribosomal RNA maturation and ribosome biogenesis in Eukarya are also oriented processes starting in the nucleolus and ending in the cytoplasm that involve exo- and endoribonucleolytic processing events coupled with nucleotide modifications. The accuracy of pre-rRNA processing is tightly monitored by quality control mechanism [for review see Fernandez-Pevida et al. (2015), Henras et al. (2015)].

Although RNA processing pathways including rRNA maturation and mRNA decay have been identified in several bacterial species, RNases and RNA degrading complexes are best studied in the model organisms *E. coli* and *B. subtilis* (Deutscher 2009; Hui et al. 2014). In *E. coli*, the canonical RNA degradosome is composed of the essential endoRNase RNase E which serves as a scaffold of the complex by interacting with the exoRNase PNPase, the RNA helicase RhlB and the glycolytic enzyme enolase (Bandyra et al. 2013; Carpousis 2007) [for review, Mackie (2013)]. In *B. subtilis*, a central endoRNase, RNase Y, has been shown to interact with the exoRNases PNPase and RNase J1, the DEAD box RNA helicase CshA and the two glycolytic enzymes enolase and frutokinase.

Throughout all domains of life, biogenesis and post-transcriptional modification processes involved in tRNA biology are the most conserved ones as they are fundamental in translating the genetic code but are also involved in regulating gene expression and in modulating apoptosis and several other biological processes (for review see Huang and Hopper (2016)]. Finally many processes involving endo- and exo-RNases have been discovered in the late years, including for example RNases involved in bacterial CRISPR guide RNA maturation (Hochstrasser and Doudna 2015) or those of the eukaryotic mi/siRNA pathways (Filipowicz et al. 2008).

Recently the identification of an ever-increasing number of ribonucleolytic enzymes combined with the availability of protein sequence of unknown function in databases has allowed the grouping of RNases in defined families. According to the presence of signature motifs in their amino acid sequence for protein enzymes, bacterial and eukaryal RNases involved in numerous physiological processes have been grouped into more than 30 families [for review, Stoecklin and Muhlemann (2013)].

4.3.2 Identified Ribonuclease Families in Archaea

Performing systematic comparative analysis and searching for gene sequences and gene-order conservation in the archaeal genomes opened the prospecting of RNA biology actors in Archaea (Koonin et al. 2001). One of the main examples of RNase prediction was the detection of an archaeal exosome by comparative genomic approaches using databases of protein domains (Koonin et al. 2001). Within the reported RNase families (Arraiano et al. 2013; Stoecklin and Muhlemann 2013), only 9 includes archaeal counterparts (Table 4.1). It should be noted that genes encoding homologs of the key bacterial endoribonucleases- RNase E/G, RNase III and RNase Y families, and of the eukaryal XRN family, that are essential to maintain the integrity of the RNA pool in the cells, could not be identified in any up-to-date annotated archaeal genome. While earlier an RNase E-like activity has been previously described in haloarchaea (Franzetti et al. 1997), this assertion has

Table 4.1 Archaeal RNase families

		Family[a]	Name	Role	Bact.[a]	Euk.[a]
3′ > 5′ EXO		RNB	RNase R	*n.d.*	✓	✓
		PDX	Rrp41/ Rrp42	*n.d.*	✓	✓
5′ > 3′ EXO	β-lactamase	β-CASP	aRNase J	*n.d.*	✓	
			aCPSF2	*n.d.*		✓
ENDO/5′ > 3′ EXO			aCPSF1	*n.d.*		✓
ENDO		Elac	RNase Z	tRNA processing	✓	✓
		RNase P	RNase P	tRNA processing	✓	✓
		SEN	End A	tRNA & rRNA processing		✓
		PIN	Nob1	rRNA processing	✓	✓
		Ferredoxin-like	Cas6	CRISPR RNA processing	✓	
		RNase H	HI/HII		✓	✓

[a]Occurrence of ribonuclease families in Eukarya (Euk.) and Bacteria (Bact.) as reported in Aravind and Koonin (2001), Arraiano et al. (2013), Condon and Putzer (2002)

been called into question since no such homologs could be identified by phylogenomic analysis (Quentin and Clouet-d'Orval personal communication). Nonetheless a putative archaeal protein of the thermophilic archaeon *Pyrococcus furiosus* with short sequence similarity to the AU-binding domain of RNase G, so-called FAU-1, has been shown to have RNA binding properties (Kanai et al. 2003).

The enzymatic activities of the archaeal ribonucleases identified have been characterized *in vitro* and are described one by one in the following sections (Table 4.1; Fig. 4.2). While tRNA and rRNA processing pathways are supported by strictly conserved endoRNases, the cellular RNA targets of most archaeal exoRNases remained to be identified (Table 4.1).

4.4 EndoRNases Identified in rRNA, tRNA, crRNA Processing and DNA Maintenance Pathways in Archaea

4.4.1 RNase P

In the different lifeforms, the majority of tRNA transcripts are synthesized as precursors containing 5′ leader and 3′ trailer sequences that should be accurately excised. The ribonucleoprotein RNase P is an endoRNase responsible for the removal of 5′ leader sequences from tRNA precursors and, therefore, has a key role in cellular homeostasis and survival in most organisms. It is interesting to

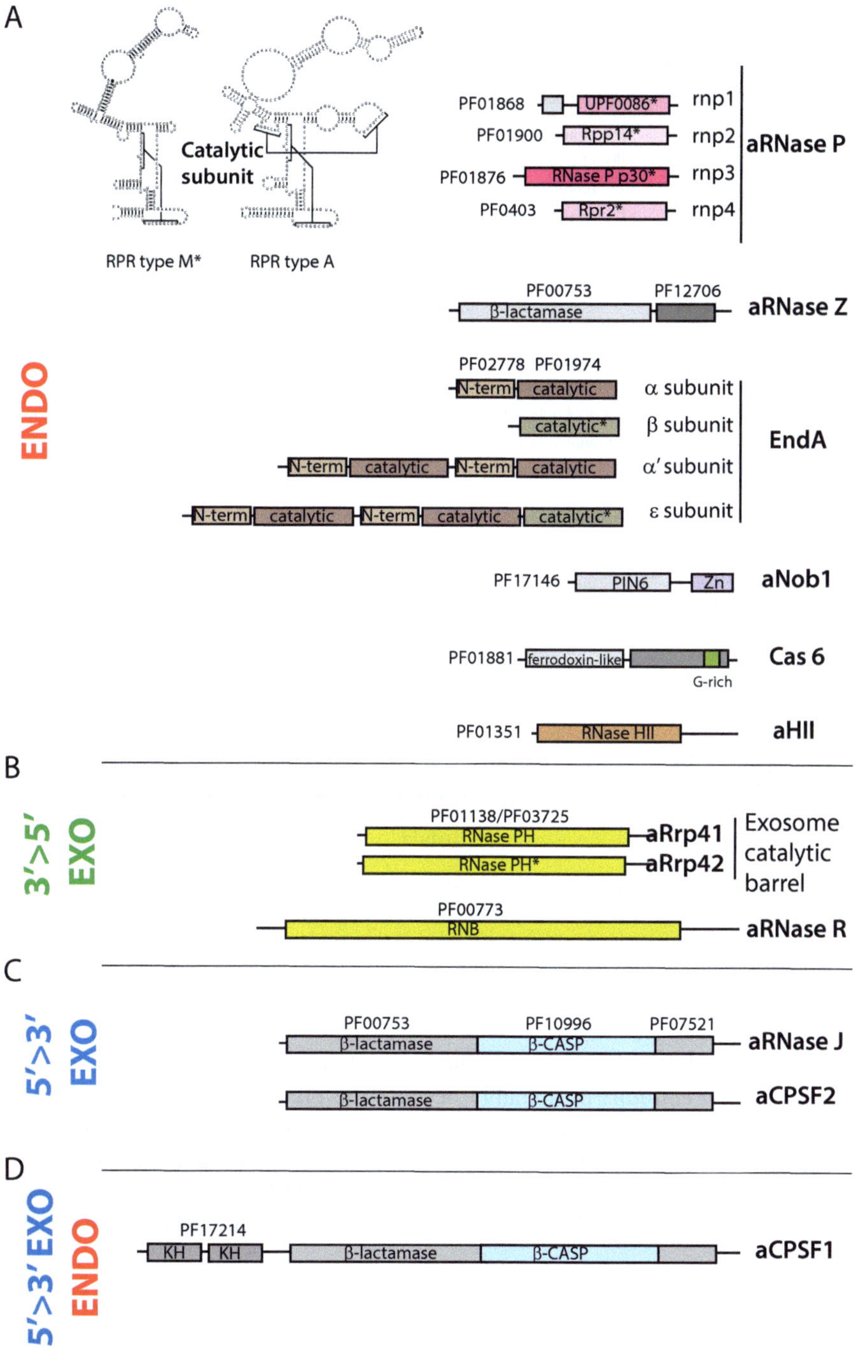

Fig. 4.2 Domain architecture of archaeal RNases. Pfam annotation (http://pfam.xfam.org/) is indicated for each enzyme domain. (**a**) EndoRNases with identified RNA targets. The two types of structure for RNA of RNase P (RPR) are shown. Archaeal RNase P RNA subunits are classified as

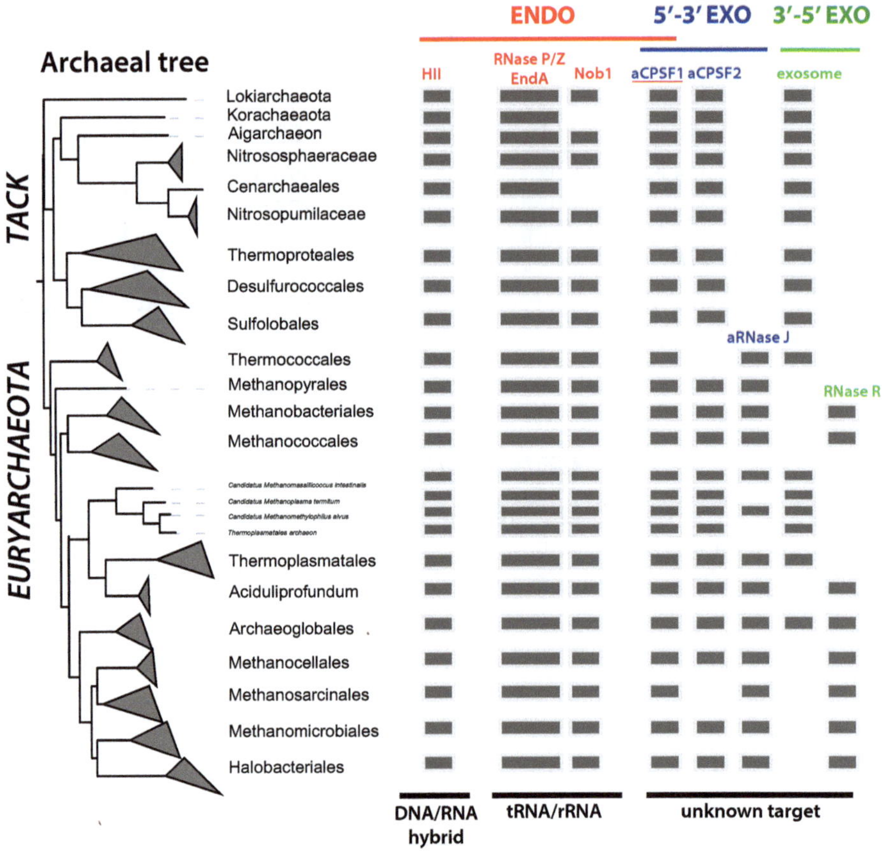

Fig. 4.3 Phylogenetic distribution of reported RNases. Archaeal phylogenetic was constructed as in Clouet-d'Orval et al. (2015). Note that the fast evolving parasitic DPANN clan hood (comprising Nanoarchaea and its relatives) is not included in this phylogeny. *Grey rectangle* indicates that at least one member of the clade possess in its genome a gene encoding the RNases mentioned in *red*, *blue* and *green* upon their activity

mention that the only know organism without RNase P in the archaeon *Nanoarchaeum equitans* presumably because the pre-tRNAs in this organism are transcribed without leader sequence (Randau et al. 2008) (Fig. 4.3).

RNase P which has divergent scaffolds across the three domains of life is an Mg^{2+}-dependent endonucleolytic ribonucleoprotein (RNP) complex which catalyzes phosphodiester bond hydrolysis of tRNA precursors generating tRNAs with mature 5′

Fig. 4.2 (continued) type M (*Methanocaldococcus jannaschii*) and type A (*P. furiosus*) (Pannucci et al. 1999) (The RNase P database, http://www.mbio.ncsu.edu/RNaseP/seqs&structures.html). (**b**) 3′-5′ ExoRNases with uncharacterized RNA target. (**c**) 5′-3′ ExoRNases with uncharacterized RNA target. (**d**) RNase with dual activity

phosphate-ends and 5′ leaders with a 3′ hydroxyl group. RNase P holoenzyme is shaped around an essential catalytic RNA subunit (termed RPR) with an average length of 400-nt, and respectively, one, at least four or 10 protein subunits (termed RPPs) in Bacteria, Archaea and Eukarya [for review see Klemm et al. (2016), Lai et al. (2010), Samanta et al. (2016)].

Unlike the bacterial RPPs that are capable of efficient catalysis in the absence of protein, the archaeal and eukaryal RPRs depend fundamentally on protein for activity. Many structures have provided significant insight into structural features that are key for catalysis and substrate recognition (Gopalan 2007). Archaeal RPR subunits are composed of two functional RNA domains, the substrate-binding domain (S-domain) and the catalytic domain (C-domain), which commonly resemble to the ancestral bacterial type A (Fig. 4.2a) (Jarrous and Gopalan 2010). However, a type M RPR is encoded by *Methanococci* and *Archaeoglobulus* (Pannucci et al. 1999) (Fig. 4.2a). The biochemical properties of the archaeal RPRs are similar to those of synthetic minimal bacterial RPRs, suggesting that the archaeal RNAs contain all of the elements required for substrate recognition and catalysis but are structurally defective in the absence of protein.

Archaeal RNase P contains at least 4 proteins—RPP21, RPP29, RPP30 and POP5—(Fig. 4.2a), with a fifth—L7Ae—associating with type M RNA, all of which are homologous to yeast and human nuclear RNase P proteins (Samanta et al. 2016). While there is no atomic resolution of full-archaeal RNase P enzymes, structures have been solved for each of the RPP subunits (Numata et al. 2004) (discussed in Chap. 7). Each subunit is briefly described in the following text.

The RPP21 subunit has a zinc ribbon in the C-terminal domain, a motif found in other nucleic-acid-binding proteins with four invariant cysteines. The RPP29 is characterized by a twisted barrel of seven antiparallel β-strands (Samanta et al. 2016). These two subunits associate to form the RRP21: RPP29 binary complex that functions in enhancing substrate specificity (Sinapah et al. 2011). This sub complex whose structure has been solved contacts the S-domain of the RNA moiety. The POP5 subunit is structurally similar to the unique bacterial RPP with β-strands forming a central cleft and near-universally conserved Arg/Lys and Phe or Tyr residues. The RPP30 shows the least sequence conservation but mutations at specific Arg residues exhibit decreases in activity (Samanta et al. 2016). The POP5: RPP30 binary complex with a hydrophobic interface contacts with the C-domain of the RNA moiety. As previously described for RPP21:RPP29, the POP5:RPP30 binary complex has been shown to affect cleavage site selection (Sinapah et al. 2011).

More recently, it has been shown that the 50S ribosomal protein L7Ae is a subunit of the type M archaeal RNase P (Cho et al. 2010). L7Ae belongs to the PF01248 family of proteins that specifically recognizes and binds Kink-turn structural motif. Interestingly this protein is also part of other macromolecular machines like box C/D and H/ACA guide ncRNAs that coordinate some aspect of translation.

4.4.2 RNase Z

The universally conserved RNase Z with representatives in all three domains of life as well as in mitochondria and chloroplasts, is the major enzyme participating in the 3′ end processing of most tRNAs and potentially in rRNA maturation pathways. RNase Z members belong to the so-called Elac family which is part of the metal dependent β-lactamase superfamily, a group of metalloproteins which perform a variety of functions (Aravind 1999) (Table 4.1). Extensive sequence analysis of the members of the metallo-β-lactamase superfamily allowed the identification of several sub-families among which are, RNase Z and β-CASP, two nucleic acid hydrolases (described hereafter in Sect. 5.2) (Aravind 1999; Daiyasu et al. 2001).

The 3′ end maturation of tRNA is far more complex than its 5′ end processing. Removal of 3′ extensions, also referred as 3′ trailers, from pre-tRNA requires an endonucleolytic cleavage that can be followed by an exonucleolytic trimming depending on the presence or absence of the terminal CCA sequence. Briefly RNase Z enzymes contain a metallo-β-lactamase domain followed by a specific pre-tRNA binding domain called exosite (Fig. 4.2a). In Archaea these enzymes come only in short-length versions (280–360 amino acids) that act as a dimer. RNase Z carries an endoribonucleolytic activity that generates the mature 3′-end of tRNA molecules by removal of the 3′-trailer elements of precursor tRNAs (Fig. 4.1). This cleavage, which comes immediately after the discriminator nucleotide (first unpaired nucleotide located 3′ of the acceptor stem) and leaves a 3′ hydroxyl group on the tRNA 3′ end and a 5′ phosphate on the trailer element requires two Zn^{2+} ions [for review see Dominski (2007), Redko et al. (2007), Spath et al. (2007), Vogel et al. (2005)].

Recombinant RNase Z from archaeal species *Haloferax volcanii, Methanococcus janaschii,* and *Pyrococcus furiosus* has been shown to efficiently and accurately cleave tRNA precursors *in vitro* (Schierling et al. 2002; Schiffer et al. 2002; Spath et al. 2008). Archaeal RNase Z enzymes require the entire structure of pre-tRNA substrates for optimal activity (Schierling et al. 2002). However, the substrate specificity of haloarchaeal and thermococcal RNase Z extend to tRNA-like structures which are present 5′ of the 5S rRNA in ribosomal operons (Holzle et al. 2008). Therefore RNase Z could also be a major actor of archaeal ribosomal RNA maturation pathway. By all these means it is not surprising that in *H. volcanni*, RNase Z gene is an essential gene (Holzle et al. 2008).

4.4.3 End A

Archaeal tRNA-splicing endoRNases (EndA) catalyze the endonucleolytic cleavage of pre-tRNAs at the 5′ and 3′ splice sites to release the intron and produces two half RNA molecules bearing 5′ hydroxyl and 2′, 3′-cyclic phosphate termini. In yeast, plants, and mammals, EndA is formed by the four non-identical subunits

Sen15p, Sen34p, Sen2p, and Sen54p [for review see in Lopes et al. (2015)]. The genes encoding the archaeal α, α′, β and ε subunits of EndA which are homologous to their eukaryotic counterparts are ubiquitous in Archaea (Fig. 4.3).

Classically archaeal EndA enzymes were classified into three types according to subunit composition: a homotetramer (α4) in some Euryarchaea like-*M. jannaschii*, a homodimer (α′2) in other Euryarchaea-like *H. volcanii)*, and a heterotetramer (α2β2) in the Crenarchaea and the Nanoarchaea (Fig. 4.2a) [for review see Lopes et al. (2015)]. Recently, a fourth type from an uncultivated archaeon *Candidatus Micrarchaeum acidiphilum*, deeply branched within Euryarchaea and referred to as ARMAN-2, which is composed of ε2 homodimer and has broad substrate specificity like the crenarchaeal and nanoarchaeal α2β2 type (Hirata et al. 2012) (Fig. 4.2a).

In all configurations, the overall folding of EndA that resembles the homotetramer one permits an efficient cleavage of the bulge-helix-bulge (BHB) motif, where two three-nucleotide bulges are separated by a four-nucleotide helix. Archaeal introns are found in different positions in the tRNA in addition to the canonical position found in Eukarya (Marck and Grosjean 2003). As long as the BHB motif is formed, the archaeal endoRNase will recognize and cleave the intron.

Finally, it was shown that various Archaea use EndA to excise pre-rRNA spacers containing BHB motifs, which remove pre-16S and pre-23S rRNA from primary transcripts (Kjems and Garrett 1988; Lykke-Andersen and Garrett 1997) (discussed in Chap. 6).

4.4.4 Nob1

In Archaea, very few are known on ribosomal assembly factors which transiently bind to and act on the nascent ribosome in a temporally and spatially well-defined and highly regulated manner. However, several archaeal homologs of eukaryotic ribosome assembly factors served as valuable models for structural, functional, and biophysical studies (Hellmich et al. 2013; Veith et al. 2012). The recent identification of eukaryotic endoRNase Nob1 homologs in archaeal genomes suggests that the 3′ terminal maturation of the archaeal pre-rRNA of the small ribosomal subunit is also processed at the D-cleavage site as in eukaryotes (discussed in Chap. 6). The Nob1-like proteins have been identified in the all the major branches of Archaea with the exception of Korarchaea (Veith et al. 2012) (Fig. 4.3). *In vitro*, the full-length Nob1 protein from the archaeon *Pyrococcus horikoshii* (PhNob1), efficiently cleaves RNA-substrates containing the D-site of the pre-16S RNA in a manganese-dependent manner. The structure of PhNob1 solved by nuclear magnetic resonance spectroscopy revealed a PIN (PilT N-terminus) domain common in many nucleases and a zinc ribbon domain, which are structurally connected by a flexible linker (Veith et al. 2012) (Fig. 4.2a).

4.4.5 Cas6

A hallmark of defense mechanisms based on clustered regularly interspaced short palindromic repeats (CRISPR) and associated sequences (Cas) are the crRNAs that guide these complexes in the destruction of invading DNA or RNA (discussed in Chap. 11). The repeat-spacer array of CRISPR gives rise to a long precursor named pre-crRNA that is, in Type I and III systems, processed by an endoRNase into crRNA intermediates (70–80 nt). Mature crRNAs are integrated into large ribonucleoprotein complexes with their cognate Cas proteins to guide them to the invading foreign RNA or DNA sequences (for review Sorek et al. (2013), van der Oost et al. (2014), Westra et al. (2012), Wiedenheft et al. (2012)].

Together with the host bacterial RNase III endoRNases, the Cas6 members are responsible for producing crRNAs from the CRISPR precursor (Fig. 4.1).

Cas6 from *P. furiosus* (PfCas6) was the first metal-independent endoRNase to be characterized as involved in CRISPR RNA processing within the repeat sequences (Carte et al. 2008). This enzyme was shown to interact with a specific sequence motif in the 5′ region of the CRISPR repeat element and to cleave at a defined site within the 3′ region of the repeat (Carte et al. 2008). Remarkably *cas6* is one of the most widely distributed CRISPR-associated genes. The Cas6 protein superfamily that has representative in Archaea and Bacteria are known as primary endoRNases in CRISPR system subtypes I-A, I-B, I-E, I-F and Type III [for review see Hochstrasser and Doudna (2015)]. The amino acid sequence of archaeal Cas6 sequence show limited conservation with only two common motifs: the ferodoxin fold and a glycine rich motif, that are also found in other RNA-binding proteins (Li 2015) (Table 4.1, Fig. 4.2a). It is only poorly understood how different Cas6 endoRNases, present in organisms with multiple CRISPR systems, differentiate between their targets [for review see Hochstrasser and Doudna (2015)].

4.4.6 RNase HII

RNase H is a ubiquitous enzyme found in all domains of life that cleaves the RNA strand embedded in RNA/DNA hybrid. In archaeal cells, RNA-primed replication intermediates, similar to those formed during eukaryotic DNA replication, have been observed in Euryarchaea and Crenarchaea, arguing for either RNA or RNA–DNA priming at the archaeal replication fork. RNase H enzymes vary greatly in domain structures and substrate specificities [for review see Tadokoro and Kanaya (2009)]. RNase H cleaves the P-O3′ bond of the substrates with a two-metal-ion catalysis mechanism, in which two metal ions are directly involved in the catalytic function. RNases H from Bacteria and Archaea have been classified into type 1 (RNase HI) and type 2 RNase H (RNase HII) based on differences in their amino acid sequences (Kochiwa et al. 2007). RNase H in its two types is a key component for the growth and survival of all organisms by playing a crucial role in

DNA replication by removing the RNA primer of Okazaki fragments and in DNA repair by removing the single ribonucleotides incorporated in the DNA [for review see Tadokoro and Kanaya (2009)]. Most of the archaeal genomes only contain RNase HII genes (Figs. 4.2a and 4.3). The crystal structure of RNase HII from *Archaeoglobus fulgidus* in complex with PCNA revealed three binding modes as the enzyme rotates around a flexible hinge while anchored to PCNA by its PIP-box motif (Bubeck et al. 2011). PCNA binding was shown to promote RNase HII activity in a hinge-dependent manner by enhancing both cleavage of misincorporated ribonucleotides in DNA duplexes, and the comprehensive hydrolysis of RNA primers formed during Okazaki fragment maturation. These findings provide insights into how RNase HII activity is directed during genome replication and repair (Bubeck et al. 2011). Interestingly, *P. furiosus* Pf-RNase HII has been shown to also be able to digest RNA-RNA duplexes in the presence of Mn^{2+} ions (Kitamura et al. 2010). Remarkably, the three-dimensional structure of Pf-RNase H is similar to that of the PIWI domain of the Pf-Ago, an argonaute protein known to act in RNA-induced silencing complex in eukaryotes, although the two enzymes share almost no similarity in their amino acid sequences. In contrast to eukaryotic Ago proteins, archaeal Ago proteins show greater affinity for RNA-DNA hybrids than for RNA-RNA hybrids (Ma et al. 2005).

As an exception, the Halobacterum sp. NRC-1, *Sulfolobus tokodaii* and *Pyrobaculum aerophilum* genomes contains additional RNase HI encoding genes. The Halobacterium sp. NRC-1 RNase HI, whose folding is induced by the binding of divalent metal ions can cleave an RNA-DNA junction (Ohtani et al. 2004b; Tannous and Kanaya 2014). Interestingly, *S. tokodaii* RNase HI enables degradation from RNA/RNA duplex (Ohtani et al. 2004a).

4.5 Ribonucleases with Uncharacterized Cellular RNA Substrates

4.5.1 3′-5′ ExoRNases: The Archaeal Exosome and RNase R

3′ to 5′ exoribonucleolytic activities have been reported in the three domains of life. These activities are performed by PNPase, RNase II and RNase R in Bacteria and by the exosome complex in Eukarya. In Archaea, only two archaeal 3′ to 5′ exoribonucleolytic activities have been described so far.

4.5.1.1 Archaeal Exosome

Prediction of the existence of a eukaryotic-like archaeal exosome was initially based on the identification of genes encoding orthologs of eukaryotic exosome subunits, the ribosomal-RNA-processing proteins Rrp4, Rrp41 and Rrp42. These components are organized in conserved super operon that were identified in most Archaea with the exception of the Halophiles and some Methanococcales (Evguenieva-Hackenberg et al. 2014; Evguenieva-Hackenberg and Klug 2009; Koonin et al. 2001) (discussed in Chap. 11).

The first experimental evidence for the existence of an exosome-like protein complex in Archaea was obtained by the purification of a 250-kDa protein complex that co-immunoprecipitate with Rrp41 from *Sulfolobus solfataricus* cellular extract. Half of this complex was also shown to co-sediment with ribosomal subunits. Orthologues to Rrp4, Rrp41, Rrp42 and Csl4, as well as a component annotated as a DnaG homologue, were found to tightly associate in a complex. A minimal core of the *S. solfataricus* and *M. thermautotrophicus* exosome was shown to consist of at least six phosphate-dependent ribonuclease PH homologues, along with Rrp4- and Csl4-like subunits (Evguenieva-Hackenberg et al. 2003; Farhoud et al. 2005) (Fig. 4.2b).

Subsequently, crystal structures of the *S. solfataricus* and *A. fulgidus* exosome core revealed a hexameric ring-like arrangement of three Rrp41-Rrp42 heterodimers in which both subunits adopt the RNase PH fold common to phosphorolytic exoRNases. Structure-guided mutagenesis revealed that the activity of the complex resides within the active sites of the Rrp41 subunits, whereas the Rrp42 subunits are inactive but contribute to the structuring of the three Rrp41 active sites (discussed in Chap. 11).

The archaeal exosome subunits share high sequence similarity with their eukaryotic counterparts, even if none of the eukaryotic subunits are active, and assemble in a structure highly similar to the bacterial trimeric PNPase ring-like structure (Dziembowski et al. 2007; Liu et al. 2006) These features support a common basis for a 3′ to 5′ RNA-degrading machineries in all three domains of life (Chlebowski et al. 2013). Nevertheless a specific distinctive of the archaeal exosome, experimentally shown for the exosome of *Sulfolobus solfataricus, Pyrococcus abyssi and Methanopyrus kandleri*, is the ability to work as an RNA-tailing complex with a heteropolymeric polyadenylation activity. Consistently heteropolymeric polyadenylation was only observed in archaeal species containing an exosome complex but not in Halophiles and the methanogens which do not encode Rrp41–Rrp42 homologues (Portnoy et al. 2005; Portnoy and Schuster 2006; Ramos et al. 2006) (Fig. 4.3).

4.5.1.2 RNase R

Only halophiles and some methanogens, devoid of exosome, possess in their genome an *rnr* gene encoding an RNase R-like enzyme (Portnoy and Schuster 2006) (Fig. 4.3). Shortly, bacterial RNase R has been described as the first ribonuclease possessing a 3′ to 5′ hydrolytic exoRNase activity with an unusual ability to digest highly structured RNA without the aid of helicase activity (Vincent and Deutscher 2009).

RNase R homologs were initially identified by amino acid sequence comparison search. Only one gene encoding RNase R-like was identified in *H. volcanii* with 34% sequence identity with *E. coli* RNase R (Portnoy and Schuster 2006). In Halophiles and in methanogens with no exosome, archaeal RNase R seems to be the unique 3′ to 5′ exoRNase. Conversely to bacterial RNase R, archaeal RNase R are not able to degrade structured RNA by themselves as they are restricted to the RNB central core domain and lack the N- (CSD1&2) and C-terminal (S1) domains specific (Matos et al. 2012) (Fig. 4.2d). *In vitro* enzymatic studies showed that recombinant HvRNase R demonstrates an optimal exoribonucleolytic activity on tRNA substrates at low salt concentration either at 25 °C and 37 °C suggesting that HvRNase R is not adapted to high salinity. However, the residual activity at high salt concentration is sufficient since *rnr* is known to be essential for *H. volcanii* viability (Matos et al. 2012). Overall the studies carried out with HvRNase R suggest that archaeal RNase R plays a critical role in RNA decay in archaeal species devoid of an exosome complex (Matos et al. 2012; Portnoy and Schuster 2006).

4.5.2 5′-3′ Exo and EndoRNases: Archaeal β-CASP Ribonucleases

The β-CASP family members that are widespread in Bacteria and Eukarya, encompass RNA cleaving (CPSF73 and RNase J) and DNA repair (Artemis, SNM1 and PSO2) enzymes. In common with RNase Z belonging to the Elac family, members of the β-CASP family are characterized by a metallo-β-lactamase domain additionally with a specific β-CASP domain. Highly conserved aspartic acid, histidine and valine residues form a signature motif that plays key roles in the enzymatic function of the β-CASP enzymes by coordinating directly or indirectly coordination with two metal ions (usually Zn^{2+}) (for review see Dominski et al. (2013)]. Notably β-CASP enzymes have been shown or suggested to share dual endonucleolytic and 5′-3′ exonucleolytic activities catalyzed by a unique catalytic site [for review see Newman et al. (2011)]. These fascinating enzymes are key players in 3′ end processing in Eukarya and in RNA decay and ribosomal RNA maturation in Bacteria.

Recently β-CASP ribonucleases have also been identified in Archaea. As their bacterial and eukaryal counterparts, some also exhibit a dual (5′-3′ exo- and endo-) ribonucleolytic activity [for review see Clouet-d'Orval et al. (2015), Dominski et al. (2013)]. The recent recognition of β-CASP ribonucleases as major players in Archaea is an important contribution towards identifying RNA-degrading mechanisms in the third domain of life (Clouet-d'Orval et al. 2015) (Table 4.1, Fig. 4.2c, d). In-depth phylogenomic analyses and structural studies revealed three major β-CASP orthologous groups in Archaea, aCPSF1 and aCPSF2 which are closely related to eukaryal CPSF73 (Cleavage Polyadenylation Specific Factor) termination factor and, aRNase J which is orthologous to bacterial RNase J (Phung et al. 2013) [for review see Dominski et al. (2013)]. The strictly conservation of archaeal β-CASP enzymes throughout archaeal phylogeny strongly suggests their essential roles in maturation and/or degradation of RNA (Clouet-d'Orval et al. 2015) (Fig. 4.3). Features of archaeal β-CASP ribonucleases are described hereafter by their enzymatic activities.

4.5.2.1 5′-3′ Exo β-CASP Ribonucleases: aRNase J & aCPSF2

aRNase J

Phylogenetic analysis showed that the members of the aRNase J group which are widespread in Euryarchaea (Fig. 4.3) have been inherited vertically, suggesting an ancient origin of the RNase J group of enzymes predating the separation of the Bacteria and the Archaea (Clouet-d'Orval et al. 2010). In contrast to bacterial RNase J which harbor an additional N- or C-terminal domain, archaeal aRNase J are restricted to the β-lactamase β-CASP core domain (Fig. 4.2c) (Clouet-d'Orval et al. 2010; Dominski et al. 2013). However, euryarchaeal aRNase J contains two small highly conserved domains, denoted Loop1 and Loop2, that are absent in bacterial RNase J (Clouet-d'Orval et al. 2010).

The exoribonucleolytic activity with a 5′ to 3′ directionality of aRNase J was first identified for recombinant proteins from Thermococcales (Clouet-d'Orval et al. 2010). Subsequent work with aRNase J of *M. jannaschii* also identified a similar 5′ to 3′ trimming activity (Levy et al. 2011). The euryarchaeal aRNase J which are highly processive enzymes do not have shown any detectable endoribonucleolytic activity under the *in vitro* conditions tested. However, they exhibit reduced *in vitro* activity on 5′ triphosphate-end RNA substrates, indicating that aRNase J exonucleases are 5′ end-dependent exoRNases and suggesting that a subset of cellular RNAs with 5′ triphosphate ends could be refractory to degradation by aRNase J *in vivo* (Clouet-d'Orval et al. 2010; Levy et al. 2011). Mutational analysis showed that residues of the highly conserved motifs implicated in the coordination of the two zinc ions and the specific Loop1 and Loop2 are critical for *in vitro* ribonucleolytic activity (Clouet-d'Orval et al. 2010). In addition, it is interesting to note that except for the difference in temperature optima, the characteristics of the exoRNase activity of euryarchaeal aRNase J are very similar to the exoRNase activity of RNase J1 from Bacteria (Li de la Sierra-Gallay et al. 2008; Mathy et al. 2007) strongly suggesting that euryarchaeal aRNase J might have similar roles in rRNA processing and mRNA decay. However the capacity of some euryarchaeal aRNase J to degrade single-stranded DNA *in vitro* suggests versatile roles of aRNase J *in vivo*. At least two other exonucleases from the β-CASP family, hSNM1 and Artemis, are known to degrade DNA [for review see Dominski et al. (2013)]. Whether these observations are informative in terms of aRNase J physiological properties and substrate specificity and, bring knowledge on aRNase J singularities remain to be seen.

aCPSF2

aCPSF2 members are strictly conserved in Crenarcheota and widely spread but more divergent in Euryarchaeota (Clouet-d'Orval et al. 2015; Phung et al. 2013) (Fig. 4.3). Much is still not known on the activity and structure of members of aCPSF2 group which is phylogenetically closely related to aCPSF1 group (Dominski et al. 2013; Phung et al. 2013). Like aRNase J, aCPSF2 enzymes are commonly restricted to the β-CASP and metallo-β-lactamase core domains with no additional N- or C-terminal extensions (Fig. 4.2c). One study reported 5′-3′ exonucleolytic activity *in vitro* activity for the crenarchaeon *S. solfataricus* (Sso) aCPSF2 that is modulated by the phosphorylation state of the 5′ end of transcript as seen for aRNase J members (Hasenohrl et al. 2011). However, in contrast to the other archaeal β-CASP proteins (Phung et al. 2013), the activity of Sso-aCPSF2 was only observed in the presence of Mg^{2+} ions (Hasenohrl et al. 2011). Furthermore, the 5′-3′ exoribonucleolytic activity of Sso-aCPSF2 has been shown to be blocked by the binding of the translation initiation factor a/eIF2 δ that protects the 5′ end of mRNA from degradation *in vivo* and *in vitro* (Hasenohrl et al. 2008). Further understanding of the enzymatic and functional properties of aCPSF2, in particular the role of magnesium, must await structural studies.

A first clue towards understanding the biological relevance of aCPSF2 was obtained in *Sulfolobus acidocaldarius* (Martens et al. 2013). Comparison of transcriptomes from wild type and aCPSF2-deleted strains revealed differential transcript abundance for 560 genes. This global effect suggests that aCPSF2 plays a major role in 5′-3′ directed mRNA decay. Since an aCPSF2-deleted strain exhibits no growth defect, the observed differences in transcripts abundance does not affect the overall cellular fitness. At this point, the processing pathways recruiting aCPSF2 remain to be identified.

To conclude, each archaeal genome encodes at least one β-CASP enzyme with a 5′-3′ exoribonucleolytic activity which could be carried out either by an aRNase J and/or an aCPSF2 group member (Clouet-d'Orval et al. 2015) (Fig. 4.3). This strongly suggests that 5′-3′ processing pathways exist in Archaea and are somehow important for the fate of cellular RNA biotypes.

4.5.3 Dual 5′-3′ Exo/Endo- β-CASP Ribonuclease: aCPSF1

Phylogenomic studies searching for β-CASP proteins among archaeal genomes have identified an outstanding group, called aCPSF1, that is strictly conserved and orthologous to the eukaryal cleavage and polyadenylation specific factor group defined by human CPSF73 and yeast CPSF100 (Clouet-d'Orval et al. 2015; Phung et al. 2013) (Fig. 4.3). Thermococcales aCPSF1 have been shown to harbor *in vitro* 5′-3′ exo- and endoribonucleolytic activities with a preference for single-stranded CA dinucleotide as described for eukaryal CPSF73 factor within the machinery required for termination of RNAP II transcription [for review see Clouet-d'Orval et al. (2015), Dominski et al. (2013)]. In Archaea, the relevance of the cleavage specificity at single-stranded CA dinucleotide remains to be elucidated (Phung et al. 2013). In contrast, *M. jannaschii* aCPSF1 [initially named mjRNAse J2 in Levy et al. (2011)] was shown to only display an endoribonucleolytic activity with no detectable exoribonucleolytic activity. aCPSF1 is dimeric in solution with a dimer interface composed of 12 amino acids that is conserved among all aCPSF1 members. Interestingly, disruption of this interface impairs the exoribonucleolytic activity of *P. abyssi* aCPSF1 but not its endoribonucleolytic activity (Phung et al. 2013).

The aCPSF1 members are distinguished by their N-terminal domain composed of a tandem repeat of two type-II KH motifs predicted to bind nucleic acids that extends the β-CASP and metallo-β-lactamase core domains (Fig. 4.2d). Crystal structures of dimeric aCPSF1 from *Pyrococcus horikoshii*, *Methanosarcina mazei* and *Methanothermobacter thermoautotrophicus* show a tripartite architecture consisting of the β-CASP and β-lactamase domains folded apart from both KH motifs (Mir-Montazeri et al. 2011; Nishida et al. 2010; Silva et al. 2011). In view of their wide phylogenetic distribution, aCPSF1 dual ribonuclease likely performs essential functions in archaeal metabolism. Interestingly, a study identified *Haloferax volcanii* aCPSF1 as an interacting components of the proteasome, suggesting a connection between RNA processing and protein degradation (Chavarria et al. 2014).

4.6 Concluding Remarks

Although we highlighted different ribonuclease families potentially involved in modulating RNA pools in Archaea cells, an integrated view of how these ribonucleases regulate RNA dynamics, levels and modification is still missing. The tight cooperation between different RNA processing systems needs also to be discovered. Furthermore ancillary RNA-modifying enzymes such as RNA helicases, poly (A) polymerases, pyrophosphohydrolases and RNA binding proteins such as Sm-like proteins, are undoubtedly important in mediating archaeal RNA processing and/or turnover. Some of these factors are beginning to be described. First, as mentioned above in Sect. 5.1.1, the archaeal exosome can work as an RNA-tailing complex of polyadenylation. More recently the Sm-like proteins in sulfolobus, SmAP1 and SmAP2, with affinity for different classes of ncRNAs and mRNAs have been shown to co-purify with proteins involved in RNA processing, RNA modification, translation and RNA turnover as well as with components of the exosome (Martens et al. 2017).

Besides, the many archaeal proteins of unknown function, for which a ribonucleolytic or related activity cannot be predicted from their sequence, makes us foresee that the discovery of archaeal ribonucleases will remain a dynamic area of research for the years to come.

In essence, deciphering RNA biology processes in Archaea will give us key paradigms for understanding fundamentally conserved processes including RNA maturation and decay which dictate post-transcriptional regulation of gene expression. In addition, examination of the distribution of the individual ribonuclease families among diverse organisms will provide a phylogenetic framework with interesting evolutionary implications. The identification of conserved pathways will retrace the history of RNA metabolism across the domains of life.

Acknowledgements BCO and MB are supported by the Centre National de la Recherche Scientifique (CNRS) with additional funding from the Agence Nationale de la Recherche (ANR-16-CE12-0016-01) and from the Université de Toulouse (IDEX « Emergence » call 2014).

References

Aravind L (1999) An evolutionary classification of the metallo-beta-lactamase fold proteins. In Silico Biol 1:69–91

Aravind L, Koonin EV (2001) A natural classification of Ribonucleases. Methods Enzymol 341:1–28

Arkhipova V, Stolboushkina E, Kravchenko O, Kljashtorny V, Gabdulkhakov A, Garber M, Nikonov S, Martens B, Blasi U, Nikonov O (2015) Binding of the 5′-triphosphate end of mRNA to the gamma-subunit of translation initiation factor 2 of the crenarchaeon Sulfolobus solfataricus. J Mol Biol 427:3086–3095

Arraiano CM, Mauxion F, Viegas SC, Matos RG, Seraphin B (2013) Intracellular ribonucleases involved in transcript processing and decay: precision tools for RNA. Biochim Biophys Acta 1829:491–513

Bandyra KJ, Bouvier M, Carpousis AJ, Luisi BF (2013) The social fabric of the RNA degradosome. Biochim Biophys Acta 1829:514–522

Barrangou R, Coute-Monvoisin AC, Stahl B, Chavichvily I, Damange F, Romero DA, Boyaval P, Fremaux C, Horvath P (2013) Genomic impact of CRISPR immunization against bacteriophages. Biochem Soc Trans 41:1383–1391

Bini E, Dikshit V, Dirksen K, Drozda M, Blum P (2002) Stability of mRNA in the hyperthermophilic archaeon Sulfolobus solfataricus. RNA 8:1129–1136

Brochier-Armanet C, Forterre P, Gribaldo S (2011) Phylogeny and evolution of the Archaea: one hundred genomes later. Current Opin Microbiol 14:274–281

Brooks SA (2010) Functional interactions between mRNA turnover and surveillance and the ubiquitin proteasome system. Wiley Interdiscip Rev RNA 1:240–252

Bubeck D, Reijns MA, Graham SC, Astell KR, Jones EY, Jackson AP (2011) PCNA directs type 2 RNase H activity on DNA replication and repair substrates. Nucleic Acids Res 39:3652–3666

Carpousis AJ (2007) The RNA degradosome of Escherichia coli: an mRNA-degrading machine assembled on RNase E. Annu Rev Microbiol 61:71–87

Carte J, Wang R, Li H, Terns RM, Terns MP (2008) Cas6 is an endoribonuclease that generates guide RNAs for invader defense in prokaryotes. Genes Dev 22:3489–3496

Charpentier B, Muller S, Branlant C (2005) Reconstitution of archaeal H/ACA small ribonucleoprotein complexes active in pseudouridylation. Nucleic Acids Res 33:3133–3144

Chavarria NE, Hwang S, Cao S, Fu X, Holman M, Elbanna D, Rodriguez S, Arrington D, Englert M, Uthandi S, Soll D, Maupin-Furlow JA (2014) Archaeal Tuc1/Ncs6 homolog required for wobble uridine tRNA thiolation is associated with ubiquitin-proteasome, translation, and RNA processing system homologs. PLoS One 9:e99104

Chlebowski A, Lubas M, Jensen TH, Dziembowski A (2013) RNA decay machines: the exosome. Biochim Biophys Acta 1829:552–560

Cho HD, Verlinde CL, Weiner AM (2005) Archaeal CCA-adding enzymes: central role of a highly conserved beta-turn motif in RNA polymerization without translocation. J Biol Chem 280:9555–9566
Cho IM, Lai LB, Susanti D, Mukhopadhyay B, Gopalan V (2010) Ribosomal protein L7Ae is a subunit of archaeal RNase P. Proc Natl Acad Sci U S A 107:14573–14578
Clouet-d'Orval B, Rinaldi D, Quentin Y, Carpousis AJ (2010) Euryarchaeal beta-CASP proteins with homology to bacterial RNase J Have 5′- to 3′-exoribonuclease activity. J Biol Chem 285:17574–17583
Clouet-d'Orval B, Phung DK, Langendijk-Genevaux PS, Quentin Y (2015) Universal RNA-degrading enzymes in Archaea: prevalence, activities and functions of beta-CASP ribonucleases. Biochimie 118:278–285
Condon C, Putzer H (2002) The phylogenetic distribution of bacterial ribonucleases. Nucleic Acids Res 30:5339–5346
Daiyasu H, Osaka K, Ishino Y, Toh H (2001) Expansion of the zinc metallo-hydrolase family of the beta-lactamase fold. FEBS Lett 503:1–6
Dar D, Prasse D, Schmitz RA, Sorek R (2016) Widespread formation of alternative 3′ UTR isoforms via transcription termination in archaea. Nat Microbiol 1:16143
Dennis PP, Omer A (2005) Small non-coding RNAs in Archaea. Curr Opin Microbiol 8:685–694
Dennis PP, Omer A, Lowe T (2001) A guided tour: small RNA function in Archaea. Mol Microbiol 40:509–519
Deutscher MP (2009) Maturation and degradation of ribosomal RNA in bacteria. Prog Mol Biol Transl Sci 85:369–391
Dominski Z (2007) Nucleases of the metallo-beta-lactamase family and their role in DNA and RNA metabolism. Crit Rev Biochem Mol Biol 42:67–93
Dominski Z, Carpousis AJ, Clouet-d'Orval B (2013) Emergence of the beta-CASP ribonucleases: highly conserved and ubiquitous metallo-enzymes involved in messenger RNA maturation and degradation. Biochim Biophys Acta 1829:532–551
Dziembowski A, Lorentzen E, Conti E, Seraphin B (2007) A single subunit, Dis3, is essentially responsible for yeast exosome core activity. Nat Struct Mol Biol 14:15–22
Evguenieva-Hackenberg E, Klug G (2009) RNA degradation in Archaea and Gram-negative bacteria different from Escherichia coli. Prog Mol Biol Transl Sci 85:275–317
Evguenieva-Hackenberg E, Walter P, Hochleitner E, Lottspeich F, Klug G (2003) An exosome-like complex in Sulfolobus solfataricus. EMBO Rep 4:889–893
Evguenieva-Hackenberg E, Hou L, Glaeser S, Klug G (2014) Structure and function of the archaeal exosome. Wiley Interdiscip Rev RNA 5:623–635
Farhoud MH, Wessels HJ, Steenbakkers PJ, Mattijssen S, Wevers RA, van Engelen BG, Jetten MS, Smeitink JA, van den Heuvel LP, Keltjens JT (2005) Protein complexes in the archaeon Methanothermobacter thermautotrophicus analyzed by blue native/SDS-PAGE and mass spectrometry. Mol Cell Proteomics 4:1653–1663
Fernandez-Pevida A, Kressler D, de la Cruz J (2015) Processing of preribosomal RNA in Saccharomyces cerevisiae. Wiley Interdiscip Rev RNA 6:191–209
Filipowicz W, Bhattacharyya SN, Sonenberg N (2008) Mechanisms of post-transcriptional regulation by microRNAs: are the answers in sight? Nat Rev Genet 9:102–114
Frank DN, Pace NR (1998) Ribonuclease P: unity and diversity in a tRNA processing ribozyme. Annu Rev Biochem 67:153–180
Franzetti B, Sohlberg B, Zaccai G, von Gabain A (1997) Biochemical and serological evidence for an RNase E-like activity in halophilic Archaea. J Bacteriol 179:1180–1185
Garrett RA (2016) Archaeal physiology: The secrets of termination. Nat Microbiol 1:16159
Gopalan V (2007) Uniformity amid diversity in RNase P. Proc Natl Acad Sci U S A 104:2031–2032
Hasenohrl D, Lombo T, Kaberdin V, Londei P, Blasi U (2008) Translation initiation factor a/eIF2 (-gamma) counteracts 5′ to 3′ mRNA decay in the archaeon Sulfolobus solfataricus. Proc Natl Acad Sci U S A 105:2146–2150

Hasenohrl D, Konrat R, Blasi U (2011) Identification of an RNase J ortholog in Sulfolobus solfataricus: implications for 5′-to-3′ directional decay and 5′-end protection of mRNA in Crenarchaeota. RNA 17:99–107

Hellmich UA, Weis BL, Lioutikov A, Wurm JP, Kaiser M, Christ NA, Hantke K, Kotter P, Entian KD, Schleiff E, Wohnert J (2013) Essential ribosome assembly factor Fap7 regulates a hierarchy of RNA-protein interactions during small ribosomal subunit biogenesis. Proc Natl Acad Sci U S A 110:15253–15258

Hennigan AN, Reeve JN (1994) mRNAs in the methanogenic archaeon Methanococcus vannielii: numbers, half-lives and processing. Mol Microbiol 11:655–670

Henras AK, Plisson-Chastang C, O'Donohue MF, Chakraborty A, Gleizes PE (2015) An overview of pre-ribosomal RNA processing in eukaryotes. Wiley Interdiscip Rev RNA 6:225–242

Hirata A, Fujishima K, Yamagami R, Kawamura T, Banfield JF, Kanai A, Hori H (2012) X-ray structure of the fourth type of archaeal tRNA splicing endonuclease: insights into the evolution of a novel three-unit composition and a unique loop involved in broad substrate specificity. Nucleic Acids Res 40:10554–10566

Hochstrasser ML, Doudna JA (2015) Cutting it close: CRISPR-associated endoribonuclease structure and function. Trends Biochem Sci 40:58–66

Holzle A, Fischer S, Heyer R, Schutz S, Zacharias M, Walther P, Allers T, Marchfelder A (2008) Maturation of the 5S rRNA 5′ end is catalyzed in vitro by the endonuclease tRNase Z in the archaeon H. volcanii. RNA 14:928–937

Huang HY, Hopper AK (2016) Multiple layers of stress-induced regulation in tRNA biology. Life (Basel) 6(2). pii: E16. doi:10.3390/life6020016

Hui MP, Foley PL, Belasco JG (2014) Messenger RNA degradation in bacterial cells. Annu Rev Genet 48:537–559

Jager A, Samorski R, Pfeifer F, Klug G (2002) Individual gvp transcript segments in Haloferax mediterranei exhibit varying half-lives, which are differentially affected by salt concentration and growth phase. Nucleic Acids Res 30:5436–5443

Jarrous N, Gopalan V (2010) Archaeal/eukaryal RNase P: subunits, functions and RNA diversification. Nucleic Acids Res 38:7885–7894

Kanai A, Oida H, Matsuura N, Doi H (2003) Expression cloning and characterization of a novel gene that encodes the RNA-binding protein FAU-1 from Pyrococcus furiosus. Biochem J 372:253–261

Kitamura S, Fujishima K, Sato A, Tsuchiya D, Tomita M, Kanai A (2010) Characterization of RNase HII substrate recognition using RNase HII-argonaute chimaeric enzymes from Pyrococcus furiosus. Biochem J 426:337–344

Kjems J, Garrett RA (1988) Novel splicing mechanism for the ribosomal RNA intron in the archaebacterium Desulfurococcus mobilis. Cell 54:693–703

Klemm BP, Wu N, Chen Y, Liu X, Kaitany KJ, Howard MJ, Fierke CA (2016) The diversity of ribonuclease P: protein and RNA catalysts with analogous biological functions. Biomolecules 6(2). pii: E27. doi:10.3390/biom6020027

Klug G, Evguenieva-Hackenberg E, Omer A, Dennis PP, Marchfelder A (2007) RNA processing. In: Cavicchioli R (ed) Archaea: molecular and cellular biology. ASM Press, Washington, DC, pp 158–174

Kochiwa H, Tomita M, Kanai A (2007) Evolution of ribonuclease H genes in prokaryotes to avoid inheritance of redundant genes. BMC Evol Biol 7:128

Koonin EV, Wolf YI, Aravind L (2001) Prediction of the archaeal exosome and its connections with the proteasome and the translation and transcription machineries by a comparative-genomic approach. Genome Res 11:240–252

Lai LB, Vioque A, Kirsebom LA, Gopalan V (2010) Unexpected diversity of RNase P, an ancient tRNA processing enzyme: challenges and prospects. FEBS Lett 584:287–296

Lapinaite A, Simon B, Skjaerven L, Rakwalska-Bange M, Gabel F, Carlomagno T (2013) The structure of the box C/D enzyme reveals regulation of RNA methylation. Nature 502:519–523

Levy S, Portnoy V, Admon J, Schuster G (2011) Distinct activities of several RNase J proteins in methanogenic archaea. RNA Biol 8:1073–1083

Li H (2015) Structural principles of CRISPR RNA processing. Structure 23:13–20

Li de la Sierra-Gallay I, Zig L, Jamalli A, Putzer H (2008) Structural insights into the dual activity of RNase J. Nat Struct Mol Biol 15:206–212

Li L, Ye K (2006) Crystal structure of an H/ACA box ribonucleoprotein particle. Nature 443:302–307

Li H, Trotta CR, Abelson J (1998) Crystal structure and evolution of a transfer RNA splicing enzyme. Science 280:279–284

Liu Q, Greimann JC, Lima CD (2006) Reconstitution, activities, and structure of the eukaryotic RNA exosome. Cell 127:1223–1237

Lopes RR, Kessler AC, Polycarpo C, Alfonzo JD (2015) Cutting, dicing, healing and sealing: the molecular surgery of tRNA. Wiley Interdiscip Rev RNA 6:337–349

Lykke-Andersen J, Garrett RA (1997) RNA-protein interactions of an archaeal homotetrameric splicing endoribonuclease with an exceptional evolutionary history. EMBO J 16:6290–6300

Ma JB, Yuan YR, Meister G, Pei Y, Tuschl T, Patel DJ (2005) Structural basis for 5′-end-specific recognition of guide RNA by the A. fulgidus Piwi protein. Nature 434:666–670

Mackie GA (2013) RNase E: at the interface of bacterial RNA processing and decay. Nat Rev Microbiol 11:45–57

Makarova KS, Wolf YI, Alkhnbashi OS, Costa F, Shah SA, Saunders SJ, Barrangou R, Brouns SJ, Charpentier E, Haft DH, Horvath P, Moineau S, Mojica FJ, Terns RM, Terns MP, White MF, Yakunin AF, Garrett RA, van der Oost J, Backofen R, Koonin EV (2015) An updated evolutionary classification of CRISPR-Cas systems. Nat Rev Microbiol 13:722–736

Marck C, Grosjean H (2002) tRNomics: analysis of tRNA genes from 50 genomes of Eukarya, Archaea, and Bacteria reveals anticodon-sparing strategies and domain-specific features. RNA 8:1189–1232

Marck C, Grosjean H (2003) Identification of BHB splicing motifs in intron-containing tRNAs from 18 archaea: evolutionary implications. RNA 9:1516–1531

Martens B, Amman F, Manoharadas S, Zeichen L, Orell A, Albers SV, Hofacker I, Blasi U (2013) Alterations of the transcriptome of Sulfolobus acidocaldarius by exoribonuclease aCPSF2. PLoS One 8:e76569

Martens B, Hou L, Amman F, Wolfinger MT, Evguenieva-Hackenberg E, Blasi U (2017) The SmAP1/2 proteins of the crenarchaeon Sulfolobus solfataricus interact with the exosome and stimulate A-rich tailing of transcripts. Nucleic Acids Res 45(13):7938–7949. doi:10.1093/nar/gkx437

Mathy N, Benard L, Pellegrini O, Daou R, Wen T, Condon C (2007) 5′-to-3′ exoribonuclease activity in bacteria: role of RNase J1 in rRNA maturation and 5′ stability of mRNA. Cell 129:681–692

Matos RG, Lopez-Vinas E, Gomez-Puertas P, Arraiano CM (2012) The only exoribonuclease present in Haloferax volcanii has an unique response to temperature changes. Biochim Biophys Acta 1820:1543–1552

Mir-Montazeri B, Ammelburg M, Forouzan D, Lupas AN, Hartmann MD (2011) Crystal structure of a dimeric archaeal cleavage and polyadenylation specificity factor. J Struct Biol 173:191–195

Nagarajan VK, Jones CI, Newbury SF, Green PJ (2013) XRN 5′-->3′ exoribonucleases: structure, mechanisms and functions. Biochim Biophys Acta 1829:590–603

Newman JA, Hewitt L, Rodrigues C, Solovyova A, Harwood CR, Lewis RJ (2011) Unusual, dual endo- and exonuclease activity in the degradosome explained by crystal structure analysis of RNase J1. Structure 19:1241–1251

Nishida Y, Ishikawa H, Baba S, Nakagawa N, Kuramitsu S, Masui R (2010) Crystal structure of an archaeal cleavage and polyadenylation specificity factor subunit from Pyrococcus horikoshii. Proteins 78:2395–2398

Nolivos S, Carpousis AJ, Clouet-d'Orval B (2005) The K-loop, a general feature of the Pyrococcus C/D guide RNAs, is an RNA structural motif related to the K-turn. Nucleic Acids Res 33:6507–6514

Numata T, Ishimatsu I, Kakuta Y, Tanaka I, Kimura M (2004) Crystal structure of archaeal ribonuclease P protein Ph1771p from Pyrococcus horikoshii OT3: an archaeal homolog of eukaryotic ribonuclease P protein Rpp29. RNA 10:1423–1432

Ohtani N, Yanagawa H, Tomita M, Itaya M (2004a) Cleavage of double-stranded RNA by RNase HI from a thermoacidophilic archaeon, Sulfolobus tokodaii 7. Nucleic Acids Res 32:5809–5819

Ohtani N, Yanagawa H, Tomita M, Itaya M (2004b) Identification of the first archaeal Type 1 RNase H gene from Halobacterium sp. NRC-1: archaeal RNase HI can cleave an RNA-DNA junction. Biochem J 381:795–802

Pannucci JA, Haas ES, Hall TA, Harris JK, Brown JW (1999) RNase P RNAs from some Archaea are catalytically active. Proc Natl Acad Sci U S A 96:7803–7808

Peeters E, Peixeiro N, Sezonov G (2013) Cis-regulatory logic in archaeal transcription. Biochem Soc Trans 41:326–331

Peeters E, Driessen RP, Werner F, Dame RT (2015) The interplay between nucleoid organization and transcription in archaeal genomes. Nat Rev Microbiol 13:333–341

Phung DK, Rinaldi D, Langendijk-Genevaux PS, Quentin Y, Carpousis AJ, Clouet-d'Orval B (2013) Archaeal beta-CASP ribonucleases of the aCPSF1 family are orthologs of the eukaryal CPSF-73 factor. Nucleic Acids Res 41:1091–1103

Portnoy V, Schuster G (2006) RNA polyadenylation and degradation in different Archaea; roles of the exosome and RNase R. Nucleic Acids Res 34:5923–5931

Portnoy V, Evguenieva-Hackenberg E, Klein F, Walter P, Lorentzen E, Klug G, Schuster G (2005) RNA polyadenylation in Archaea: not observed in Haloferax while the exosome polynucleotidylates RNA in Sulfolobus. EMBO Rep 6:1188–1193

Proudfoot NJ (2011) Ending the message: poly(A) signals then and now. Genes Dev 25:1770–1782

Ramos CR, Oliveira CL, Torriani IL, Oliveira CC (2006) The Pyrococcus exosome complex: structural and functional characterization. J Biol Chem 281:6751–6759

Randau L, Schroder I, Soll D (2008) Life without RNase P. Nature 453:120–123

Redko Y, Li de Lasierra-Gallay I, Condon C (2007) When all's zed and done: the structure and function of RNase Z in prokaryotes. Nat Rev Microbiol 5:278–286

Samanta MP, Lai SM, Daniels CJ, Gopalan V (2016) Sequence analysis and comparative study of the protein subunits of archaeal RNase P. Biomolecules 6

Santangelo TJ, Cubonova L, Skinner KM, Reeve JN (2009) Archaeal intrinsic transcription termination in vivo. J Bacteriol 191:7102–7108

Schierling K, Rosch S, Rupprecht R, Schiffer S, Marchfelder A (2002) tRNA 3′ end maturation in archaea has eukaryotic features: the RNase Z from Haloferax volcanii. J Mol Biol 316:895–902

Schiffer S, Rosch S, Marchfelder A (2002) Assigning a function to a conserved group of proteins: the tRNA 3′-processing enzymes. EMBO J 21:2769–2777

Silva AP, Chechik M, Byrne RT, Waterman DG, Ng CL, Dodson EJ, Koonin EV, Antson AA, Smits C (2011) Structure and activity of a novel archaeal beta-CASP protein with N-terminal KH domains. Structure 19:622–632

Sinapah S, Wu S, Chen Y, Pettersson BM, Gopalan V, Kirsebom LA (2011) Cleavage of model substrates by archaeal RNase P: role of protein cofactors in cleavage-site selection. Nucleic Acids Res 39:1105–1116

Sorek R, Lawrence CM, Wiedenheft B (2013) CRISPR-mediated adaptive immune systems in bacteria and archaea. Annu Rev Biochem 82:237–266

Spang A, Saw JH, Jorgensen SL, Zaremba-Niedzwiedzka K, Martijn J, Lind AE, van Eijk R, Schleper C, Guy L, Ettema TJ (2015) Complex archaea that bridge the gap between prokaryotes and eukaryotes. Nature 521:173–179

Spath B, Canino G, Marchfelder A (2007) tRNase Z: the end is not in sight. Cell Mol Life Sci 64:2404–2412

Spath B, Schubert S, Lieberoth A, Settele F, Schutz S, Fischer S, Marchfelder A (2008) Two archaeal tRNase Z enzymes: similar but different. Arch Microbiol 190:301–308

Stoecklin G, Muhlemann O (2013) RNA decay mechanisms: specificity through diversity. Biochim Biophys Acta 1829:487–490

Tadokoro T, Kanaya S (2009) Ribonuclease H: molecular diversities, substrate binding domains, and catalytic mechanism of the prokaryotic enzymes. FEBS J 276:1482–1493

Tang TH, Rozhdestvensky TS, Clouet-d'Orval B, Bortolin ML, Huber H, Charpentier B, Branlant C, Bachellerie JP, Brosius J, Huttenhofer A (2002) RNomics in Archaea reveals a further link between splicing of archaeal introns and rRNA processing. Nucleic Acids Res 30:921–930

Tannous E, Kanaya S (2014) Divalent metal ion-induced folding mechanism of RNase H1 from extreme halophilic archaeon Halobacterium sp. NRC-1. PLoS One 9:e109016

Thompson LD, Daniels CJ (1988) A tRNA(Trp) intron endonuclease from Halobacterium volcanii. Unique substrate recognition properties. J Biol Chem 263:17951–17959

van der Oost J, Westra ER, Jackson RN, Wiedenheft B (2014) Unravelling the structural and mechanistic basis of CRISPR-Cas systems. Nat Rev Microbiol 12:479–492

Veith T, Martin R, Wurm JP, Weis BL, Duchardt-Ferner E, Safferthal C, Hennig R, Mirus O, Bohnsack MT, Wohnert J, Schleiff E (2012) Structural and functional analysis of the archaeal endonuclease Nob1. Nucleic Acids Res 40:3259–3274

Vincent HA, Deutscher MP (2009) The roles of individual domains of RNase R in substrate binding and exoribonuclease activity. The nuclease domain is sufficient for digestion of structured RNA. J Biol Chem 284:486–494

Vogel A, Schilling O, Spath B, Marchfelder A (2005) The tRNase Z family of proteins: physiological functions, substrate specificity and structural properties. Biol Chem 386:1253–1264

Weiner AM (2004) tRNA maturation: RNA polymerization without a nucleic acid template. Curr Biol 14:R883–R885

Werner F (2007) Structure and function of archaeal RNA polymerases. Mol Microbiol 65:1395–1404

Werner F, Grohmann D (2011) Evolution of multisubunit RNA polymerases in the three domains of life. Nat Rev Microbiol 9:85–98

Westra ER, Swarts DC, Staals RH, Jore MM, Brouns SJ, van der Oost J (2012) The CRISPRs, they are a-changin': how prokaryotes generate adaptive immunity. Annu Rev Genet 46:311–339

Wiedenheft B, Sternberg SH, Doudna JA (2012) RNA-guided genetic silencing systems in bacteria and archaea. Nature 482:331–338

Yip WS, Shigematsu H, Taylor DW, Baserga SJ (2016) Box C/D sRNA stem ends act as stabilizing anchors for box C/D di-sRNPs. Nucleic Acids Res 44:8976–8989

Zago MA, Dennis PP, Omer AD (2005) The expanding world of small RNAs in the hyperthermophilic archaeon Sulfolobus solfataricus. Mol Microbiol 55:1812–1828

Chapter 5
The Archaeal Exosome: Degradation and Tailing at the 3′-End of RNA

Elena Evguenieva-Hackenberg, Susann Gauernack, and Gabriele Klug

Abstract Processing of many nascent RNAs into functional molecules includes ribonucleolytic trimming at the 3′-end. Additionally, ribonucleases are needed for removal of non-functional RNAs and for mRNA degradation adjusting mRNA levels to physiological needs of the cell. In most Archaea RNA processing and degradation at the 3′-end are performed by a protein complex named exosome. The archaeal exosome degrades RNA phosphorolytically from the 3′-end releasing nucleoside 5′-diphosphates (NDPs). In a reverse reaction, it uses NDPs to synthesize heteropolymeric, adenine-rich tails at the 3′-end of RNAs dedicated to degradation. The exosome consists of a hexameric ring structure composed of the archaeal proteins aRrp41 and aRrp42, and of a multimeric RNA-binding cap containing three different proteins (aRrp4, aCsl4 and aDnaG) with totally four different RNA binding domains (S1, KH, Zn-ribbon and the N-terminal domain of aDnaG). The hexameric ring and a variable aRrp4-aCsl4-heterotrimer form the nine-subunit core of the archaeal exosome, a structure that is evolutionary preserved in the bacterial polynucleotide phosphorylase (PNPase) and in the eukaryotic exosome. However, while in most eukaryotic exosomes the nine-subunit core is catalytically inactive, both bacterial PNPase and the archaeal exosome contain three phosphorolytic sites in the channels of their hexameric rings and function as exoribonucleases and RNA tailing enzymes. aDnaG is the archaea-specific subunit of the archaeal exosome and probably increases its substrate versatility, contributes to the regulation of its functions and determines its subcellular localization.

5.1 Introduction

Most RNAs are transcribed as precursors, which need to undergo ribonucleolytic trimming and sometimes enzymatic modifications to become functional. Essentially all RNAs in Eukarya and at least all rRNAs and tRNAs in Prokarya (Bacteria

E. Evguenieva-Hackenberg (✉) • S. Gauernack • G. Klug
Institut für Mikrobiologie und Molekularbiologie, Heinrich-Buff-Ring 26-32, 35392 Giessen, Germany
e-mail: Elena.Evguenieva-Hackenberg@mikro.bio.uni-giessen.de; Angelika.S.Gauernack@bio.uni-giessen.de; Gabriele.Klug@mikro.bio.uni-giessen.de

B. Clouet-d'Orval (ed.), *RNA Metabolism and Gene Expression in Archaea*, Nucleic Acids and Molecular Biology 32, DOI 10.1007/978-3-319-65795-0_5

and Archaea) undergo such maturation, for which ribonucleases (RNases) and RNase-containing protein complexes are of fundamental importance (Mörl and Marchfelder 2001; Deutscher 2009; Arraiano et al. 2010; Shepherd and Ibba 2015). Furthermore, RNases are necessary for degradation of non-functional RNAs in the processes of quality control and surveillance (Li et al. 2002; Cheng and Deutscher 2003; Houseley and Tollervey 2009; Wilusz et al. 2011), and for mRNA turnover (Arraiano et al. 2010; Belasco 2010). Generally, the mechanisms for degradation of stable (tRNA, rRNA) and unstable (mRNA) RNAs are similar (Deutscher 2009; Houseley and Tollervey 2009). Like mRNA in bacteria, archaeal mRNA is short-lived with typical half-lives between 0.5 and 10 min (Evguenieva-Hackenberg and Klug 2011), and the longevity of individual mRNA species can vary under different environmental conditions (Klug 1991; Goldenberg et al. 1996; Jäger et al. 2002). Thus, regulated RNA decay probably plays an important role for fitness and survival not only in bacteria (Condon and Bechhofer 2011; Deutscher 2015), but also in Archaea.

RNA is processed or degraded by endoribonucleases and exoribonucleases with the help of accessory factors like RNA-binding proteins and helicases. Endoribonucleases cleave RNA internally, while exoribonucleases release mononucleotides from the 5′-end or the 3′-end of transcripts. The underlying mechanisms are much better understood in Eukarya and Bacteria than in Archaea (Houseley and Tollervey 2009; Belasco 2010). Examples for archaeal endoribonucleases are the splicing endoribonuclease, RNase P and RNase Z involved in tRNA maturation (Nieuwlandt et al. 1991; Kleman-Leyer et al. 1997; Schierling et al. 2002), the β-CASP1 RNase, which probably represents the long sought major endoribonuclease in Archaea (Levy et al. 2011; Clouet-d'Orval et al. 2015), and CRISPR-Cas enzymes, which are parts of RNA-mediated adaptive immune systems (Carte et al. 2008; Richter et al. 2012; Sheppard et al. 2016; Charpentier et al. 2015). In Archaea exoribonucleolytic degradation in 5′ to 3′ direction is performed by other β-CASP RNases including homologs of bacterial RNase J (Clouet-d'Orval et al. 2010, 2015; Hasenöhrl et al. 2011). Finally, exoribonucleolytic degradation from the 3′-end is conducted by the archaeal exosome (Koonin et al. 2001; Evguenieva-Hackenberg et al. 2003, 2014) or the homolog of bacterial RNase R (Portnoy and Schuster 2006). This review is focused on the archaeal exosome, a protein complex with phosphorolytic activity found in all Archaea with the exception of halophiles and some methanogens. Much of our understanding of the archaeal exosome is based on its similarity to enzymes from Bacteria and Eukarya.

5.2 Conservation of Exoribonucleolytic Machineries Acting at the 3′-End

The key cellular roles of the RNases are reflected by their high conservation. Prime example is the similarity between the three-dimensional structures of bacterial polynucleotide phosphorylase (PNPase) and the exosome cores in Eukarya and Archaea (Fig. 5.1).

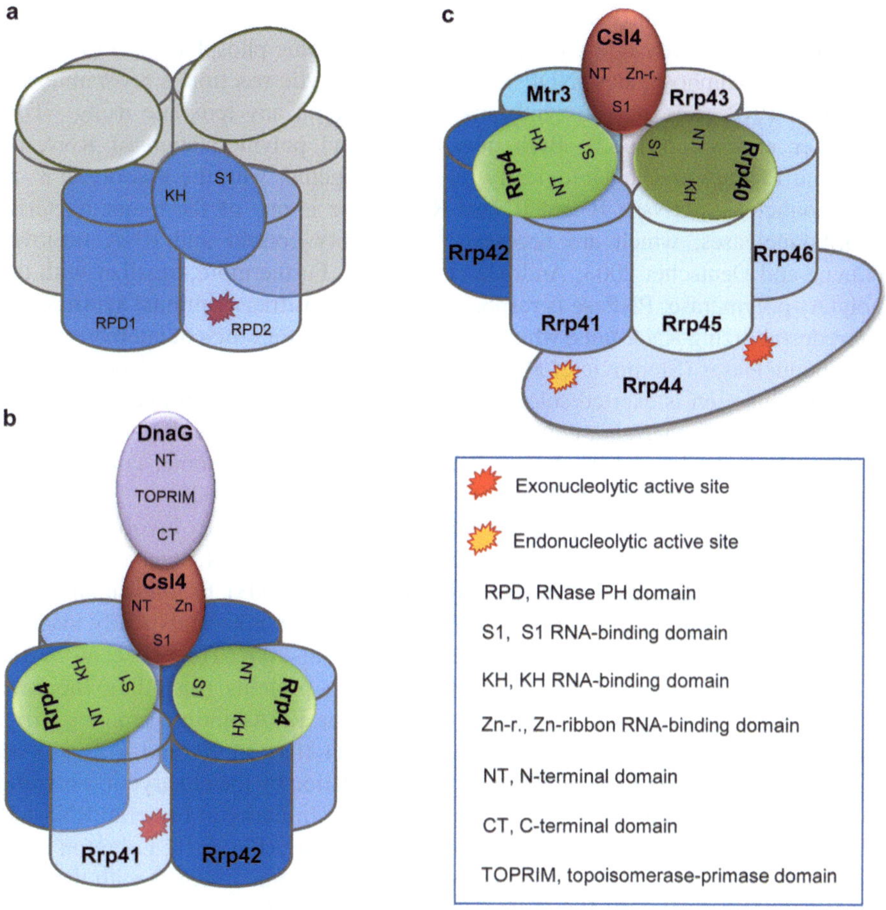

Fig. 5.1 Conservation of machineries acting at the 3′-end of RNA in the three domains of life. (**a**) Schematic representation of the homotrimeric bacterial PNPase: The domains in one of the monomers are indicated. Each monomer contains two RPD domains, an S1 domain, and a KH-domain. The six RPDs of the homotrimer build a hexameric ring, in which each RPD2 harbors a phosphorolytic active site located near the *bottom* of the central channel. The S1 and KH domains build an RNA-binding cap on the *top* of PNPase (Symmons et al. 2000). (**b**) Schematic representation of the Rrp44-containing eukaryotic exosome. Proteins and their domains are indicated. (**c**) Schematic representation of the archaeal exosome. Three aRrp41-aRrp42 dimers build a hexameric ring with three active sites located in aRrp41 (Lorentzen et al. 2005). The RNA binding cap consists of Rrp4, Csl4 and DnaG (Büttner et al. 2005; Hou et al. 2014). As shown by Witharana et al. (2012), different isoforms of the exosome have a different stoichiometry of Rrp4 and Csl4, and thus also different amount of DnaG

Bacterial PNPase is a homotrimer, with each monomer having two RNase PH-like domains (RPDs), only one of them with an active site, and two RNA-binding domains, an S1- and a KH-domain. The six RPDs build a ring with three active sites, on the top of which the S1 and KH-domains form an

RNA-binding platform (Fig. 5.1a; Symmons et al. 2000). PNPase degrades RNA exoribonucleolytically in 3′–5′ direction using inorganic phosphate (Pi) to release nucleoside 5′-diphosphates (NDPs). The phosphorolytic reaction is reversible and therefore PNPase is able to synthesize RNA without any template using NDPs (Littauer and Soreq 1982). Both degradation and polymerization of RNA by bacterial PNPase are physiologically relevant. Together with the hydrolytic 3′–5′ exoribonucleases RNase II and RNase R, PNPase is one of the major bacterial exoribonucleases, which are necessary for quality control and RNA turnover (Cheng and Deutscher 2003; Andrade et al. 2009). Furthermore, together with the poly(A)-polymerase, PNPase is responsible for the posttranscriptional synthesis of short destabilizing RNA-tails, which serve as scaffolds for efficient interaction with exoribonucleases (Slomovic et al. 2008). In bacteria lacking poly(A)-polymerase, the latter function is exerted solely by PNPase (Mohanty and Kushner 2000; Rott et al. 2003). In *Escherichia coli* PNPase is in a complex with an RNA helicase or binds to the endoribonuclease RNase E as a part of the degradosome, a multiprotein RNA-degrading complex (Py et al. 1996; Lin and Lin-Chao 2005).

The eukaryotic exosome is an essential multisubunit complex involved 3′–5′ in exoribonucleolytic processing and degradation of all major RNA classes (Mitchell et al. 1997; Allmang et al. 1999; Mitchell and Tollervey 2003). Its nine-subunit core shows striking structural similarities to PNPase, although it is catalytically inactive in human and yeast (Liu et al. 2006; Dziembowski et al. 2007; Tomecki et al. 2010). It comprises a hexameric ring of RPD-harboring proteins (including the here relevant Rrp41 and Rrp42), on the top of which the RNA-binding proteins Rrp4, Rrp40 (both have S1 and KH domains) and Csl4 (has an S1 and a Zn-ribbon domain) are located (Liu et al. 2006). RNA degradation by the eukaryotic exosome relies on its tenth subunit Rrp44/Dis3, which shows similarity to bacterial RNase R and has hydrolytic endo- and exoribonuclease activities (Fig. 5.1b; Lebreton et al. 2008; Schaeffer et al. 2009). In the cytoplasm the exosome interacts with the Ski7 complex to degrade mRNA (van Hoof et al. 2000; Mitchell and Tollervey 2003), while in the nucleus it binds Rrp6, an exoribonuclease with similarity to bacterial RNase D (Briggs et al. 1998; Milligan et al. 2005; Januszyk et al. 2011). Ski7 and Rrp6 use similar surfaces and recognition motifs for binding to the exosome (Kowalinski et al. 2016). The eukaryotic exosome is activated by addition of short destabilizing poly(A) tails, which are synthesized by non-canonical poly (A)-polymerases like the TRAMP complex at the 3′-end of substrates dedicated for degradation (LaCava et al. 2005; Vanácová et al. 2005).

The presence of genes with homology to Rrp41, Rrp42, Rrp4 and Csl4 in archaeal genomes led to the prediction of the archaeal exosome (Koonin et al. 2001), which was first isolated from *Sulfolobus solfataricus* (Evguenieva-Hackenberg et al. 2003). In addition to the archaeal proteins aRrp41, aRrp42, aRrp4 and aCsl4, a protein annotated as a bacterial-type primase DnaG (aDnaG) was shown to be an integral part of the native archaeal exosome (Fig. 5.1c; Walter et al. 2006; Witharana et al. 2012). In the nine-subunit archaeal exosome, aRrp41 and aRrp42 form a hexameric ring with three active sites located in aRrp41, while aRrp4 and aCsl4 constitute a heterotrimeric RNA-binding platform at the top of the

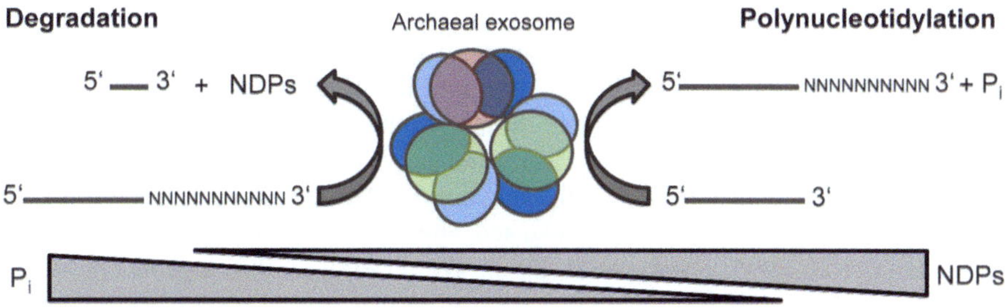

Fig. 5.2 Two functions of the archaeal exosome at the 3′-end of RNA. On the *left* side, phosphorolytic RNA degradation: RNA is degraded exoribonucleolytically using P_i and releasing NDPs. On the *right* side, RNA synthesis (polynucleotidylation): RNA is synthesized without any template using NDPs and releasing Pi. The regulation of the two functions in vivo is unclear. In vitro, high P_i concentrations promote RNA degradation, while at high NDP concentrations RNA polynucleotidylation (tailing) takes place (Portnoy et al. 2005; Evguenieva-Hackenberg et al. 2008), as indicated at the *bottom*. *Solid line* transcribed RNA; *NNNNNNNN* posttranscriptionally added heteropolymeric RNA tail

hexamer. This complex is structurally very similar to PNPase and the nine-subunit core of the eukaryotic exosome (Fig. 5.1; Lorentzen et al. 2005; Büttner et al. 2005). Functionally it strongly resembles PNPase, since it phosphorolytically degrades RNA in the presence of Pi, but also synthesizes heteropolymeric, purine-rich tails at high NDP concentrations (Fig. 5.2; Lorentzen et al. 2005; Portnoy et al. 2005; Evguenieva-Hackenberg et al. 2008).

The high similarity between PNPase and the cores of the eukaryotic and archaeal exosomes indicates evolutionary pressure for structure maintenance despite the loss of active sites in the eukaryotic RPD-hexamer. In the cases of PNPase and the archaeal exosome, RNA interacts with the RNA-binding cap on the top of the enzyme and the 3′-end is threaded through the central channel of the hexameric ring. The narrowest part of the channel (called "neck") can be passed by single-strand RNA only, to finally reach one of the three active sites near the bottom (Lorentzen et al. 2007; Navarro et al. 2008). Similarly, the 3′-end of a transcript is threaded through the channel of the yeast RPD-hexamer to reach the active site of Rrp44, which is at the bottom of the eukaryotic exosome (Makino et al. 2013). Thus, a feature of this structural arrangement is that it restricts the substrate range to single-stranded RNA. Additionally, strong interactions between the substrate and the neck ensure processive RNA degradation (Audin et al. 2016).

5.3 Subunits of the Archaeal Exosome and Their Roles

The catalytically active hexameric ring of the archaeal exosome was intensely studied and several crystal structures with and without RNA are available (Lorentzen et al. 2005; Lorentzen and Conti 2005; Büttner et al. 2005; Navarro et al. 2008; Ng et al. 2010). It was already a subject of several reviews (Evguenieva-Hackenberg 2010; Malet and Lorentzen 2011; Evguenieva-Hackenberg and Bläsi 2013; Evguenieva-Hackenberg et al. 2014) and therefore is described very briefly here. It is built of three aRrp41-aRrp42 dimers, each harboring an active site located in aRrp41. However, aRrp41 alone does not have RNase activity, because both proteins are needed for binding of the nucleotides N1–N4 (N1 being the ultimate nucleotide at the 3′-end, which is released after an exoribonucleolytic cleavage) (Lorentzen et al. 2005; Ramos et al. 2006; Walter et al. 2006). In a recent study, Audin et al. (2016) have shown that the 3′-end of the substrate is highly flexible in the lumen of the hexamer and each of the three active sites equally contributes to catalysis. The strongest binding between a substrate and the hexamer is at the neck, where the nucleotide N10 interacts with conserved loops of all three aRrp41 subunits of the hexameric ring (Navarro et al. 2008; Audin et al. 2016). This binding with nanomolar substrate affinity prevents RNA release between the catalytic steps and is the basis for fast and processive degradation (Audin et al. 2016). Ribooligonucleotides shorter than 10 nt are not fixed at the neck and their degradation is slower and seems to be not processive (Hartung et al. 2010; Audin et al. 2016). Final products of degradation are NDPs and ribooligonucleotides with a length of 4–5 nt (Evguenieva-Hackenberg et al. 2008) or 2–3 nt (Hartung et al. 2010). Besides Pi, Mg^{2+} ions are needed for the phosphorolytic degradation of RNA by the archaeal exosome (Evguenieva-Hackenberg et al. 2008; Lorentzen and Conti 2012).

In vivo, the nine-subunit archaeal exosome contains aRrp4-aCsl4 heterotrimers with different stoichiometries (Witharana et al. 2012). To analyze each of these two proteins separately in the context of a nine-subunit exosome, complexes with homotrimeric RNA-binding caps have been reconstituted in vitro. In the crystal structures of Rrp4-exosomes and Csl4-exosomes, the N-terminal domains of both Rrp4 and Csl4 are tightly bound to the hexameric ring, their middle RNA-binding S1-domains form an entry pore for RNA above the central channel of the hexamer and the C-terminal KH (of Rrp4) and Zn-ribbon (of Csl4) domains are at the periphery (Büttner et al. 2005; Lorentzen et al. 2007; Lu et al. 2010). Biochemical analyses revealed that archaeal nine-subunit exosomes degrade RNA faster than the hexameric ring alone due to more efficient substrate binding (Büttner et al. 2005; Walter et al. 2006; Luz et al. 2010). Furthermore, Rrp4 from *S. solfataricus* (SsoRrp4) confers preference for poly(A) to the exosome (Roppelt et al. 2010a), while SsoCsl4 is necessary for the interaction with aDnaG, the archaea-specific component of the exosome (Fig. 5.1c; Hou et al. 2013).

Despite its annotation as a bacterial type primase (She et al. 2001) and biochemical data suggesting such a function (Bauer et al. 2013), aDnaG is an integral part of the archaeal exosome (Walter et al. 2006; Witharana et al. 2012; reviewed in Evguenieva-Hackenberg et al. 2014). Recently we revealed that the N-terminal domain of aDnaG is a novel archaeal RNA-binding domain (Hou et al. 2014). Thus, the composite RNA-binding cap of the archaeal exosome is comprised of aRrp4, aCsl4 and aDnaG. In *S. solfataricus* two of these proteins (SsoRrp4 and SsoDnaG) preferentially bind poly(A) RNA and enhance degradation of adenine-rich RNA by the exosome (Roppelt et al. 2010a; Hou et al. 2013, 2014). Effective binding of adenine-rich RNA by the RNA degrading machinery of *S. solfataricus* is probably beneficial for this organism with an AT-rich genome (She et al. 2001) and A-rich posttranscriptionally added RNA-tails (Portnoy et al. 2005). The presence of three different RNA-binding proteins with totally four different RNA-binding domains in the cap of the archaeal exosome probably not only ensures efficient substrate binding, but may also help to resolve complex RNA structures, thus enabling processing and degradation of structured RNAs. Indeed, DnaG was necessary for in vitro polyadenylation of ribosomal RNA by the reconstituted Csl4-exosome of *S. solfataricus* (Hou et al. 2014).

Interestingly, the archaeal *dnaG* gene is more conserved in Archaea than the genes *rrp4, rrp41, rrp42* and *csl4*. The *dnaG* gene is present in all so far sequenced archaeal genomes, besides genes encoding the PriS and PriL subunits of the eukaryotic-type primase (Evguenieva-Hackenberg et al. 2014). This conservation of aDnaG in exosome-less Archaea suggests an exosome-independent function. However, the conserved RNA-binding domain at the N-terminus of the protein and its strong binding to the exosome in Archaea containing this protein complex argue for a major role of aDnaG in RNA metabolism.

5.4 Functions and Regulation of the Archaeal Exosome

The phosphorolytic mechanism of the archaeal exosome predestinates it for being both an exoribonuclease and a polyadenylating enzyme like bacterial PNPase. Tests with reconstituted exosome and with exosome in archaeal cell-free lysates confirmed that the protein complex can degrade RNA using Pi and that in a reverse reaction it can use NDPs to synthesize RNA (Lorentzen et al. 2005; Portnoy et al. 2005; Walter et al. 2006). RNA synthesis was most efficient with ADP and GDP resulting in very long RNA tails (larger than 700 nt), while CDP or UDP generated only short tails (approximately 20 nt) (Evguenieva-Hackenberg et al. 2008). In line with these results, posttranscriptionally added RNA-tails were detected in exosome-containing Archaea, but are completely missing in halophiles and methanogens lacking this protein complex (Portnoy et al. 2005; Portnoy and Schuster 2006). This also strongly supports the physiological role of the archaeal exosome as an RNA polyadenylating machine in addition to being a 3′–5′ exoribonuclease (Fig. 5.2). The tails detected in *S. solfataricus* were AG-rich and were found at 3′-ends of rRNA and mRNA degradation products (Portnoy et al. 2005). Most probably, like in bacteria, these tails destabilize RNA serving as platforms for efficient RNase binding (Hui et al. 2014). Indeed, a tailed rRNA transcript was degraded faster by the nine-subunit exosome than the same transcript without a tail (Evguenieva-Hackenberg et al. 2008). Unfortunately, our knowledge on the exosome functions still relies on biochemical and molecular analyses but genetic analyses are lacking. Similarly to eukaryotes, the genes for exosomal subunits seem to be essential in Archaea and so far this hampered deeper investigation of their functions in the cell. Despite the big progress in developing tools for genetic modifications and gene silencing in Archaea (Albers and Driessen 2008; Zebec et al. 2016), the analysis of essential genes still remains a challenge.

In Archaea containing exosomal genes, the exosome is a major 3′–5′ exoribonuclease and the only RNA-tailing enzyme (Portnoy et al. 2005; Walter et al. 2006; Portnoy and Schuster 2006). Thus, similarly to the eukaryotic exosome, it probably interacts with virtually all RNAs in the cell. However, how different functions (complete RNA degradation, 3′-end trimming or 3′-tailing) on specific RNA molecules are regulated remains unknown. The direction of the phosphorolytic reaction in vitro depends on the Pi and NDP concentrations and on the concentration of Mg^{2+}, suggesting that variations in their local concentrations in vivo may modulate the activity of the archaeal exosome (Evguenieva-Hackenberg et al. 2008). Further, it can be assumed that the accessibility of an RNA substrate (protection by a higher-order structure and/or by RNA-binding proteins) determines 3′-trimming (for example, during maturation of ribosomal RNA). Similarly, the accessibility of RNA-binding sites at the exosome may determine substrate selection. The last is probably influenced by the composition of different isoforms of the exosome and by interaction with other proteins as described for its eukaryotic counterpart (LaCava et al. 2005; Makino et al. 2015; Kowalinski et al. 2016). The archaeal exosome was found to interact with the splicing endoribonuclease in *Methanotermobacter thermoauthotrophicus* (Farhoud et al. 2005), the S1-domain containing, RNA binding protein TK2227 in *Thermococcus kodakarensis* (Li et al. 2010) and with EF1-alpha in *S. solfataricus* (Witharana et al. 2012).

In *S. solfataricus*, the exosome is found in a light, soluble form (270 kDa) and in a heavy form co-sedimenting with ribosomal subunits and membranes (Evguenieva-Hackenberg et al. 2003; Roppelt et al. 2010b). More DnaG and Csl4 were found in the heavy than in the light form, but the functional implications of these differences are still not clear (Witharana et al. 2012). The heavy form seems to be associated with the cytoplasmic membrane and therefore it was proposed that DnaG may participate in the subcellular localization of the exosome in the prokaryotic, archaeal cell (Roppelt et al. 2010b; Evguenieva-Hackenberg et al. 2011). The subcellular localization of the archaeal exosome was also revealed by immunofluorescence microscopy showing that Rrp41 and DnaG are localized at the periphery of the *S. solfataricus* cell (Roppelt et al. 2010b). This localization may also have a regulatory role, for example keeping the destructive exosome away from potential mRNA targets in the cytoplasm. Examples for prokaryotic RNases with specific and even regulated localization are found in bacteria (Khemici et al. 2008; Deutscher 2015; Cascante-Estepa et al. 2016).

Considering the multilayered regulation of bacterial PNPase, probably much remains to be discovered about the regulation of the exosome in Archaea. Indeed, bacterial PNPase is not only autoregulated at the level of mRNA stability and translatability, but is also part of distinct protein complexes and its activity is influenced by small molecules reflecting the metabolic and energetic status of the cell (recently reviewed by Briani et al. 2016). Furthermore, PNPase does not only degrade RNA or synthesize destabilizing RNA tails, but also influences the polyadenylation status of certain mRNAs and is even found in ribonucleoprotein particles protecting small regulatory RNAs (Bandyra et al. 2016; Mildenhall et al. 2016). Given the high structural and functional similarity between bacterial PNPase and the archaeal exosome, such mechanisms should be considered also for Archaea.

It is noteworthy that in contrast to the higher similarities between the archaeal and eukaryotic replication and translation machineries compared to the bacterial ones, the archaeal mechanisms for RNA degradation are more similar to those in bacteria. This seems reasonable when we consider the similarities between archaeal and bacterial mRNAs including their polycistronic nature and the general lack of introns, methylguanosine caps and long poly(A) tails (Brown and Reeve 1985). It can be speculated that these characteristics are pivotal for the prokaryotic way of life, which demands short mRNA half-lives for prompt response to rapidly changing conditions. In addition, the phosphorolytic mechanism of the archaeal exosome and bacterial PNPase saves energy in the NDPs, which then can be used for RNA tailing and for synthesis of ATP, GTP, and their signaling derivatives. Despite the challenge to regulate the reversible phosphorolytic reaction, this could confer an important advantage for the survival of prokaryotic cells in nature.

5.5 Outlook

More than 10 years research on the archaeal exosome led to a thorough characterization of the structure and mechanisms of action of its nine-subunit core. However, a structural analysis of its DnaG-containing, physiologically relevant form is still lacking. Structural information about the DnaG-exosome is needed to understand the mechanisms of substrate selection by the composite RNA-binding cap of the archaeal exosome. Identification of additional protein interaction partners and global analyses of RNA substrates of the exosome in the future may contribute to further understanding of the role and the regulation of this fascinating protein complex. Last but not least, construction of conditional exosome mutants still remains a major goal in the exosome research with Archaea, because this will ultimately open opportunities to study its role in the archaeal cell.

Acknowledgements Our work on the archaeal exosome was supported by Deutsche Forschungsgemeinschaft.

References

Albers SV, Driessen AJ (2008) Conditions for gene disruption by homologous recombination of exogenous DNA into the *Sulfolobus solfataricus* genome. Archaea 2:145–149

Allmang C, Kufel J, Chanfreau G, Mitchell P, Petfalski E, Tollervey D (1999) Functions of the exosome in rRNA, snoRNA and snRNA synthesis. EMBO J 18(19):5399–5410

Andrade JM, Pobre V, Silva IJ, Domingues S, Arraiano CM (2009) The role of 3′-5′ exoribonucleases in RNA degradation. Prog Mol Biol Transl Sci 85:187–229. doi:10.1016/S0079-6603(08)00805-2

Arraiano CM, Andrade JM, Domingues S, Guinote IB, Malecki M, Matos RG, Moreira RN, Pobre V, Reis FP, Saramago M, Silva IJ, Viegas SC (2010) The critical role of RNA processing and degradation in the control of gene expression. FEMS Microbiol Rev 34(5):883–923. doi:10.1111/j.1574-6976.2010.00242.x

Audin MJ, Wurm JP, Cvetkovic MA, Sprangers R (2016) The oligomeric architecture of the archaeal exosome is important for processive and efficient RNA degradation. Nucleic Acids Res 44(6):2962–2973. doi:10.1093/nar/gkw062

Bandyra KJ, Sinha D, Syrjanen J, Luisi BF, De Lay NR (2016) The ribonuclease polynucleotide phosphorylase can interact with small regulatory RNAs in both protective and degradative modes. RNA 22(3):360–372. doi:10.1261/rna.052886.115

Bauer RJ, Graham BW, Trakselis MA (2013) Novel interaction of the bacterial-Like DnaG primase with the MCM helicase in archaea. J Mol Biol 425(8):1259–1273. doi:10.1016/j.jmb.2013.01.025

Belasco JG (2010) All things must pass: contrasts and commonalities in eukaryotic and bacterial mRNA decay. Nat Rev Mol Cell Biol 11(7):467–478. doi:10.1038/nrm2917

Briani F, Carzaniga T, Dehò G (2016) Regulation and functions of bacterial PNPase. Wiley Interdiscip Rev RNA 7(2):241–258. doi:10.1002/wrna.1328

Briggs MW, Burkard KT, Butler JS (1998) Rrp6p, the yeast homologue of the human PM-Scl 100-kDa autoantigen, is essential for efficient 5.8 S rRNA 3′ end formation. J Biol Chem 273 (21):13255–13263

Brown JW, Reeve JN (1985) Polyadenylated, noncapped RNA from the archaebacterium *Methanococcus vannielii*. J Bacteriol 162:909–917

Büttner K, Wenig K, Hopfner KP (2005) Structural framework for the mechanism of archaeal exosomes in RNA processing. Mol Cell 20:461–471

Carte J, Wang R, Li H, Terns RM, Terns MP (2008) Cas6 is an endoribonuclease that generates guide RNAs for invader defense in prokaryotes. Genes Dev 22:3489–3496. doi:10.1101/gad.1742908

Cascante-Estepa N, Gunka K, Stülke J (2016) Localization of components of the RNA-degrading machine in *Bacillus subtilis*. Front Microbiol 7:1492

Charpentier E, Richter H, van der Oost J, White MF (2015) Biogenesis pathways of RNA guides in archaeal and bacterial CRISPR-Cas adaptive immunity. FEMS Microbiol Rev 39:428–441. doi:10.1093/femsre/fuv023

Cheng ZF, Deutscher MP (2003) Quality control of ribosomal RNA mediated by polynucleotide phosphorylase and RNase R. Proc Natl Acad Sci USA 100(11):6388–6393

Clouet-d'Orval B, Rinaldi D, Quentin Y, Carpousis AJ (2010) Euryarchaeal beta-CASP proteins with homology to bacterial RNase J Have 5′- to 3′-exoribonuclease activity. J Biol Chem 285 (23):17574–17583. doi:10.1074/jbc.M109.095117

Clouet-d'Orval B, Phung DK, Langendijk-Genevaux PS, Quentin Y (2015) Universal RNA-degrading enzymes in Archaea: prevalence, activities and functions of β-CASP ribonucleases. Biochimie 118:278–285. doi:10.1016/j.biochi.2015.05.021

Condon C, Bechhofer DH (2011) Regulated RNA stability in the Gram positives. Curr Opin Microbiol 14(2):148–154. doi:10.1016/j.mib.2011.01.010

Deutscher MP (2009) Maturation and degradation of ribosomal RNA in bacteria. Prog Mol Biol Transl Sci 85:369–391. doi:10.1016/S0079-6603(08)00809-X

Deutscher MP (2015) How bacterial cells keep ribonucleases under control. FEMS Microbiol Rev 39(3):350–361. doi:10.1093/femsre/fuv012

Dziembowski A, Lorentzen E, Conti E, Séraphin B (2007) A single subunit, Dis3, is essentially responsible for yeast exosome core activity. Nat Struct Mol Biol 14:15–22

Evguenieva-Hackenberg E (2010) The archaeal exosome. Adv Exp Med Biol 702:29–38

Evguenieva-Hackenberg E, Bläsi U (2013) Attack from both ends: mRNA degradation in the crenarchaeon *Sulfolobus solfataricus*. Biochem Soc Trans 41(1):379–383. doi:10.1042/BST20120282
Evguenieva-Hackenberg E, Klug G (2011) New aspects of RNA processing in prokaryotes. Curr Opin Microbiol 14(5):587–592. doi:10.1016/j.mib.2011.07.025
Evguenieva-Hackenberg E, Walter P, Hochleitner E, Lottspeich F, Klug G (2003) An exosome-like complex in *Sulfolobus solfataricus*. EMBO Rep 4:889–893
Evguenieva-Hackenberg E, Roppelt V, Finsterseifer P, Klug G (2008) Rrp4 and Csl4 are needed for efficient degradation but not for polyadenylation of synthetic and natural RNA by the archaeal exosome. Biochemistry 47:13158–13168. doi:10.1021/bi8012214
Evguenieva-Hackenberg E, Roppelt V, Lassek C, Klug G (2011) Subcellular localization of RNA degrading proteins and protein complexes in prokaryotes. RNA Biol 8:49–54
Evguenieva-Hackenberg E, Hou L, Glaeser S, Klug G (2014) Structure and function of the archaeal exosome. Wiley Interdiscip Rev RNA 5(5):623–635. doi:10.1002/wrna.1234
Farhoud MH, Wessels HJ, Steenbakkers PJ, Mattijssen S, Wevers RA, van Engelen BG, Jetten MS, Smeitink JA, van den Heuvel LP, Keltjens JT (2005) Protein complexes in the archaeon *Methanothermobacter thermautotrophicus* analyzed by blue native/SDS-PAGE and mass spectrometry. Mol Cell Proteomics 4:1653–1663
Goldenberg D, Azar I, Oppenheim AB (1996) Differential mRNA stability of the cspA gene in the cold-shock response of *Escherichia coli*. Mol Microbiol 19:241–248
Hartung S, Niederberger T, Hartung M, Tresch A, Hopfner KP (2010) Quantitative analysis of processive RNA degradation by the archaeal RNA exosome. Nucleic Acids Res 38:5166–5176. doi:10.1093/nar/gkq238
Hasenöhrl D, Konrat R, Bläsi U (2011) Identification of an RNase J ortholog in *Sulfolobus solfataricus*: implications for 5′-to-3′ directional decay and 5′-end protection of mRNA in Crenarchaeota. RNA 17(1):99–107. doi:10.1261/rna.2418211
Hou L, Klug G, Evguenieva-Hackenberg E (2013) The archaeal DnaG protein needs Csl4 for binding to the exosome and enhances its interaction with adenine-rich RNAs. RNA Biol 10:415–424. doi:10.4161/rna.23450
Hou L, Klug G, Evguenieva-Hackenberg E (2014) Archaeal DnaG contains a conserved N-terminal RNA-binding domain and enables tailing of rRNA by the exosome. Nucleic Acids Res 42(20):12691–12706. doi:10.1093/nar/gku969
Houseley J, Tollervey D (2009) The many pathways of RNA degradation. Cell 136(4):763–776. doi:10.1016/j.cell.2009.01.019
Hui MP, Foley PL, Belasco JG (2014) Messenger RNA degradation in bacterial cells. Annu Rev Genet 48:537–559. doi:10.1146/annurev-genet-120213-092340
Jäger A, Samorski R, Pfeifer F, Klug G (2002) Individual gvp transcript segments in *Haloferax mediterranei* exhibit varying half-lives, which are differentially affected by salt concentration and growth phase. Nucleic Acids Res 30(24):5436–5443
Januszyk K, Liu Q, Lima CD (2011) Activities of human RRP6 and structure of the human RRP6 catalytic domain. RNA 17(8):1566–1577
Khemici V, Poljak L, Luisi BF, Carpousis AJ (2008) The RNase E of *Escherichia coli* is a membrane-binding protein. Mol Microbiol 70:799–813. doi:10.1111/j.1365-2958.2008.06454
Kleman-Leyer K, Armbruster DW, Daniels CJ (1997) Properties of *H. volcanii* tRNA intron endonuclease reveal a relationship between the archaeal and eucaryal tRNA intron processing systems. Cell 89(6):839–847
Klug G (1991) Endonucleolytic degradation of puf mRNA in *Rhodobacter capsulatus* is influenced by oxygen. Proc Natl Acad Sci USA 88(5):1765–1769
Koonin EV, Wolf YI, Aravind L (2001) Prediction of the archaeal exosome and its connections with the proteasome and the translation and transcription machineries by a comparative-genomic approach. Genome Res 11:240–252
Kowalinski E, Kögel A, Ebert J, Reichelt P, Stegmann E, Habermann B, Conti E (2016) Structure of a cytoplasmic 11-subunit RNA exosome complex. Mol Cell 63(1):125–134. doi:10.1016/j.molcel.2016

LaCava J, Houseley J, Saveanu C, Petfalski E, Thompson E, Jacquier A, Tollervey D (2005) RNA degradation by the exosome is promoted by a nuclear polyadenylation complex. Cell 121:713–724
Lebreton A, Tomecki R, Dziembowski A, Séraphin B (2008) Endonucleolytic RNA cleavage by a eukaryotic exosome. Nature 456(7224):993–996. doi:10.1038/nature07480
Levy S, Portnoy V, Admon J, Schuster G (2011) Distinct activities of several RNase J proteins in methanogenic archaea. RNA Biol 8:1073–1083. doi:10.4161/rna.8.6.16604
Li Z, Reimers S, Pandit S, Deutscher MP (2002) RNA quality control: degradation of defective transfer RNA. EMBO J 21(5):1132–1138
Li Z, Santangelo TJ, Cuboňová L, Reeve JN, Kelman Z (2010) Affinity purification of an archaeal DNA replication protein network. MBio 1(5):e00221-10. doi:10.1128/mBio.00221-10
Lin PH, Lin-Chao S (2005) RhlB helicase rather than enolase is the beta-subunit of the *Escherichia coli* polynucleotide phosphorylase (PNPase)-exoribonucleolytic complex. Proc Natl Acad Sci USA 102(46):16590–16595
Littauer U, Soreq H (1982) Polynucleotide phosphorylase. In: Boyer PD (ed) The enzymes, Vol 15, Part B. Academic Press, New York, pp 517–533
Liu Q, Greimann JC, Lima CD (2006) Reconstitution, activities, and structure of the eukaryotic RNA exosome. Cell 127:1223–1237. Erratum in: Cell 2007, 131:188–189
Lorentzen E, Conti E (2005) Structural basis of 3′ end RNA recognition and exoribonucleolytic cleavage by an exosome RNase PH core. Mol Cell 20:473–481
Lorentzen E, Conti E (2012) Crystal structure of a 9-subunit archaeal exosome in pre-catalytic states of the phosphorolytic reaction. Archaea 2012:721869. doi:10.1155/2012/721869
Lorentzen E, Walter P, Fribourg S, Evguenieva-Hackenberg E, Klug G, Conti E (2005) The archaeal exosome core is a hexameric ring structure with three catalytic subunits. Nat Struct Mol Biol 12:575–581
Lorentzen E, Dziembowski A, Lindner D, Seraphin B, Conti E (2007) RNA channelling by the archaeal exosome. EMBO Rep 8:470–476
Lu C, Ding F, Ke A (2010) Crystal structure of the *S. solfataricus* archaeal exosome reveals conformational flexibility in the RNA-binding ring. PLoS One 5:e8739. doi:10.1371/journal.pone.0008739
Luz JS, Ramos CR, Santos MC, Coltri PP, Palhano FL, Foguel D, Zanchin NI, Oliveira CC (2010) Identification of archaeal proteins that affect the exosome function in vitro. BMC Biochem 11:22. doi:10.1186/1471-2091-11-22
Makino DL, Baumgärtner M, Conti E (2013) Crystal structure of an RNA-bound 11-subunit eukaryotic exosome complex. Nature 495(7439):70–75. doi:10.1038/nature11870
Makino DL, Schuch B, Stegmann E, Baumgärtner M, Basquin C, Conti E (2015) RNA degradation paths in a 12-subunit nuclear exosome complex. Nature 524(7563):54–58. doi:10.1038/nature14865
Malet H, Lorentzen E (2011) Mechanisms of RNA recruitment by the exosome. RNA Biol 8 (3):398–403. Epub 2011 May 1
Mildenhall KB, Wiese N, Chung D, Maples VF, Mohanty BK, Kushner SR (2016) RNase E-based degradosome modulates polyadenylation of mRNAs after Rho-independent transcription terminators in *Escherichia coli*. Mol Microbiol 101(4):645–655. doi:10.1111/mmi.13413
Milligan L, Torchet C, Allmang C, Shipman T, Tollervey D (2005) A nuclear surveillance pathway for mRNAs with defective polyadenylation. Mol Cell Biol 25(22):9996–10004
Mitchell P, Tollervey D (2003) An NMD pathway in yeast involving accelerated deadenylation and exosome-mediated 3′-5′ degradation. Mol Cell 11(5):1405–1413
Mitchell P, Petfalski E, Shevchenko A, Mann M, Tollervey D (1997) The exosome: a conserved eukaryotic RNA processing complex containing multiple 3′ to 5′ exoribonucleases. Cell 91:457–466
Mohanty BK, Kushner SR (2000) Polynucleotide phosphorylase functions both as a 3′ right-arrow 5′ exonuclease and a poly(A) polymerase in *Escherichia coli*. Proc Natl Acad Sci USA 97:11966–11971

Mörl M, Marchfelder A (2001) The final cut. The importance of tRNA 3′-processing. EMBO Rep 2(1):17–20

Navarro MV, Oliveira CC, Zanchin NI, Guimarães BG (2008) Insights into the mechanism of progressive RNA degradation by the archaeal exosome. J Biol Chem 283:14120–14131. doi:10.1074/jbc.M801005200

Ng CL, Waterman DG, Antson AA, Ortiz-Lombardía M (2010) Structure of the *Methanothermobacter thermautotrophicus* exosome RNase PH ring. Acta Crystallogr D Biol Crystallogr 66:522–528. doi:10.1107/S0907444910002908

Nieuwlandt DT, Haas ES, Daniels CJ (1991) The RNA component of RNase P from the archaebacterium *Haloferax volcanii*. J Biol Chem 266(9):5689–5695

Portnoy V, Schuster G (2006) RNA polyadenylation and degradation in different Archaea; roles of the exosome and RNase R. Nucleic Acids Res 34:5923–5931

Portnoy V, Evguenieva-Hackenberg E, Klein F, Walter P, Lorentzen E, Klug G, Schuster G (2005) RNA polyadenylation in Archaea: not observed in *Haloferax* while the exosome polynucleotidylates RNA in *Sulfolobus*. EMBO Rep 6:1188–1193

Py B, Higgins CF, Krisch HM, Carpousis AJ (1996) A DEAD-box RNA helicase in the *Escherichia coli* RNA degradosome. Nature 381(6578):169–172

Ramos CR, Oliveira CL, Torriani IL, Oliveira CC (2006) The *Pyrococcus* exosome complex: structural and functional characterization. J Biol Chem 281:6751–6759

Richter H, Zoephel J, Schermuly J, Maticzka D, Backofen R, Randau L (2012) Characterization of CRISPR RNA processing in *Clostridium thermocellum* and *Methanococcus maripaludis*. Nucleic Acids Res 40(19):9887–9896. doi:10.1093/nar/gks737

Roppelt V, Hobel CF, Albers SV, Lassek C, Schwarz H, Klug G, Evguenieva-Hackenberg E (2010a) The archaeal exosome localizes to the membrane. FEBS Lett 4:2791–2795. doi:10.1016/j.febslet.2010.05.013

Roppelt V, Klug G, Evguenieva-Hackenberg E (2010b) The evolutionarily conserved subunits Rrp4 and Csl4 confer different substrate specificities to the archaeal exosome. FEBS Lett 584:2931–2936. doi:10.1016/j.febslet.2010.05.014

Rott R, Zipor G, Portnoy V, Liveanu V, Schuster G (2003) RNA polyadenylation and degradation in cyanobacteria are similar to the chloroplast but different from *Escherichia coli*. J Biol Chem 278:15771–15777

Schaeffer D, Tsanova B, Barbas A, Reis FP, Dastidar EG, Sanchez-Rotunno M, Arraiano CM, van Hoof A (2009) The exosome contains domains with specific endoribonuclease, exoribonuclease and cytoplasmic mRNA decay activities. Nat Struct Mol Biol 16(1):56–62. doi:10.1038/nsmb

Schierling K, Rösch S, Rupprecht R, Schiffer S, Marchfelder A (2002) tRNA 3′ end maturation in archaea has eukaryotic features: the RNase Z from *Haloferax volcanii*. J Mol Biol 316 (4):895–902

She Q, Singh RK, Confalonieri F, Zivanovic Y, Allard G, Awayez MJ, Chan-Weiher CC, Clausen IG, Curtis BA, De Moors A, Erauso G, Fletcher C, Gordon PM, Heikamp-de Jong I, Jeffries AC, Kozera CJ, Medina N, Peng X, Thi-Ngoc HP, Redder P, Schenk ME, Theriault C, Tolstrup N, Charlebois RL, Doolittle WF, Duguet M, Gaasterland T, Garrett RA, Ragan MA, Sensen CW, Van der Oost J (2001) The complete genome of the crenarchaeon *Sulfolobus solfataricus* P2. Proc Natl Acad Sci USA 98:7835–7840

Shepherd J, Ibba M (2015) Bacterial transfer RNAs. FEMS Microbiol Rev 39(3):280–300. doi:10.1093/femsre/fuv004

Sheppard NF, Glover CV III, Terns RM, Terns MP (2016) The CRISPR-associated Csx1 protein of *Pyrococcus furiosus* is an adenosine-specific endoribonuclease. RNA 22:216–224. doi:10.1261/rna.039842

Slomovic S, Portnoy V, Yehudai-Resheff S, Bronshtein E, Schuster G (2008) Polynucleotide phosphorylase and the archaeal exosome as poly(A)-polymerases. Biochim Biophys Acta 1779:247–255. doi:10.1016/j.bbagrm.2007.12.004

Symmons MF, Jones GH, Luisi BF (2000) A duplicated fold is the structural basis for polynucleotide phosphorylase catalytic activity, processivity, and regulation. Structure 8:1215–1226

Tomecki R, Kristiansen MS, Lykke-Andersen S, Chlebowski A, Larsen KM, Szczesny RJ, Drazkowska K, Pastula A, Andersen JS, Stepien PP, Dziembowski A, Jensen TH (2010) The human core exosome interacts with differentially localized processive RNases: hDIS3 and hDIS3L. EMBO J. 29(14):2342–2357. doi:10.1038/emboj.2010.121

van Hoof A, Staples RR, Baker RE, Parker R (2000) Function of the ski4p (Csl4p) and Ski7p proteins in 3′-to-5′ degradation of mRNA. Mol Cell Biol 20(21):8230–8243

Vanácová S, Wolf J, Martin G, Blank D, Dettwiler S, Friedlein A, Langen H, Keith G, Keller W (2005) A new yeast poly(A) polymerase complex involved in RNA quality control. PLoS Biol 3:e189

Walter P, Klein F, Lorentzen E, Ilchmann A, Klug G, Evguenieva-Hackenberg E (2006) Characterization of native and reconstituted exosome complexes from the hyperthermophilic archaeon *Sulfolobus solfataricus*. Mol Microbiol 62(4):1076–1089

Wilusz JE, Whipple JM, Phizicky EM, Sharp PA (2011) tRNAs marked with CCACCA are targeted for degradation. Science 334(6057):817–821. doi:10.1126/science.1213671

Witharana C, Roppelt V, Lochnit G, Klug G, Evguenieva-Hackenberg E (2012) Heterogeneous complexes of the RNA exosome in *Sulfolobus solfataricus*. Biochimie 94:1578–1587. doi:10.1016/j.biochi.2012.03.026

Zebec Z, Zink IA, Kerou M, Schleper C (2016) Efficient CRISPR-mediated post-transcriptional gene silencing in a hyperthermophilic archaeon using multiplexed crRNA expression. G3 (Bethesda) 6:3161–3168. doi:10.1534/g3.116.032482

Chapter 6
Life and Death of Ribosomes in Archaea

Sébastien Ferreira-Cerca

Abstract Ribosomes are universally conserved large ribonucleoprotein particles ensuring protein synthesis in every cell. The universal conservation of ribosome function and structure offers a unique paradigm for understanding how RNP assembly mechanisms and function have evolved. Consequently, deciphering the general principles and differences of ribosome synthesis and function can contribute to a better understanding of the evolution history of these fundamental processes.

However, to achieve such a level of understanding, it is necessary to define conserved and specific principles of the ribosome life cycle in model organisms representative of all domains of life. Whereas, ribosome synthesis has been well characterized in both bacteria and eukarya, the archaeal ribosome biogenesis pathway is, in contrast, still largely unexplored.

In the following chapter, I provide a general survey of selected known and/or putative key features of the archaeal ribosome biogenesis pathway and highlight examples of functional convergence shared in the different domains of life.

Altogether, the archaeal ribosome life cycle appears to proceed via a mixture of bacterial-like and eukaryotic-like features to which archaeal specific features have been eventually implemented. In addition, it also suggests that several aspects of the eukaryotic ribosome life cycle have evolved, to some extent, on the basis of a "simplified" *archaeal-like* ribosome biogenesis pathway.

6.1 Introduction

Ribosomes are universally conserved macromolecular machineries essential for the life of every organism. Early pioneering work from Carl Woese and colleagues took advantage of this fascinating feature to develop the foundation of modern molecular phylogenetic analysis (Albers et al. 2013; Fox et al. 1977; Pace et al. 2012; Woese et al. 1990; Woese and Fox 1977). Using 16S ribosomal RNA (rRNA)

S. Ferreira-Cerca
Institute for Biochemistry, Genetics and Microbiology, University of Regensburg, Regensburg, Germany
e-mail: sebastien.ferreira-cerca@ur.de

B. Clouet-d'Orval (ed.), *RNA Metabolism and Gene Expression in Archaea*, Nucleic Acids and Molecular Biology 32, DOI 10.1007/978-3-319-65795-0_6

sequencing—a component of the ribosome—Woese and collaborators have not only discovered a new world—"the third domain of life" or archaea—but have also shed light on the evolutionary history of life on Earth (Albers et al. 2013; Eme and Doolittle 2015; Pace et al. 2012).

As such the history of the discovery of the third domain of life and the subsequent molecular basis of our understanding of the evolution are intimately linked to our insight of how ribosomes are made and how they function throughout the evolution. Therefore, our formal knowledge of the ribosome life cycle across the domains of life can also provide an additional "functional phylogeny" viewpoint to our comprehension of the evolution history and relationships among organisms.

Same-Same, But Different

Despite its universally conserved status, ribosomes are not all the same "from *Escherichia coli* to elephants". Whereas, the core ribosome activities, e.g., the decoding of mRNA and peptide-bond formation, are universally conserved, the structural components—i.e., the ribosomal RNA and ribosomal proteins (r-proteins)—vary both in number and size across the tree of life (Table 6.1). Indeed,

Table 6.1 Ribosome and ribosome biogenesis key numbers across the tree of life

	Bacteria	Archaea	Eukarya
rDNA repeats numbers[a]	1–16	1–4	~150–200 (>1000 in plants)
rRNAs	16S, 23S and 5S	16S, 23S and 5S	18S, 25/28S, 5.8S and 5S
rRNA expansion segments[b]	No	Yes (variable—only few)	Yes (variable amounts and length)
Ribosomal proteins[c]	32 universal r-proteins		
	≈49–59 44 ubiquitous	≈58–68 54 ubiquitous/71 described	≈78–80 78 ubiquitous
rRNA modifications[d]	Protein-based (≈36 modifications)	Protein- and sRNP-based	Protein- and sRNP-based
2′-O-Methylation/ Pseudourydilation	4/11	*Sso* (67/9); *Hv* (4/2) 7-127 C/D box sRNA	*Sc* (55/49); *Hs* (94/95)
Additional "stand-alone" base modifications	≈20	*Hv* (7)	12
Ribosome biogenesis factors[e]	≈50	>50?	>200

[a]Hadjiolov (1985); Klappenbach et al. (2001); Stoddard et al. (2015); Warner (1999)
[b]Armache et al. (2013); Parker et al. (2015); Petrov et al. (2015)
[c]Lecompte et al. (2002); Nakao et al. (2004); Yutin et al. (2012)
[d]Dennis et al. (2015); Grosjean et al. (2008); Lafontaine and Tollervey (1998); Sharma and Lafontaine (2015); Sloan et al. (2016)
[e]Ebersberger et al. (2014); Grosjean et al. (2014); Hage and Tollervey (2004); Henras et al. (2015); Thomson et al. (2013); Woolford and Baserga (2013)
Sso Sulfolobus solfataricus, *Hv Haloferax volcanii*, *Hs Homo sapiens*, *Sc Saccharomyces cerevisiae*

ribosomes have evolved around a conserved minimal universal core of ribosomal protein, which is composed of 32 r-proteins (Lecompte et al. 2002; Yutin et al. 2012). However, the total number of r-proteins vary from around 49–80, in bacteria and eukarya respectively (Ban et al. 2014; Grosjean et al. 2014; Lecompte et al. 2002; Nakao et al. 2004; Yutin et al. 2012). Similarly, the rRNA vary in size, by virtue of acquisition of additional expansion segments (Armache et al. 2013; Parker et al. 2015; Petrov et al. 2015). Finally, several aspects of the translation mechanisms itself (see Chap. 3) and ribosome biogenesis (this chapter) also vary across the tree of life.

Strikingly, the archaeal information processing machineries (like transcription, translation . . .) are generally described to be more closely related to their eukaryotic counterparts (Albers et al. 2013; Allers and Mevarech 2005; Eme and Doolittle 2015; Graham et al. 2000; Jain et al. 1999; Rivera et al. 1998). Early on and more recently, this remarkable analogy has been at the basis to suggest a common origin for archaea and eukarya, (1) either as a sister group (classical 3 domains of life representation) or (2) a eukaryotic origin from within the archael lineage (2 domains of life representation) (Cox et al. 2008; Lake 1985, 2015; Lake et al. 1984; Raymann et al. 2015; Spang et al. 2015; Williams et al. 2012; Zaremba-Niedzwiedzka et al. 2017).

Independently of these still much debated possibilities, the universal conservation of ribosome function and structure offers a unique paradigm for understanding how RNP assembly mechanisms and function have evolved. Consequently, deciphering the general principles and differences of ribosome synthesis and function can contribute to a better understanding of the evolution history of these fundamental processes.

However, to achieve such a level of knowledge, it is necessary to define conserved and specific principles of the ribosome life cycle in model organisms representative of the different domains of life. Whereas, ribosome synthesis has been well characterized in both bacteria and eukarya, the archaeal ribosome biogenesis pathway is, in contrast, still largely unexplored (Blombach et al. 2011; Yip et al. 2013).

Due to our current lack of knowledge of the life and death of ribosomes in archaea, the following chapter aims to provide a selected survey of known and/or putative key features of the archaeal ribosome biogenesis pathway and highlight features presumably shared between archaea and/or bacteria, and/or eukarya.

6.2 General Principles of Bacterial and Eukaryotic Ribosome Biogenesis: Common Themes and Fundamental Differences

To date, ribosome biogenesis in bacteria and eukarya has been best characterized in *Escherichia coli* and *Saccharomyces cerevisiae*, respectively. Very briefly summarized, ribosome synthesis is a complex, stepwise, multilayered and coordinated process, where the rRNAs emerge as an immature precursor transcript (pre-rRNA) which is matured in a series of relatively well documented steps (Henras et al. 2015; Nerurkar et al. 2015; Shajani et al. 2011; Thomson et al. 2013). Ribosomal assembly steps include rRNA folding, processing and chemical modifications of the rRNA. These steps require the coordinated action of endo- and exo-nuclease activities and RNA modifications machineries. Moreover, the r-proteins are simultaneously assembled with the nascent pre-rRNA in a relatively defined subdomain-dependent hierarchical order (Davis et al. 2016; de la Cruz et al. 2015; Mulder et al. 2010; Nierhaus 1991). Finally, diverse amounts of transiently acting factors also called ribosome assembly factors have been proposed to facilitate the formation of the ribosomal subunits (Fig. 6.1 and Table 6.1) (Henras et al. 2015; Nerurkar et al. 2015; Shajani et al. 2011; Thomson et al. 2013 and examples below).

However, this over simplified description of the bacterial and eukaryotic ribosome synthesis could easily provide the impression that there is a large amounts of similarities between these two pathways. Whereas the general blueprint of the ribosome biogenesis pathway are indeed very similar, the molecular complexity of the ribosome biogenesis process has been dramatically expanded in the course of evolution (Hage and Tollervey 2004) (Table 6.1). This molecular intricacy has been very well summarized by James Williamson: "*the impression is that a government defense contractor was given a fully functional working prototype bacterial ribosome that was subsequently "reengineered" to do the same job, and was ultimately delivered at a cost of one third of the nation's gross domestic product.*" [quoted from Williamson (2003)]. This increase complexity can be easily highlighted by the plethora of eukaryotic ribosome assembly factors, where more than 200 factors have been proposed to be involved in eukaryotic ribosome biogenesis. Remarkably, a minor fraction of these "eukaryotic" ribosome biogenesis factors are also already presents in most archaea (Blombach et al. 2011; Ebersberger et al. 2014; Yip et al. 2013). In contrast to eukaryotes, bacterial ribosome biogenesis apparently only requires a condensed subset of ribosome biogenesis factors (Hage and Tollervey 2004; Henras et al. 2015; Nerurkar et al. 2015; Shajani et al. 2011; Thomson et al. 2013) (Table 6.1).

Nevertheless, the excerpt mentioned above summarizes some interesting key differences of ribosomal subunits function and synthesis in one sentence. These apparent differences in terms of complexity and cellular economics, have been originally supported by early pioneering in vitro reconstitution experiments of the ribosomal subunits from their individual constituent (Mizushima and Nomura 1970; Nierhaus 1991; Nierhaus and Lafontaine 2004; Nomura and Erdmann 1970). These central studies were initially suggesting that the bacterial ribosome is a self-assembling macromolecular machine. These early major observations were therefore implying that the total information for the ribosome assembly pathway as well as the quaternary structure of the active ribosomes resides completely in the r-proteins and rRNA primary sequences, a key feature presumably shared with a primitive ribosome ancestor. Remarkably, in vitro ribosome self-assembly is a characteristic shared among prokaryotic ribosomes since it was also demonstrated that both archaeal ribosomal subunits can be assembled in vitro (Londei et al. 1986;

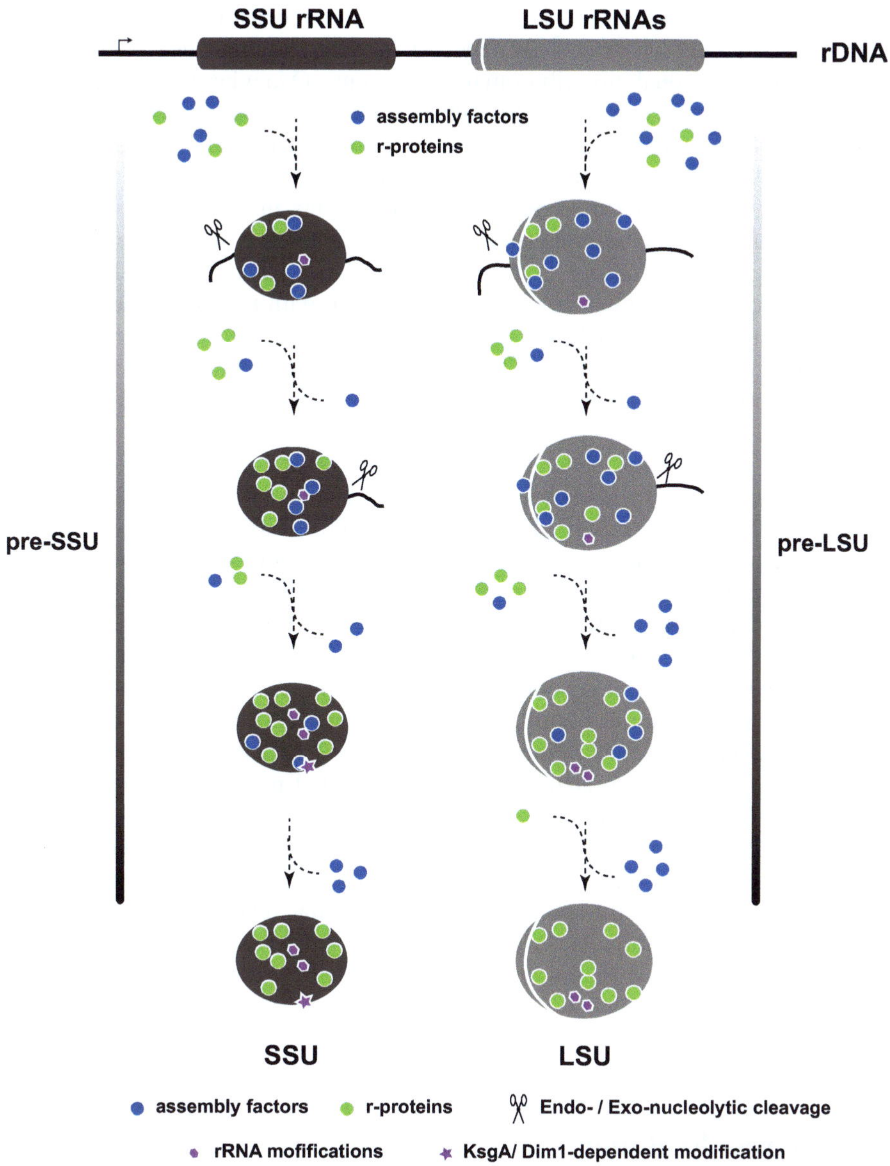

Fig. 6.1 General blueprint of ribosome biogenesis across evolution. A general and simplified schematic representation of ribosome biogenesis' key steps are depicted. In brief, co-transcriptional assembly of ribosome assembly factors (*blue dot*), r-proteins (*green dot*), and rRNA modifications (*lilac hexagon*) allow formation of the first nascent pre-ribosomal intermediates (pre-SSU and pre-LSU). The pre-ribosomal particles are further matured in a series of coordinated steps including rRNA folding, processing and chemical modifications of the rRNA. Whereas additional r-proteins are gradually assembled and stabilized, ribosome assembly factors dynamically associate and dissociate from the evolving pre-particles. The almost universal KsgA/Dim1-dependent SSU-rRNA modification is indicated by a lilac star. Note that the number of processing steps, modifications, assembly factors and r-proteins vary in an inter- and intra-domain of life specific-manner. See main text for detailed description

Sanchez et al. 1990, 1996). In contrast, in vitro reconstitution of eukaryotic ribosomal subunits has been only reported for *Dictyostelium discoideum* in the presence of additional non-ribosomal constituents (Mangiarotti and Chiaberge 1997). Finally, these in vitro reconstitution studies also suggested that r-proteins assembly follows a hierarchical order summarized into r-proteins "assembly maps" (Nierhaus 1991; Nierhaus and Lafontaine 2004).

Whereas, these in vitro reconstitution experiments initially suggested a self-assembly process, the non-physiological conditions used in such experiments (ionic-strength, temperature...) were also indicating that ribosomal assembly might be facilitated in vivo. It is only recently that, with the further development of genetic and in vivo analyses, several reports have been both supporting and challenging our view of how ribosomes are assembled in vivo in bacteria (Bubunenko et al. 2006, 2007; Davis et al. 2016; Gupta and Culver 2014; Mulder et al. 2010).

Overall, many of the general assembly features derived from these in vitro assembly studies can be also observed in vivo (Chen and Williamson 2013; Davis et al. 2016; Mulder et al. 2010; Talkington et al. 2005). However, in vivo bacterial ribosome assembly additionally proceed via the help of additional ribosome assembly factors presumably facilitating and/or providing directionality to the intrinsic self-assembling pathway (Bunner et al. 2010; Maki et al. 2002; Shajani et al. 2011). Noteworthy, the general in vivo r-proteins assembly pathway in eukaryotes strikingly follows very similar general domain-specific hierarchical assembly routes as described for the bacterial ribosomes (de la Cruz et al. 2015; Ferreira-Cerca et al. 2007; Ohmayer et al. 2013). From this point of view archaeal ribosome assembly might also follow a very analogous factor-facilitated modular hierarchical self-assembling process in vivo.

In this strikingly similar conceptual framework it is essential to disentangle the nature and function of the ribosome assembly factors involved in archaeal ribosome biogenesis, and to identify and decipher the molecular functional mimicry which have been developed and selected to facilitate very similar assembly principles with the help of apparently different molecular contributors.

6.3 Synthesis of the Ribosome Structural Components Across Evolution

The earliest step of the ribosome assembly process is the synthesis of the ribosomal constituents. It is also worth to note at this point that ribosome synthesis is one of the major metabolic pathway in actively dividing cells suggesting that highly regulated mechanisms must have evolved to ensure the precise coordinated production of the different ribosomal constituents according to cellular need (Iskakova et al. 2004; Nomura 1999; Planta 1997; Warner 1999).

Despite differences, some common principles about rRNA gene organization in pro- and eukaryotes can be highlighted (Fig. 6.2).

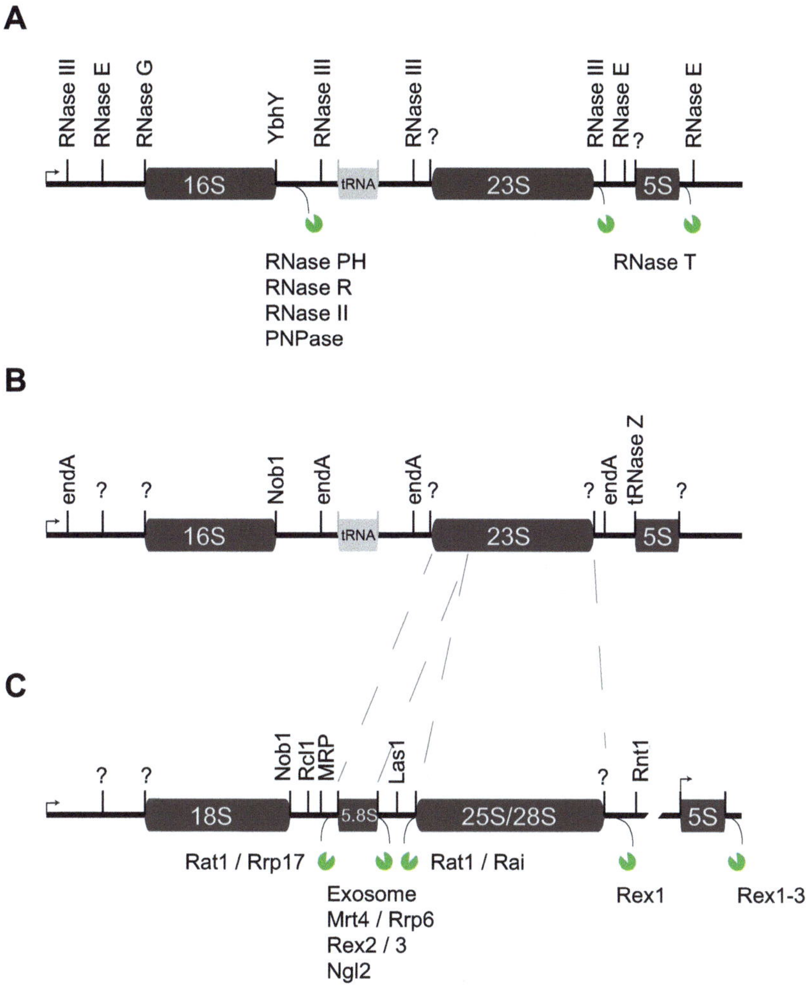

Fig. 6.2 Exemplary polycistronic organization of the rDNA across the evolution and its respective processing sites. Ribosomal DNA polycistronic organization and rRNA processing sites in (**a**) *E. coli*, representative of bacteria, (**b**) archaea and (**c**) *S. cerevisiae*, representative of eukaryotes. Schematic representation of exemplary rDNA polycistron, processing sites and the known respective endo- and exonuclease (*green pacman*) activities required for the maturation or the pre-rRNA are indicated. The *dashed lines* between panel (**a**) and (**c**) indicate the evolutionary origin of the eukaryotic 5.8S and 25S rRNA as a product of the splitting of prokaryotic 23S rRNA (see text)

Most rRNA genes are organized as operon-like structures where the different rRNAs are usually transcribed as a common rRNA precursor containing the 16S, 23S and 5S rRNA and the 18S, 5.8S, 25/28S in most of the pro-and eukaryotes respectively (Hadjiolov 1985; Hage and Tollervey 2004; Henras et al. 2015; Nomura 2001; Woolford and Baserga 2013; Yip et al. 2013). The mature rRNAs are usually flanked and separated from each other by internal and external transcribed spacers. This organization as long been thought to facilitate stoichiometric rRNA production. However, in eukaryotes the 5S rRNA is transcribed as an independent unit. In addition, in some archaea the rRNA are transcribed from independent genes (Yip et al. 2013). Finally, it was also shown, in eukaryotes, that the transcription of the small ribosomal subunit (SSU) and large ribosomal subunit (LSU) rRNAs can be uncoupled (Burman and Mauro 2012; Liang and Fournier 1997). Suggesting that such operon-like structure is not an essential prerequisite for the production of functional ribosomes.

In eukaryotes, the rRNA genes are often organized in tandemly repeated copies at one or a few chromosomal loci, whereas in bacteria the rRNA genes are scattered on the chromosome. The amount of rRNA gene repetition varies between organisms—from one gene to several thousand repeats (Hadjiolov 1985; Klappenbach et al. 2001; Stoddard et al. 2015; Yip et al. 2013). For example, *E. coli* possesses seven rRNA operons (Srivastava and Schlessinger 1990), and the yeast *S. cerevisiae* about 150 tandemly repeated copies of the rRNA genes located on chromosome XII (Goffeau et al. 1996; Nomura 2001) (Table 6.1).

In prokaryotes, all the cellular genes are transcribed with the help of a single RNA polymerase, whereas in eukaryotes there are three different kinds of RNA polymerases transcribing a different set of genes.

The RNA polymerase I (Pol I) is dedicated to the production of the rRNA contained in the operon-like structure (18S, 5.8S, 25/28S rRNA), while the RNA polymerase III, besides other cellular RNAs (i.e., tRNA), synthesizes the 5S rRNA (Sentenac and Riva 2013).

RNA Pol I-dependent gene transcription occurs in a sub-nuclear compartment, the nucleolus. The nucleolus was first described by microscopy analysis in the early nineteenth century and was found to be the location of rRNA synthesis (Hernandez-Verdun et al. 2010; Pederson 2011). The evolutionary origin of the nucleolus as an electron dense ultrastructure remains elusive. Despite rRNA synthesis, no nucleolar-like morphology (classical electron microscopy definition) has been characterized in prokaryotic cells. However, super-resolution microscopy (Jin et al. 2015, 2016) and molecular phylogenetic analyses (Staub et al. 2004) are challenging this classical view and suggest an early origin of a nucleolar-like structure which could be simply defined as the intracellular clustering of the rRNA genes/transcription and ribosome maturation machineries. Whereas, it is clear that prokaryotic cells analysed so far do not show the classical eukaryotic-like nucleolar morphology, it is likely that the eukaryotic nucleolus-ancestor originally evolved on the basis of only few comparable eukaryotic features that might be going beyond the classical ultrastructure definition of the nucleolus. In this regard, the origin of the nucleolus is presumably tightly linked to the evolution of a primordial eukaryotic-like ribosome biogenesis pathway and the onset of eukaryogenesis.

In *E. coli* cells, about half of the r-protein genes are clustered into four operons whereas the remaining r-proteins genes are scattered throughout the genome in additional operons containing 1–4 genes (Iskakova et al. 2004; Nomura 1999). Similarly, archaeal r-proteins genes are also clustered into several operons scattered throughout the genome. Whereas, several auto-regulation mechanisms have been described in bacteria (Iskakova et al. 2004; Nomura 1999), how the coordinated expression of the different operons is achieved in archaea is not well characterized.

In eukaryotes, the r-proteins genes are scattered throughout the entire genome and are independently transcribed by RNA Pol II. In the yeast *S. cerevisiae,* r-protein gene promoters often share binding sites for common transcription regulators (Lempiäinen and Shore 2009; Planta 1997; Warner 1999). In any case, cells face the challenge to accurately coordinate the expression of a large number of r-proteins with the rRNAs. How this balanced gene expression is faithfully achieved in archaea remains to be fully explored.

6.4 Ribosome Assembly in Archaea: Lessons from Bacteria and Eukarya

6.4.1 Ribosomal RNA Modifications and Their Contribution to the Ribosomal Subunit Assembly Process

Nucleoside modifications of rRNA have been already reported more than 50 years ago (Littlefield and Dunn 1958a, b; Smith and Dunn 1959). These modifications occur on the pre-rRNA during ribosome synthesis and are essentially of three types: base modifications; methylation of the 2′-hydroxyl group of sugar residues (2′-O-methylation); and conversion of uridine residues to pseudouridine (Ψ) by base rotation. Whereas these modifications are essentially clustered in the ribosomes active centers (Decatur and Fournier 2002; Piekna-Przybylska et al. 2008), the nature, extent and position of these modifications are not "fully" conserved across the domains of life (Lafontaine and Tollervey 1998; Piekna-Przybylska et al. 2008; Sloan et al. 2016). Moreover, additional intra-domain modifications variability (nature, extent and position) has been also described (Dennis et al. 2015; Lafontaine and Tollervey 1998; Sloan et al. 2016; Yip et al. 2013). A more systematic survey of the nature, extent and position of rRNA modifications in archaea still largely remains to be done (Dennis et al. 2015; Yip et al. 2013) (Table 6.1).

Moreover, the role of the numerous rRNA modifications remain in most cases elusive and have been suggested to participate in the stabilization of the ribosomes native structure, thereby fine-tuning the translation process (Lafontaine and Tollervey 1998; Sloan et al. 2016). Moreover, the absence of a subset of rRNA modification and/or their respective RNA modifier have been linked to ribosome assembly defects (Lafontaine and Tollervey 1998; Sloan et al. 2016). Recently, with the advance of high-resolution cryo-electron microscopy analysis, some rRNA modifications have been observed in their native context, thereby confirming some of their putative functions (Fischer et al. 2015; Polikanov et al. 2015). Interestingly, the number of rRNA modifications increases from prokaryotes to eukaryotes (Table 6.1). Similarly, hyperthermophile archaea generally present an increasing amounts of post-transcriptional modifications very close to those observed in higher eukaryotes (Dennis et al. 2015; Lafontaine and Tollervey 1998; Sloan et al. 2016). In contrast, Haloarchaea appears to have a reduced set of modifications when compared to those regularly observed in both archaea and bacteria (Grosjean et al. 2008; Hansen et al. 2002; Kirpekar et al. 2005). Remarkably, the functional contributions of those variations for ribosome assembly and/or function, in the different biological context aforementioned, are still unclear.

Mechanistically, substrate recognition also differs between the protein-only bacterial-type modification system and, the dual usage of protein-based and RNA-guided modification machineries present in archaea and eukarya (Omer et al. 2003; Sloan et al. 2016; Yip et al. 2013). As such, these mechanistic differences also implies that rRNA modifications must be properly coordinated and can potentially influence the proper timing of the ribosome assembly process (Siibak and Remme 2010; Sloan et al. 2016).

Archaea and eukarya share a common RNA-guided modification system allowing the recruitment of the respective enzymatic activities performing pseudourydilation or 2′-O-methylation reactions (Lafontaine and Tollervey 1998; Sloan et al. 2016; Yip et al. 2013). Whereas, in all the different domains of life, the modification reactions are performed by proteins, the emergence of an RNA-guided machinery in archaea and eukarya (hereafter sRNA), implies that most rRNA/guide-RNA complexes will occur at a time where the rRNA is likely to be kept in a locally accessible/unfolded state. In addition, this local structural conformation can potentially provide additional directionality to the ribosomal subunit assembly either by increasing the kinetic window of local unfolded state and/or by delaying the kinetic of downstream assembly steps (Lafontaine and Tollervey 1998; Sloan et al. 2016; Yip et al. 2013). In agreement with this idea most sRNA-guided modifications are believed to be established during the early steps of eukaryotic ribosome biogenesis (Lafontaine and Tollervey 1998; Sloan et al. 2016; Yip et al. 2013). Moreover, sRNAs have been suggested to serve as "RNA chaperone", whereby sRNAs function as scaffold to facilitate rRNA folding (e.g., Bachellerie et al. 1995; Dennis et al. 2015; Steitz and Tycowski 1995). The functional distribution of this feature in archaea remains to be further analyzed (Dennis et al. 2015). In contrast, in a protein-only based system the timing of those modifications is mostly dictated by the intrinsic substrate specificity of the enzyme and/or its co-factors performing the modification. Accordingly, a great variety of substrate recognition has been observed e.g., RNA sequence specificity, 2D and 3D context specificity (Siibak and Remme 2010; Sloan et al. 2016). Strikingly, whereas in eukaryotes, most, if not all, RNA-guided modifications occur during the early ("less folded") steps of ribosome biogenesis, protein-based modifications are mostly occurring during the latest steps of ribosome biogenesis (Sloan et al. 2016). In most cases, the influence of the modification itself and/or the modification reaction on ribosome assembly progression remains to be, in most cases, further explored.

Finally, despite an apparent clustering of the rRNA modifications in the ribosomal subunits functional centers, a strict universal or inter-domain (bacteria-archaea or archaea-eukarya) conservation of rRNA modifications (same modification type at the same relative residue) is, surprisingly, a relatively rare event in the course of the evolution (Lafontaine and Tollervey 1998; Piekna-Przybylska et al. 2008; Sloan et al. 2016).

The following examples will highlight some rRNA modifications having various degree of conservation between the different domains of life and for which either the modification itself and/or the modifiers have been involved in proper ribosome assembly.

6.4.1.1 The Universal KsgA/Dim1-Dependent Modification Paradigm

One of such "universally conserved" modification is the RsmA/KsgA/Dim1-dependent dimethylation of two universally conserved adenosines located at the 3′ end of the 16S/18S rRNA (Lafontaine et al. 1994; Mangat and Brown 2008). The addition of four methyl groups on two adenosines located in helix 45 (h45) of the 16S rRNA has been suggested to contribute to the necessary destabilization of h45 allowing its proper folding and incorporation into the nascent ribosomal subunit, thereby contributing to the decoding site stability (Demirci et al. 2010; Heus et al. 1983; Van Charldorp et al. 1981). Moreover, additional studies have been suggesting a central "quality control" function of the KsgA/Dim1- dependent modification during the latest steps of SSU biogenesis (Connolly et al. 2008; Strunk et al. 2011; Xu et al. 2008). Surprisingly, the modification itself does not appear to be essential for viability, whereas the KsgA/Dim1 structural fold has been shown to have, at least, an additional essential function in the early steps of eukaryotic ribosome biogenesis (Lafontaine et al. 1995). In opposition to its described universally conserved nature, early studies have already suggested that in some organelles the SSU are either not modified (Klootwijk et al. 1975) or not fully modified (Van Buul et al. 1984). In addition, a recent work on the obligate nanoarchaeal symbiont *Nanoarchaeum equitans* revealed that KsgA and its dependent modifications are not present in this organism (Seistrup et al. 2016). Remarkably, sRNA-dependent 2′-O-methylation of h45 readily takes place in this biological context, suggesting that proper maturation and incorporation of h45 might use an alternative molecular strategy which presumably still depends on the h45 modifications status (Seistrup et al. 2016). Together these results suggest that a larger survey of the (archaeal) biodiversity can potentially reveal the natural flexibility and adaptation of the ribosome assembly pathway (Seistrup et al. 2016; our own unpublished results).

6.4.1.2 The Bacterial/Archaea-Eukarya Dichotomy

2′-O-Methylation of U2552 from Protein-Based to RNA-Guided Modification

Methylation of the 23S/25-28S rRNA at position U2552 (*E. coli* numbering) is a widespread modification found in all domains of life. However, the mechanisms by which this methylation is acquired diverge in the course of evolution. In bacteria (in those containing this modification), rlmE/rrmJ performs the modification reaction (Arai et al. 2015; Bügl et al. 2000; Caldas et al. 2000; Hager et al. 2004) whereas in eukaryotes modification is guided by a sRNA (Lapeyre and Purushothaman 2004). Interestingly, only a subset of archaea does contain an rlmE/rrmJ homolog (mostly in Euryarchaeota with the exception of the Thermococcales) predicted to perform the modification reaction, whereas the presence of a guide sRNA in some Crenarchaeota, in which rlmE/rrmJ homologs are absent, has been predicted to guide the methylation [e.g., *Pyrococcus abyssi* -sR25 a C/D box sRNA (Dennis et al. 2015; Hansen et al. 2002)]. Therefore, it is likely that during the course of the evolution process an RNA-guided machinery has functionally replaced the rlmE/rrmJ-dependent modification. Interestingly, whereas rlmE/rrmJ-dependent modification occurs in the context of a well assembled 50S particle (Caldas et al. 2000), the sRNA-dependent modification in eukaryotes is believed to occur in the earliest steps of the LSU biogenesis (Lapeyre and Purushothaman 2004). Consequently, although the nature of the modification has been preserved, the alteration of the modification machinery seems to have modify the timing of occurrence of this modification in the course of the LSU assembly. Moreover, in addition to its role in fine-tuning ribosome function, the U2252-modification has been suggested to be required for proper ribosomal assembly in bacteria (Arai et al. 2015). The different timing and molecular mechanisms enabling this modification and, the molecular consequences on the ribosome assembly pathway will have to be further dissected.

Pseudourydilation U1915/U1917

Similarly to the example described above, U1915/U1917 (*E. coli* numbering) in H69 of the 23S rRNA are universally conserved pseudourydilated residues. Whereas in bacteria rluD performs the isomerization of the uridine U1915/U1917 (and also U1911), it has been proposed that both archaea and eukarya utilize sRNA-dependent modification.

Likewise, to the example described above, rluD-dependent modifications have been suggested to collectively occur during the latest steps of the LSU formation (Leppik et al. 2007). Most strikingly, these modifications have been also described to be necessary for ribosome assembly in bacteria (Gutgsell et al. 2005). In *H. volcanii*, yeast and human, a single sRNA has been suggested to guide the modification of residues U1940/U1942; U2258/U2260 and U3741/U3743, respectively (Badis et al. 2003; Blaby et al. 2011; Grosjean et al. 2008). Whereas, deletion of the respective H/ACA sRNA or the pseudouridine synthase mildly affect growth, the impact of the sRNA-mediated modification on ribosome assembly has not been unambiguously analyzed in these organisms.

(Hyper) Modification of 16S/18S rRNA h31

Helix h31 contributes to the formation of the P-site tRNA binding pocket of the SSU. Interestingly, despite sequence variability, h31 has been suggested to be universally modified at the same relative residue (*E. coli* G966/*S. cerevisiae* U1191). In bacteria, N(2)-Methylguanine (G966) is performed by rsmD (Lesnyak et al. 2007). In most archaea-eukarya the guanine is replaced by a uridine which undergo various degree of modifications requiring up to three individual steps.

The minimal h31 modification status is the addition of aminocarboxypropyl residue (acp3) (Grosjean et al. 2008; Kowalak et al. 2000) by the recently characterized Tsr3, a factor conserved in most archaea and eukarya (Meyer et al. 2016). In eukarya, the acp3 modification is normally preceded by first the sRNP-dependent isomerization of uridine (*S. cerevisiae* U1191) in pseudouridine (Ψ), which is followed by Nep1-dependent N1-methylation (m1) (Liang et al. 2009; Meyer et al. 2011; Wurm et al. 2010). Whereas, the acp3 addition is independent from the other two modifications, Nep1-dependent modification is dependent on the first isomerization of the uridine (Meyer et al. 2011, 2016; Thomas et al. 2011; Wurm et al. 2010). Interestingly, these modifications occur at different steps of the SSU assembly and could contribute to timely control the maturation of the future SSU P-site.

The full complement of rRNA pseudourydilation has not been fully characterized in archaea, therefore the extent of m1-Ψ modification found in h31 in archaea is still unclear. However, archaeal Nep1 has been shown to be able to methylate a Ψ-modified h31 in vitro (Wurm et al. 2010). Therefore, in absence of a clearly identified sRNA and/or further experimental validation, the presence of a Nep1 and Tsr3 homologs might be a good indication for the occurrence of m1-acp3-Ψ modification in archaea. Strikingly, Nep1 is absent in Methanogens class II and in Haloarchaea suggesting that these group of organisms might only contain an acp3 modified h31, as previously experimentally demonstrated for *H. volcanii* (Grosjean et al. 2008; Kowalak et al. 2000).

Finally, both eukaryotic Nep1 and Tsr3 have been shown to facilitate early and late steps of ribosome assembly, respectively (Meyer et al. 2011, 2016). Whereas, Nep1-dependent modification is not essential for proper SSU assembly, the lack of acp3 modification leads to late SSU rRNA processing defect in yeast (Meyer et al. 2011, 2016). A similar function of Nep1 and Tsr3 and their respective modifications has not been addressed in archaea.

Additional Key Modifications of the Archaeal rRNA?

As mentioned above the modifications status of archaeal rRNA show a large degree of variability. Therefore, in absence of experimental data in most archaea, it is difficult to fully appreciate the complete extent and function(s) of the rRNA modifications in this domain of life. Despite these current limitations, the role in ribosome function and/or assembly of additional (almost) archaeal specific modifications still need to be fully characterized.

Among them, adenosine methylation (m6A) of the SSU rRNA (*H. volcanii* A1432) is believed to be widespread in archaea, whereas only few eukaryotic SSU rRNA contain this modification (Grosjean et al. 2008). In addition, the chemically enigmatic C-N330 (according to its molecular mass) modification of the SSU rRNA (*H. volcanii* C1352) is also believed to be widespread in archaea (Grosjean et al. 2008) [also found in the hyperthermophile bacterium *Thermatoga maritima* (Guymon et al. 2007)]. However, neither the function, nor the enzymatic activities responsible for these modifications have been characterized so far.

6.4.2 Ribosomal RNA Processing and Ribosome Biogenesis Factors

As mentioned earlier, the rRNAs are usually transcribed as large precursors containing the mature rRNAs separated by internal and external spacers which are trimmed in a defined stepwise manner with the help of endo- and exo-nucleolytic activities (see also Chapter 4 related to archaeal ribonucleases).

In the last decades, the nearly complete set of rRNA processing and ribosome biogenesis factors have been described in both bacteria and eukarya, some of which are also found in archaeal genomes (Table 6.1) (Blombach et al. 2011; Deutscher 2009, 2015; Ebersberger et al. 2014; Henras et al. 2015; Shajani et al. 2011; Sulthana and Deutscher 2013; Thomson et al. 2013; Yip et al. 2013).

In the light of the observed differences of the nuclease counterpart involved in the rRNA processing in model bacteria (e.g., *E. coli* vs *B. subtilis*) (Britton et al. 2007; Condon 2007; Deutscher 2009), it is not surprising that the set of nucleases and ribosome biogenesis factors are apparently not conserved between bacteria and eukarya. Nevertheless, some emerging common principles shared between bacteria and eukarya, and probably archaea, can be eventually defined.

6.4.2.1 Early Circular-Like and Circular Pre-rRNA as a Common Early Step for the Efficient Maturation of SSU/LSU?

In brief, the first step of bacterial rRNA processing involves the endonucleolytic cleavage activity of RNase III within a double stranded RNA (dsRNA) stem structure which is formed by the base pairing of the 5′ leader and 3′ trailer sequences of the respective 16S and 23S rRNA precursors (Deutscher 2009; Gegenheimer et al. 1977; Gegenheimer and Apirion 1975; Young and Steitz 1978) (Figs. 6.2 and 6.3).

In a very similar way, in archaea, the 5′ leader and 3′ trailer sequences of the respective 16S and 23S pre-rRNA can form a double stranded RNA structure

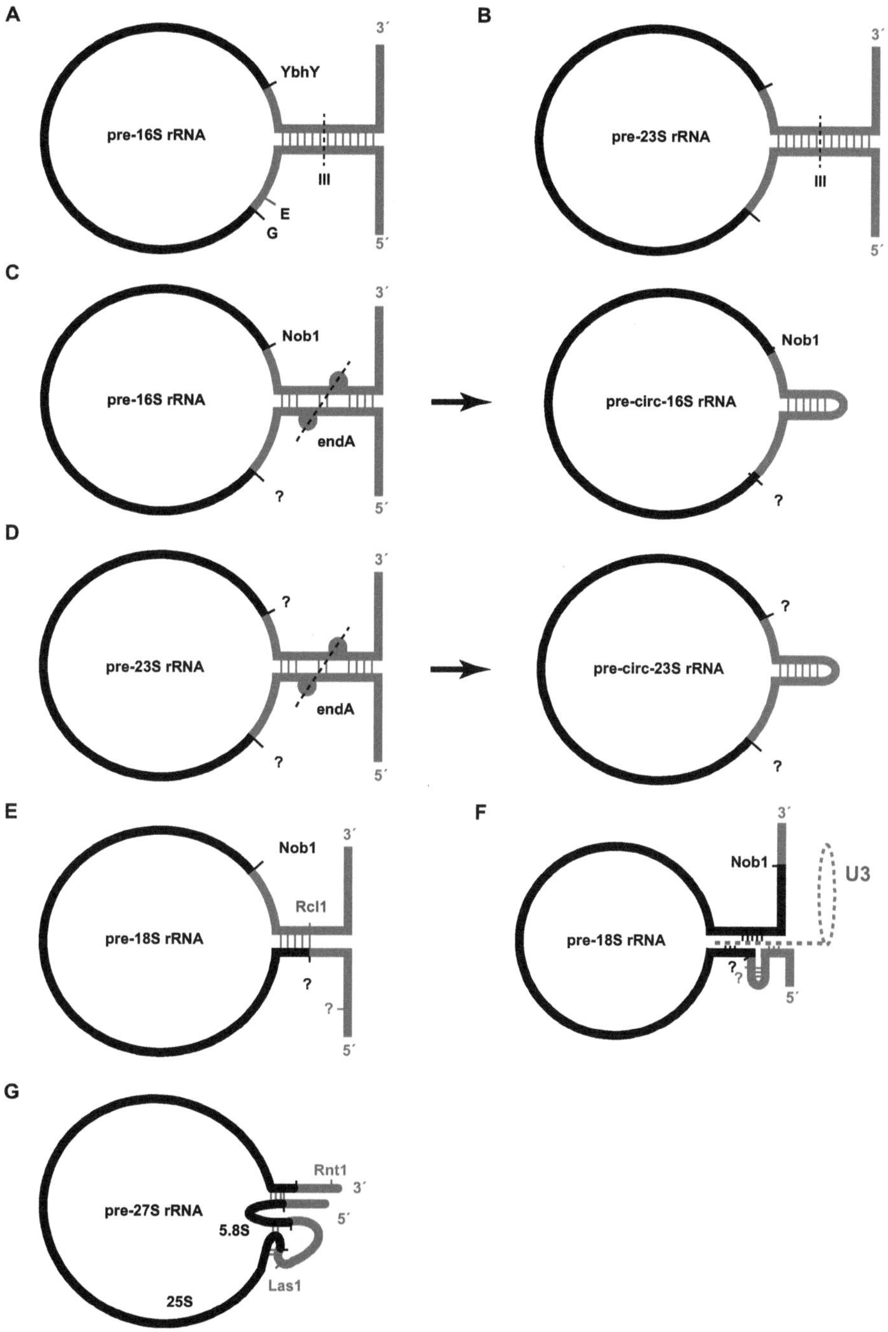

Fig. 6.3 Circular and circular-like pre-rRNA intermediates across the evolution. (**a–b**) Base pairing of pre-16S rRNA and pre-23S rRNA resulting in the formation of circular like pre-rRNA intermediates in bacteria is depicted in panel (**a**) and (**b**), respectively. (**c–d**) Base pairing of pre-16S rRNA and pre-23S rRNA resulting in the formation of circular pre-rRNA

(Figs. 6.2 and 6.3). In contrast, these dsRNA contain a bulge-helix-bulge motif, a known recognition site for the tRNA splicing endonuclease, also known as endA, which shares homology with the catalytic subunits of the eukaryotic tRNA splicing endonuclease (e.g., Lopes et al. 2015; Lykke-Andersen et al. 1997). Moreover, the 5′ and 3′ ends of the resulting 16S and 23S rRNA processing intermediates have been suggested to be respectively ligated into a circular pre-rRNAs (Danan et al. 2012; Tang et al. 2002 and our unpublished results) (Fig. 6.3).

Interestingly, in eukaryotes, such a pseudo-circularization has been previously suggested to facilitate the positioning of the different processing sites into close proximities (Fig. 6.3). The formation of these circular-like pre-rRNA were suggested to be formed by base pairing between the rRNA and its spacers (Veldman et al. 1981). Moreover, additional factors that could potentially stabilize a similar kind of structure have been also suggested. Among them, the sRNA U3 performs extensive base pairing with the 5′ leader sequence of the pre-18S rRNA and with sequence located further in the 3′ end of the 18S rRNA. In this way, the sRNA U3 and its numerous associated factors could contribute to the general stabilization of a circular-like processing intermediate which also prevents the premature formation of the SSU central pseudoknot (Dutca et al. 2011; Henras et al. 2015; Phipps et al. 2011; Woolford and Baserga 2013) (Fig. 6.3). Similarly, in bacteria, it has been suggested that the Nus transcription elongation factors that piggy backs with the RNA polymerase could hold the 5′ end of the nascent transcript, thereby facilitating the initial 5′ leader and 3′ trailer base pairing required for the RNAse III mediated processing, and stabilizing the formation of a pseudo-circular pre-rRNA (Bubunenko et al. 2013) (Fig. 6.3). In contrast to the U3 sRNA and most of the early eukaryotic ribosome biogenesis factors, some of the bacterial Nus transcription elongation factors are also conserved in archaea and could also potentially facilitate the initial base pairing of the nascent rRNA.

Together, these observations suggest that pseudo-circular pre-rRNA formation could be a common feature during the initial steps of ribosome biogenesis. Interestingly, the formation of circular-like or circular-pre-rRNA could facilitates the early steps of pre-rRNA assembly, by (1) providing molecular constraints stabilizing an rRNA scaffold facilitating early maturation events; (2) providing initial topological constraints bringing the 5′ and 3′ ends of the mature rRNA into a relative close proximity as observed in the final mature ribosomal subunits, (3) providing additional protection against unspecific nuclease activities.

Fig. 6.3 (continued) intermediates in archaea is depicted in panel (**c**) and (**d**), respectively. After cleavage at the Bulge Helix Bulge motif by the tRNA splicing endonuclease (endA) the resulting cleavage products are ligated into a circular pre-rRNA (*right panels*). (**e-g**) Putative circular-like pre-rRNA intermediates in eukaryotic cells are depicted. (**e**) Circular-like pre-18S rRNA formation established by base pairing between the 18S rRNA and the rRNA 3′ spacer sequence (ITS1) as described in yeast (Veldman et al. 1981). (**f**) Circular-like pre-18S rRNA formation established by base pairing between the 18S rRNA and the U3 sRNA is depicted (see text for details). (**g**) Circular-like pre-27S rRNA formation established by base pairing between the 25S, 5.8S rRNA and RNA spacer elements as described in yeast (Veldman et al. 1981). Endonucleolytic cleavages of the pre-rRNA and the known/putative enzymes are indicated

Whereas, intramolecular ligation of the pre-16S and pre-23S rRNA emerges as one of the unique features of archaeal ribosome biogenesis, the exact functional relevance for archaea of both the pre-rRNA spacers base pairing and the formation of circular pre-rRNA intermediates for the synthesis of functional ribosomal subunits, and how these pre-circular rRNA are further matured into linear rRNA awaits in-depth characterization.

6.4.2.2 Late 16S/18S rRNA Maturation: An Example of Evolutionary Converging Functional Mimicry?

After RNase III cleavages, the 5′- and 3′-extended pre-rRNA is further processed in its 5′ end by the sequential action of RNase E and G in *E. coli* (Deutscher 2009). However, in bacteria lacking RNAse G/E, like *Bacillus subtilis*, RNase J1 has been shown to produce the mature 5′ end of the 16S rRNA (Britton et al. 2007). Finally, the 3′ end processing in *E. coli* has been recently resolved. Whereas, the action of several exonuclease activities have been described to initiate the maturation of the16S rRNA 3′ end (Sulthana and Deutscher 2013), YbeY, an endonuclease, has been described to perform the final maturation of the 16S rRNA 3′ end (Vercruysse et al. 2016). Finally, additional ribosome biogenesis factors, notably the Era GTPase, RsgA, RbfA have been described to facilitate the late maturation steps of the SSU (Clatterbuck Soper et al. 2013; Datta et al. 2007; Guo et al. 2011; Jomaa et al. 2011; Sharma et al. 2005; Tu et al. 2009, 2011).

None of the nucleases and/or ribosome biogenesis factors described above are involved in the maturation of the 18S rRNA in eukaryotes. However, it is striking to notice several similarities during the late assembly steps of bacterial and eukaryotic SSU. First, the processing pathway requires a similar amount of processing steps. Second, two of the processing sites are located in the 5′ leader sequence. Third, the 3′ end processing of the 18S rRNA also depends on the Nob1 endonuclease. Finally, in human cells, exonucleolytic activities have been also suggested to initiate the 3′ end processing of the 18S rRNA (Ishikawa et al. 2016; Preti et al. 2013) (Figs. 6.2 and 6.3).

Moreover, from a structural point of view, the recent identification of the relative binding sites of key ribosome biogenesis factors involved in the late steps of the SSU biogenesis in bacteria and eukarya are strikingly comparable (Clatterbuck Soper et al. 2013; Datta et al. 2007; Granneman et al. 2010; Guo et al. 2011; Jomaa et al. 2011; Sharma et al. 2005; Strunk et al. 2011; Tu et al. 2009, 2011; Vercruysse et al. 2016; Xu et al. 2008) (Fig. 6.4). In summary, most of these factors cluster at the interface of the nascent SSU and binds to/or in close proximities of the future active centers of the SSU. Accordingly, these features have been interpreted to avoid premature engagement of the nascent ribosomal subunit into the translation pool (Granneman et al. 2010; Guo et al. 2011; Sharma et al. 2005; Strunk et al. 2011; Vercruysse et al. 2016; Xu et al. 2008). An alternative, non-mutually exclusive possibility, is that these factors could also shield the functional centers

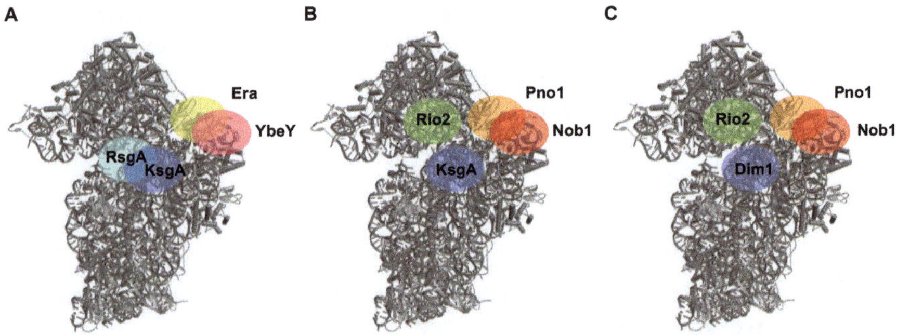

Fig. 6.4 Relative topological positioning of selected SSU ribosome biogenesis factors involved in late SSU biogenesis steps. The indicated ribosome biogenesis factors (putatively) involved in late SSU maturation in (**a**) *E. coli*, (**b**) in most archaea, and (**c**) *S.cerevisiae* were positioned on a model 30S ribosomal subunit from *Pyroccocus furiosus* (PDB: 4V6U) (Armache et al. 2013) (see text for detail references and discussion)

from unspecific ribonuclease cleavage. Finally, general principles of r-proteins assembly are very well conserved from bacteria to eukarya (see above).

Previous analysis of rRNA processing in Sulfolobales and Haloarchaea suggest some similarity with the processing of the pre-16S/18S rRNA in bacteria/eukarya (Chant and Dennis 1986; Ciammaruconi and Londei 2001; Durovic and Dennis 1994). The latest steps of maturation presumably requires the linearization of the circular-pre-rRNA into a pre-16S rRNA, either containing a 5′ and 3′ extension which needs to be removed, or the 5′ end will be directly mature out of the circular-pre-rRNA leaving out a 3′ extended pre-rRNA. The ribonucleases responsible for the maturation of the 5′ end of 16S rRNA in archaea is unknown. In contrast, the 3′ end maturation has been suggested to be achieved by the action of the PIN domain endonuclease Nob1, which is also responsible for the eukaryotic 3′ end maturation of the 18S rRNA (Fatica et al. 2003, 2004; Lamanna and Karbstein 2009; Lebaron et al. 2012; Pertschy et al. 2009; Veith et al. 2012). Interestingly, this suggests that both archaea and eukarya use a similar 3′ end maturation of the SSU rRNA (Fig. 6.4).

Additional factors have been involved in facilitating 3′ end processing of the 16S/18S rRNA. In *E.coli*, factor like the Era GTPase interacts with a conserved RNA sequence located in the 3′ end of the 16S rRNA via its type II KH-domain (Sharma et al. 2005; Tu et al. 2009, 2011). Moreover, Era, YbeY and the r-proteins uS11 have been shown to interact (Vercruysse et al. 2016). Phylogenetic analysis has been previously suggesting that the Era GTPase is not present in archaeal genome (Mittenhuber 2001). However, several ORF in different archaeal genome have been annotated as ERA-like GTPases. Interestingly all these putative homologs apparently lack the KH-domain essential for SSU synthesis (our own observation).

In eukarya Pno1/Dim2, a type I KH-domain containing protein, similarly interact with Nob1 (Woolls et al. 2011) and are both located in close proximity of uS11 and the 18S 3′ end (Fig. 6.4). Noteworthy, mutational analysis have demonstrated the requirement of uS11, Nob1 and Pno1/Dim2 for 3′ end 18S rRNA processing (Jakovljevic et al. 2004; Lamanna and Karbstein 2009; Lebaron et al. 2012; Pertschy et al. 2009; Woolls et al. 2011). Furthermore, structural analysis of an archaeal Pno1/Dim2 homologue have demonstrated that aPno1/aDim2 KH-domain binds a very similar 16S rRNA sequence as the Era KH-domain (Jia et al. 2010).

Together these findings remarkably suggest a very high degree of converging functional mimicry at the basis of the maturation of the 16S/18S rRNA 3′ end. In this scenario, evolutionary constraints have, surprisingly, selected very similar *modus operandi* whereby a KH-domain containing protein is used to presumably "hold" in place the 3′ end of the SSU rRNA prior to endonucleolytic cleavage.

Finally, in addition to these ribosome biogenesis factors, the presence of additional homologs of the eukaryotic ribosome biogenesis factors, like Rio1/Rio2 and Fap7, in most archaeal genome also suggest a possible function of these factors in archaeal ribosome biogenesis (Ebersberger et al. 2014; Hellmich et al. 2013; Loc'h et al. 2014).

6.4.2.3 Final 23S rRNA Maturation in Archaea: Alone in the Dark?

In *E coli*, RNase III-dependent 23S pre-rRNA intermediate formation is a prerequisite for final maturation of 23S rRNA and precedes final maturation of the 23S rRNA 5′ end which is performed by an unknown endonuclease and the 3′ end formation mainly involving the exonuclease RNase T (Deutscher 2009; King et al. 1984; Li et al. 1999; Sirdeshmukh and Schlessinger 1985; Srivastava and Schlessinger 1990). RNase T is also involved in the maturation of various cellular RNAs (i.e., 5S rRNA, tRNA) [see for review (Nicholson 1999)]. Alternatively, it has been shown, in *B. subtilis* that mini-RNase III can directly generate the matured 5′- 3′ ends of the 23S rRNA (Redko et al. 2008; Redko and Condon 2010).

In eukaryotes, the rRNA processing situation of the LSU rRNAs is more complex by virtue of the presence of an additional rRNA, the 5.8S rRNA, which has originally evolved from the splitting of domain I of the 23S rRNA (Henras et al. 2015; Venema and Tollervey 1995) (Fig. 6.2).

In archaea, the processing situation is also poorly characterized and no obvious similarity can be inferred from the existing literature. Presumably, the circular-pre-23S rRNA will be generated by the actions of several endo-/exo-nucleolytic steps. Moreover, in contrast to the SSU ribosomes biogenesis factors, only very few putative LSU biogenesis factors could be inferred from sequence homology analysis (Ebersberger et al. 2014; our own unpublished results).

6.5 Death of Ribosomes in Archaea?

In the last years, several studies have shown that, wrongly assembled or non-functional ribosomes are sorted out and targeted for degradation (Deutscher 2009; Lafontaine 2010; Luidalepp et al. 2016; Maiväli et al. 2013; Paier et al. 2015; Piir et al. 2011; Sulthana et al. 2016). Similarly, stress-induced degradation of ribosomal subunits has been also proposed. Whereas, the "danger signals" inducing ribosomal subunit decay appear in some cases to be conserved, the molecular basis of recognition and degradation owing to their respective molecular machineries differs between bacteria and eukarya. Non-functional ribosome decay (NRD) and/or stress-induced degradation is also likely to occur in archaea, however it is unclear which signalization or degradation machinery will carry out this function.

In bacteria, ribosome clearance (quality control/starvation-induced degradation) is initiated by cleavage of the rRNA backbone by RNAse E thereby generating an "entry" for RNase R/PH/II -mediated exonuclease digestions (Luidalepp et al. 2016; Sulthana et al. 2016). Interestingly, in the case of non-functional ribosome decay, point mutations affecting SSU function are not degraded, whereas some mutations affecting LSU function leads to degradation of both the mutated 50S and its associated none-mutated 30S particles (Paier et al. 2015).

In eukaryotes, two main degradation pathways for the degradation on non-functional/misassembled ribosomal subunits can be distinguished. A nuclear degradation pathway, which mainly target wrongly assembled pre-ribosomal particles depending on the TRAMP complex and nuclear exosome (Lafontaine 2010; Maiväli et al. 2013). A cytosolic degradation pathway well described for the clearance of non-functional ribosomal subunit where the cytoplasmic exosome play a crucial role (Cole et al. 2009; LaRiviere et al. 2006). Moreover, additional factors like Dom34 (Pelota), Hbs1, Ski7 and Xrn1 have been suggested to participate in the SSU-NRD, whereas the ubiquitin/proteasome system have been involved in the LSU-NRD pathway (Lafontaine 2010; Maiväli et al. 2013). Interestingly, in contrast to bacterial NRD, only the mutated ribosomal subunit is targeted to degradation (Cole et al. 2009; LaRiviere et al. 2006).

Finally, starvation induced ribosomal subunit largely relies on vacuolar-dependent degradation: ribophagy (Lafontaine 2010; Maiväli et al. 2013).

What are the degradation rules and pathway(s) in archaea? How do they relate to the bacterial and eukaryotic principles? Owing to the lack of experimental data, it is difficult to predict a general molecular principle of archaeal (pre-) ribosomal subunits degradation. However, the exosome or the RNase R and the β-CASP protein family (Evguenieva-Hackenberg and Klug 2009; Phung et al. 2013; Portnoy et al. 2005; Portnoy and Schuster 2006) are likely to play an important role in this degradation pathway.

Further studies will be needed to define the molecular mechanisms, including the RNA degradation machinery, and the specific regulatory mechanisms responsible for ribosome degradation in archaea.

6.6 Conclusion: Perspectives

Understanding the life and death of ribosomes in archaea is still in its early days. Whereas, comparative genomics and early studies have been providing some insights into the archaeal ribosomes life cycle, most of the detailed experimental work needs to be perform in order to provide an accurate and unbiased view of the ribosomal subunit metabolism(s) in archaea. Based on the current available data, some of which have been summarized in this chapter, archaeal ribosome metabolism likely take advantage of a mixture of bacterial- and eukaryote-like features to which archaeal specific features have been implemented. A comprehensive molecular analysis of the ribosomal subunit metabolism in the different domain of life will be crucial to further understand the evolution history of the fascinating translation machinery life cycle.

Acknowledgements I am grateful to Daniela Strauß, Robert Knüppel and Béatrice Clouet d´ Orval for discussions, suggestions and proof-reading. This work is supported by the chair of Biochemistry III "House of the Ribosome"—University of Regensburg and by the DFG-funded collaborative research centre CRC/SFB960-AP1 "Ribosome formation: principles of RNP biogenesis and control of their function".

References

Albers S-V, Forterre P, Prangishvili D, Schleper C (2013) The legacy of Carl Woese and Wolfram Zillig: from phylogeny to landmark discoveries. Nat Rev Microbiol 11:713–719

Allers T, Mevarech M (2005) Archaeal genetics—the third way. Nat Rev Genet 6:58–73. doi:10.1038/nrg1504

Arai T, Ishiguro K, Kimura S, Sakaguchi Y, Suzuki T, Suzuki T (2015) Single methylation of 23S rRNA triggers late steps of 50S ribosomal subunit assembly. Proc Natl Acad Sci U S A 112: E4707–E4716. doi:10.1073/pnas.1506749112

Armache J-P, Anger AM, Márquez V, Franckenberg S, Fröhlich T, Villa E, Berninghausen O, Thomm M, Arnold GJ, Beckmann R, Wilson DN (2013) Promiscuous behaviour of archaeal ribosomal proteins: implications for eukaryotic ribosome evolution. Nucleic Acids Res 41:1284–1293. doi:10.1093/nar/gks1259

Bachellerie J-P, Michot B, Nicoloso M, Balakin A, Ni J, Fournier MJ (1995) Antisense snoRNAs: a family of nucleolar RNAs with long complementarities to rRNA. Trends Biochem Sci 20:261–264. doi:10.1016/S0968-0004(00)89039-8

Badis G, Fromont-Racine M, Jacquier A (2003) A snoRNA that guides the two most conserved pseudouridine modifications within rRNA confers a growth advantage in yeast. RNA 9:771–779. doi:10.1261/rna.5240503

Ban N, Beckmann R, Cate JH, Dinman JD, Dragon F, Ellis SR, Lafontaine DL, Lindahl L, Liljas A, Lipton JM, McAlear MA, Moore PB, Noller HF, Ortega J, Panse VG, Ramakrishnan V, Spahn CM, Steitz TA, Tchorzewski M, Tollervey D, Warren AJ, Williamson JR, Wilson D, Yonath A, Yusupov M (2014) A new system for naming ribosomal proteins. Curr Opin Struct Biol 24:165–169. doi:10.1016/j.sbi.2014.01.002

Blaby IK, Majumder M, Chatterjee K, Jana S, Grosjean H, de Crécy-Lagard V, Gupta R (2011) Pseudouridine formation in archaeal RNAs: the case of Haloferax volcanii. RNA 17:1367–1380. doi:10.1261/rna.2712811

Blombach F, Brouns SJJ, van der Oost J (2011) Assembling the archaeal ribosome: roles for translation-factor-related GTPases. Biochem Soc Trans 39:45. doi:10.1042/BST0390045

Britton RA, Wen T, Schaefer L, Pellegrini O, Uicker WC, Mathy N, Tobin C, Daou R, Szyk J, Condon C (2007) Maturation of the 5′ end of Bacillus subtilis 16S rRNA by the essential ribonuclease YkqC/RNase J1. Mol Microbiol 63:127–138. doi:10.1111/j.1365-2958.2006.05499.x

Bubunenko M, Baker T, Court DL (2007) Essentiality of ribosomal and transcription antitermination proteins analyzed by systematic gene replacement in Escherichia coli. J Bacteriol 189:2844–2853. doi:10.1128/JB.01713-06

Bubunenko M, Court DL, Refaii AA, Saxena S, Korepanov A, Friedman DI, Gottesman ME, Alix J-H (2013) Nus transcription elongation factors and RNase III modulate small ribosome subunit biogenesis in E. coli. Mol Microbiol 87:382–393. doi:10.1111/mmi.12105

Bubunenko M, Korepanov A, Court DL, Jagannathan I, Dickinson D, Chaudhuri BR, Garber MB, Culver GM (2006) 30S ribosomal subunits can be assembled in vivo without primary binding ribosomal protein S15. RNA 12:1229–1239. doi:10.1261/rna.2262106

Bügl H, Fauman EB, Staker BL, Zheng F, Kushner SR, Saper MA, Bardwell JCA, Jakob U (2000) RNA methylation under heat shock control. Mol Cell 6:349–360. doi:10.1016/S1097-2765(00)00035-6

Bunner AE, Nord S, Wikström PM, Williamson JR (2010) The effect of ribosome assembly cofactors on in vitro 30S subunit reconstitution. J Mol Biol 398:1–7. doi:10.1016/j.jmb.2010.02.036

Burman LG, Mauro VP (2012) Analysis of rRNA processing and translation in mammalian cells using a synthetic 18S rRNA expression system. Nucleic Acids Res 40:8085–8098. doi:10.1093/nar/gks530

Caldas T, Binet E, Bouloc P, Costa A, Desgres J, Richarme G (2000) The FtsJ/RrmJ heat shock protein of Escherichia coli is a 23 S ribosomal RNA methyltransferase. J Biol Chem 275:16414–16419. doi:10.1074/jbc.M001854200

Chant J, Dennis P (1986) Archaebacteria: transcription and processing of ribosomal RNA sequences in Halobacterium cutirubrum. EMBO J 5:1091–1097

Chen SS, Williamson JR (2013) Characterization of the ribosome biogenesis landscape in E. coli using quantitative mass spectrometry. J Mol Biol 425:767–779. doi:10.1016/j.jmb.2012.11.040

Ciammaruconi A, Londei P (2001) In vitro processing of the 16S rRNA of the thermophilic archaeon Sulfolobus solfataricus. J Bacteriol 183:3866–3874. doi:10.1128/JB.183.13.3866-3874.2001

Clatterbuck Soper SF, Dator RP, Limbach PA, Woodson SA (2013) In vivo X-Ray footprinting of Pre-30S ribosomes reveals chaperone-dependent remodeling of late assembly intermediates. Mol Cell 52:506–516. doi:10.1016/j.molcel.2013.09.020

Cole SE, LaRiviere FJ, Merrikh C, Moore MJ (2009) A convergence of rRNA and mRNA quality control pathways revealed by mechanistic analysis of nonfunctional rRNA decay. Mol Cell 34:440–450. doi:10.1016/j.molcel.2009.04.017

Condon C (2007) Maturation and degradation of RNA in bacteria. Ecol Ind Microbiol RNA Tech. Margaret McFall-Ngai Victor Lorenzo Emmanuelle Charpentier Renée Schroeder 10:271–278. doi:10.1016/j.mib.2007.05.008

Connolly K, Rife JP, Culver G (2008) Mechanistic insight into the ribosome biogenesis functions of the ancient protein KsgA. Mol Microbiol 70:1062–1075. doi:10.1111/j.1365-2958.2008.06485.x

Cox CJ, Foster PG, Hirt RP, Harris SR, Embley TM (2008) The archaebacterial origin of eukaryotes. Proc Natl Acad Sci 105:20356–20361. doi:10.1073/pnas.0810647105

Danan M, Schwartz S, Edelheit S, Sorek R (2012) Transcriptome-wide discovery of circular RNAs in archaea. Nucleic Acids Res 40:3131–3142. doi:10.1093/nar/gkr1009

Datta PP, Wilson DN, Kawazoe M, Swami NK, Kaminishi T, Sharma MR, Booth TM, Takemoto C, Fucini P, Yokoyama S, Agrawal RK (2007) Structural aspects of RbfA action

during small ribosomal subunit assembly. Mol Cell 28:434–445. doi:10.1016/j.molcel.2007.08.026
Davis JH, Tan YZ, Carragher B, Potter CS, Lyumkis D, Williamson JR (2016) Modular assembly of the bacterial large ribosomal subunit. Cell 167:1610–1622.e15. doi:10.1016/j.cell.2016.11.020
de la Cruz J, Karbstein K, Woolford JL (2015) Functions of ribosomal proteins in assembly of eukaryotic ribosomes in vivo. Annu Rev Biochem 84:93–129. doi:10.1146/annurev-biochem-060614-033917
Decatur WA, Fournier MJ (2002) rRNA modifications and ribosome function. Trends Biochem Sci 27:344–351. doi:10.1016/S0968-0004(02)02109-6
Demirci H, Murphy F, Belardinelli R, Kelley AC, Ramakrishnan V, Gregory ST, Dahlberg AE, Jogl G (2010) Modification of 16S ribosomal RNA by the KsgA methyltransferase restructures the 30S subunit to optimize ribosome function. RNA 16:2319–2324. doi:10.1261/rna.2357210
Dennis PP, Tripp V, Lui L, Lowe T, Randau L (2015) C/D box sRNA-guided 2′-O-methylation patterns of archaeal rRNA molecules. BMC Genomics 16:632. doi:10.1186/s12864-015-1839-z
Deutscher MP (2009) Chapter 9: Maturation and degradation of ribosomal RNA in bacteria. In: Progress in molecular biology and translational science. Academic Press, Amsterdam, pp 369–391
Deutscher MP (2015) Twenty years of bacterial RNases and RNA processing: how we've matured. RNA 21:597–600. doi:10.1261/rna.049692.115
Durovic P, Dennis PP (1994) Separate pathways for excision and processing of 16S and 23S rRNA from the primary rRNA operon transcript from the hyperthermophilic archaebacterium Sulfolobus acidocaldarius: similarities to eukaryotic rRNA processing. Mol Microbiol 13:229–242. doi:10.1111/j.1365-2958.1994.tb00418.x
Dutca LM, Gallagher JEG, Baserga SJ (2011) The initial U3 snoRNA:pre-rRNA base pairing interaction required for pre-18S rRNA folding revealed by in vivo chemical probing. Nucleic Acids Res 39:5164–5180. doi:10.1093/nar/gkr044
Ebersberger I, Simm S, Leisegang MS, Schmitzberger P, Mirus O, von Haeseler A, Bohnsack MT, Schleiff E (2014) The evolution of the ribosome biogenesis pathway from a yeast perspective. Nucleic Acids Res 42:1509–1523. doi:10.1093/nar/gkt1137
Eme L, Doolittle WF (2015) Archaea. Curr Biol 25:R851–R855. doi:10.1016/j.cub.2015.05.025
Evguenieva-Hackenberg E, Klug G (2009) Chapter 7: RNA degradation in archaea and gram-negative bacteria different from Escherichia coli. In: Progress in molecular biology and translational science. Academic Press, Amsterdam, pp 275–317
Fatica A, Oeffinger M, Dlakić M, Tollervey D (2003) Nob1p is required for cleavage of the 3′ end of 18S rRNA. Mol Cell Biol 23:1798–1807. doi:10.1128/MCB.23.5.1798-1807.2003
Fatica A, Tollervey D, Dlakić M (2004) PIN domain of Nob1p is required for D-site cleavage in 20S pre-rRNA. RNA 10:1698–1701. doi:10.1261/rna.7123504
Ferreira-Cerca S, Pöll G, Kühn H, Neueder A, Jakob S, Tschochner H, Milkereit P (2007) Analysis of the in vivo assembly pathway of eukaryotic 40S ribosomal proteins. Mol Cell 28:446–457. doi:10.1016/j.molcel.2007.09.029
Fischer N, Neumann P, Konevega AL, Bock LV, Ficner R, Rodnina MV, Stark H (2015) Structure of the E. coli ribosome-EF-Tu complex at <3 A resolution by Cs-corrected cryo-EM. Nature 520:567–570
Fox GE, Magrum LJ, Balch WE, Wolfe RS, Woese CR (1977) Classification of methanogenic bacteria by 16S ribosomal RNA characterization. Proc Natl Acad Sci 74:4537–4541
Gegenheimer P, Apirion D (1975) Escherichia coli ribosomal ribonucleic acids are not cut from an intact precursor molecule. J Biol Chem 250:2407–2409
Gegenheimer P, Watson N, Apirion D (1977) Multiple pathways for primary processing of ribosomal RNA in Escherichia coli. J Biol Chem 252:3064–3073
Goffeau A, Barrell BG, Bussey H, Davis RW, Dujon B, Feldmann H, Galibert F, Hoheisel JD, Jacq C, Johnston M, Louis EJ, Mewes HW, Murakami Y, Philippsen P, Tettelin H, Oliver SG (1996) Life with 6000 genes. Science 274:546. doi:10.1126/science.274.5287.546

Graham DE, Overbeek R, Olsen GJ, Woese CR (2000) An archaeal genomic signature. Proc Natl Acad Sci 97:3304–3308. doi:10.1073/pnas.97.7.3304
Granneman S, Petfalski E, Swiatkowska A, Tollervey D (2010) Cracking pre-40S ribosomal subunit structure by systematic analyses of RNA–protein cross-linking. EMBO J 29:2026–2036. doi:10.1038/emboj.2010.86
Grosjean H, Breton M, Sirand-Pugnet P, Tardy F, Thiaucourt F, Citti C, Barré A, Yoshizawa S, Fourmy D, de Crécy-Lagard V, Blanchard A (2014) Predicting the minimal translation apparatus: lessons from the reductive evolution of mollicutes. PLoS Genet 10:e1004363. doi:10.1371/journal.pgen.1004363
Grosjean H, Gaspin C, Marck C, Decatur WA, de Crécy-Lagard V (2008) RNomics and modomics in the halophilic archaea Haloferax volcanii: identification of RNA modification genes. BMC Genomics 9:470–470. doi:10.1186/1471-2164-9-470
Guo Q, Yuan Y, Xu Y, Feng B, Liu L, Chen K, Sun M, Yang Z, Lei J, Gao N (2011) Structural basis for the function of a small GTPase RsgA on the 30S ribosomal subunit maturation revealed by cryoelectron microscopy. Proc Natl Acad Sci U S A 108:13100–13105. doi:10.1073/pnas.1104645108
Gupta N, Culver GM (2014) Multiple in vivo pathways for E. coli small ribosomal subunit assembly occur on one pre-rRNA. Nat Struct Mol Biol 21:937–943. doi:10.1038/nsmb.2887
Gutgsell NS, Deutscher MP, Ofengand J (2005) The pseudouridine synthase RluD is required for normal ribosome assembly and function in Escherichia coli. RNA 11:1141–1152. doi:10.1261/rna.2550105
Guymon R, Pomerantz SC, Ison JN, Crain PF, McCloskey JA (2007) Post-transcriptional modifications in the small subunit ribosomal RNA from Thermotoga maritima, including presence of a novel modified cytidine. RNA 13:396–403. doi:10.1261/rna.361607
Hadjiolov AA (1985) The nucleolus and ribosome biogenesis. Springer, New York
Hage AE, Tollervey D (2004) A surfeit of factors: why is ribosome assembly so much more complicated in eukaryotes than bacteria? RNA Biol 1:9–14. doi:10.4161/rna.1.1.932
Hager J, Staker BL, Jakob U (2004) Substrate binding analysis of the 23S rRNA methyltransferase RrmJ. J Bacteriol 186:6634–6642. doi:10.1128/JB.186.19.6634-6642.2004
Hansen MA, Kirpekar F, Ritterbusch W, Vester B (2002) Posttranscriptional modifications in the a-loop of 23S rRNAs from selected archaea and eubacteria. RNA 8:202–213
Hellmich UA, Weis BL, Lioutikov A, Wurm JP, Kaiser M, Christ NA, Hantke K, Kötter P, Entian K-D, Schleiff E, Wöhnert J (2013) Essential ribosome assembly factor Fap7 regulates a hierarchy of RNA–protein interactions during small ribosomal subunit biogenesis. Proc Natl Acad Sci U S A 110:15253–15258. doi:10.1073/pnas.1306389110
Henras AK, Plisson-Chastang C, O'Donohue M-F, Chakraborty A, Gleizes P-E (2015) An overview of pre-ribosomal RNA processing in eukaryotes. Wiley Interdiscip Rev RNA 6:225–242. doi:10.1002/wrna.1269
Hernandez-Verdun D, Roussel P, Thiry M, Sirri V, Lafontaine DLJ (2010) The nucleolus: structure/function relationship in RNA metabolism. Wiley Interdiscip Rev RNA 1:415–431. doi:10.1002/wrna.39
Heus HA, Van Kimmenade JMA, van Knippenberg PH, Hinz H-J (1983) Calorimetric measurements of the destabilisation of a ribosomal RNA hairpin by dimethyiation of two adjacent adenosines. Nucleic Acids Res 11:203–210. doi:10.1093/nar/11.1.203
Ishikawa H, Yoshikawa H, Izumikawa K, Miura Y, Taoka M, Nobe Y, Yamauchi Y, Nakayama H, Simpson RJ, Isobe T, Takahashi N (2016) Poly(A)-specific ribonuclease regulates the processing of small-subunit rRNAs in human cells. Nucleic Acids Res. doi:10.1093/nar/gkw1047
Iskakova M, Connell SR, Nierhaus KH (2004) Regulation of ribosome biosynthesis in Escherichia coli. In: Protein synthesis and ribosome structure. Wiley, New york, pp 429–448
Jain R, Rivera MC, Lake JA (1999) Horizontal gene transfer among genomes: the complexity hypothesis. Proc Natl Acad Sci 96:3801–3806. doi:10.1073/pnas.96.7.3801

Jakovljevic J, de Mayolo PA, Miles TD, Nguyen TM-L, Léger-Silvestre I, Gas N, Woolford JL Jr (2004) The carboxy-terminal extension of yeast ribosomal protein S14 is necessary for maturation of 43S preribosomes. Mol Cell 14:331–342. doi:10.1016/S1097-2765(04)00215-1

Jia MZ, Horita S, Nagata K, Tanokura M (2010) An archaeal Dim2-Like protein, aDim2p, forms a ternary complex with a/eIF2α and the 3′ end fragment of 16S rRNA. J Mol Biol 398:774–785. doi:10.1016/j.jmb.2010.03.055

Jin DJ, Cagliero C, Martin CM, Izard J, Zhou YN (2015) The dynamic nature and territory of transcriptional machinery in the bacterial chromosome. Front Microbiol 6:497. doi:10.3389/fmicb.2015.00497

Jin DJ, Mata Martin C, Sun Z, Cagliero C, Zhou YN (2016) Nucleolus-like compartmentalization of the transcription machinery in fast-growing bacterial cells. Crit Rev Biochem Mol Biol 1–11. doi:10.1080/10409238.2016.1269717

Jomaa A, Stewart G, Mears JA, Kireeva I, Brown ED, Ortega J (2011) Cryo-electron microscopy structure of the 30S subunit in complex with the YjeQ biogenesis factor. RNA 17:2026–2038. doi:10.1261/rna.2922311

King TC, Sirdeshmukh R, Schlessinger D (1984) RNase III cleavage is obligate for maturation but not for function of Escherichia coli pre-23S rRNA. Proc Natl Acad Sci U S A 81:185–188

Kirpekar F, Hansen LH, Rasmussen A, Poehlsgaard J, Vester B (2005) The archaeon Haloarcula marismortui has few modifications in the central parts of its 23 S ribosomal RNA. J Mol Biol 348:563–573. doi:10.1016/j.jmb.2005.03.009

Klappenbach JA, Saxman PR, Cole JR, Schmidt TM (2001) rrndb: the ribosomal RNA operon copy number database. Nucleic Acids Res 29:181–184

Klootwijk J, Klein I, Grivell LA (1975) Minimal post-transcriptional modification of yeast mitochondrial ribosomal RNA. J Mol Biol 97:337–350. doi:10.1016/S0022-2836(75)80044-1

Kowalak JA, Bruenger E, Crain PF, McCloskey JA (2000) Identities and phylogenetic comparisons of posttranscriptional modifications in 16 S ribosomal RNA from Haloferax volcanii. J Biol Chem 275:24484–24489. doi:10.1074/jbc.M002153200

Lafontaine DLJ (2010) A "garbage can" for ribosomes: how eukaryotes degrade their ribosomes. Trends Biochem Sci 35:267–277. doi:10.1016/j.tibs.2009.12.006

Lafontaine D, Delcour J, Glasser A-L, Desgres J, Vandenhaute J (1994) The DIM1 gene responsible for the conserved m62Am62A dimethylation in the 3′-terminal loop of 18 S rRNA is essential in yeast. J Mol Biol 241:492–497. doi:10.1006/jmbi.1994.1525

Lafontaine DL, Tollervey D (1998) Birth of the snoRNPs: the evolution of the modification-guide snoRNAs. Trends Biochem Sci 23:383–388. doi:10.1016/S0968-0004(98)01260-2

Lafontaine D, Vandenhaute J, Tollervey D (1995) The 18S rRNA dimethylase Dim1p is required for pre-ribosomal RNA processing in yeast. Genes Dev 9:2470–2481. doi:10.1101/gad.9.20.2470

Lake JA (1985) Evolving ribosome structure: domains in archaebacteria, eubacteria, eocytes and eukaryotes. Annu Rev Biochem 54:507–530. doi:10.1146/annurev.bi.54.070185.002451

Lake JA (2015) Eukaryotic origins. Philos Trans R Soc B Biol Sci 370. doi:10.1098/rstb.2014.0321

Lake JA, Henderson E, Oakes M, Clark MW (1984) Eocytes: a new ribosome structure indicates a kingdom with a close relationship to eukaryotes. Proc Natl Acad Sci 81:3786–3790

Lamanna AC, Karbstein K (2009) Nob1 binds the single-stranded cleavage site D at the 3′-end of 18S rRNA with its PIN domain. Proc Natl Acad Sci U S A 106:14259–14264. doi:10.1073/pnas.0905403106

Lapeyre B, Purushothaman SK (2004) Spb1p-directed formation of Gm2922 in the ribosome catalytic center occurs at a late processing stage. Mol Cell 16:663–669. doi:10.1016/j.molcel.2004.10.022

LaRiviere FJ, Cole SE, Ferullo DJ, Moore MJ (2006) A late-acting quality control process for mature eukaryotic rRNAs. Mol Cell 24:619–626. doi:10.1016/j.molcel.2006.10.008

Lebaron S, Schneider C, van Nues RW, Swiatkowska A, Walsh D, Böttcher B, Granneman S, Watkins NJ, Tollervey D (2012) Proof reading of pre-40S ribosome maturation by a translation initiation factor and 60S subunits. Nat Struct Mol Biol 19:744–753. doi:10.1038/nsmb.2308

Lecompte O, Ripp R, Thierry J, Moras D, Poch O (2002) Comparative analysis of ribosomal proteins in complete genomes: an example of reductive evolution at the domain scale. Nucleic Acids Res 30:5382–5390. doi:10.1093/nar/gkf693

Lempiäinen H, Shore D (2009) Growth control and ribosome biogenesis. Curr Opin Cell Biol 21:855–863. doi:10.1016/j.ceb.2009.09.002

Leppik M, Peil L, Kipper K, Liiv A, Remme J (2007) Substrate specificity of the pseudouridine synthase RluD in Escherichia coli. FEBS J 274:5759–5766. doi:10.1111/j.1742-4658.2007.06101.x

Lesnyak DV, Osipiuk J, Skarina T, Sergiev PV, Bogdanov AA, Edwards A, Savchenko A, Joachimiak A, Dontsova OA (2007) Methyltransferase that modifies guanine 966 of the 16 S rRNA: functional identification and tertiary structure. J Biol Chem 282:5880–5887. doi:10.1074/jbc.M608214200

Li Z, Pandit S, Deutscher MP (1999) Maturation of 23S ribosomal RNA requires the exoribonuclease RNase T. RNA 5:139–146

Liang W-Q, Fournier MJ (1997) Synthesis of functional eukaryotic ribosomal RNAs in trans: development of a novel in vivo rDNA system for dissecting ribosome biogenesis. Proc Natl Acad Sci U S A 94:2864–2868

Liang X, Liu Q, Fournier MJ (2009) Loss of rRNA modifications in the decoding center of the ribosome impairs translation and strongly delays pre-rRNA processing. RNA 15:1716–1728. doi:10.1261/rna.1724409

Littlefield JW, Dunn DB (1958a) Natural occurrence of thymine and three methylated adenine bases in several ribonucleic acids. Nature 181:254–255. doi:10.1038/181254a0

Littlefield JW, Dunn DB (1958b) The occurrence and distribution of thymine and three methylated-adenine bases in ribonucleic acids from several sources. Biochem J 70:642–651

Loc'h J, Blaud M, Réty S, Lebaron S, Deschamps P, Bareille J, Jombart J, Robert-Paganin J, Delbos L, Chardon F, Zhang E, Charenton C, Tollervey D, Leulliot N (2014) RNA mimicry by the Fap7 adenylate kinase in ribosome biogenesis. PLoS Biol 12:e1001860. doi:10.1371/journal.pbio.1001860

Londei P, Teixidò J, Acca M, Cammarano P, Amils R (1986) Total reconstitution of active large ribosomal subunits of the thermoacidophilic archaebacterium Sulfolobus solfataricus. Nucleic Acids Res 14:2269–2285

Lopes RR, Kessler AC, Polycarpo C, Alfonzo JD (2015) Cutting, dicing, healing and sealing: the molecular surgery of tRNA. Wiley Interdiscip Rev RNA 6:337–349. doi:10.1002/wrna.1279

Luidalepp H, Berger S, Joss O, Tenson T, Polacek N (2016) Ribosome shut-down by 16S rRNA fragmentation in stationary-phase Escherichia coli. J Mol Biol 428:2237–2247. doi:10.1016/j.jmb.2016.01.033

Lykke-Andersen J, Aagaard C, Semionenkov M, Garrett RA (1997) Archaeal introns: splicing, intercellular mobility and evolution. Trends Biochem Sci 22:326–331. doi:10.1016/S0968-0004(97)01113-4

Maiväli Ü, Paier A, Tenson T (2013) When stable RNA becomes unstable: the degradation of ribosomes in bacteria and beyond. Biol Chem 394:845. doi:10.1515/hsz-2013-0133

Maki JA, Schnobrich DJ, Culver GM (2002) The DnaK chaperone system facilitates 30S ribosomal subunit assembly. Mol Cell 10:129–138. doi:10.1016/S1097-2765(02)00562-2

Mangat CS, Brown ED (2008) Ribosome biogenesis; the KsgA protein throws a methyl-mediated switch in ribosome assembly. Mol Microbiol 70:1051–1053. doi:10.1111/j.1365-2958.2008.06484.x

Mangiarotti G, Chiaberge S (1997) Reconstitution of functional eukaryotic ribosomes from Dictyostelium discoideum ribosomal proteins and RNA. J Biol Chem 272:19682–19687. doi:10.1074/jbc.272.32.19682

Meyer B, Wurm JP, Kötter P, Leisegang MS, Schilling V, Buchhaupt M, Held M, Bahr U, Karas M, Heckel A, Bohnsack MT, Wöhnert J, Entian K-D (2011) The bowen–conradi syndrome protein Nep1 (Emg1) has a dual role in eukaryotic ribosome biogenesis, as an essential assembly factor and in the methylation of Ψ1191 in yeast 18S rRNA. Nucleic Acids Res 39:1526–1537. doi:10.1093/nar/gkq931

Meyer B, Wurm JP, Sharma S, Immer C, Pogoryelov D, Kötter P, Lafontaine DLJ, Wöhnert J, Entian K-D (2016) Ribosome biogenesis factor Tsr3 is the aminocarboxypropyl transferase responsible for 18S rRNA hypermodification in yeast and humans. Nucleic Acids Res 44:4304–4316. doi:10.1093/nar/gkw244

Mittenhuber G (2001) Comparative genomics of prokaryotic GTP-binding proteins (the Era, Obg, EngA, ThdF (TrmE), YchF and YihA families) and their relationship to eukaryotic GTP-binding proteins (the DRG, ARF, RAB, RAN, RAS and RHO families). J Mol Microbiol Biotechnol 3:21–35

Mizushima S, Nomura M (1970) Assembly mapping of 30S ribosomal proteins from E. coli. Nature 226:1214–1218. doi:10.1038/2261214a0

Mulder AM, Yoshioka C, Beck AH, Bunner AE, Milligan RA, Potter CS, Carragher B, Williamson JR (2010) Visualizing ribosome biogenesis: parallel assembly pathways for the 30 S subunit. Science 330:673. doi:10.1126/science.1193220

Nakao A, Yoshihama M, Kenmochi N (2004) RPG: the ribosomal protein gene database. Nucleic Acids Res 32:D168–D170. doi:10.1093/nar/gkh004

Nerurkar P, Altvater M, Gerhardy S, Schütz S, Fischer U, Weirich C, Panse VG (2015) Chapter 3 - Eukaryotic ribosome assembly and nuclear export. In: Jeon KW (ed) International review of cell and molecular biology. Academic Press, London, pp 107–140

Nicholson AW (1999) Function, mechanism and regulation of bacterial ribonucleases. FEMS Microbiol Rev 23:371–390. doi:10.1111/j.1574-6976.1999.tb00405.x

Nierhaus KH (1991) The assembly of prokaryotic ribosomes. Biochimie 73:739–755. doi:10.1016/0300-9084(91)90054-5

Nierhaus KH, Lafontaine DL (2004) Ribosome assembly. In: Protein synthesis and ribosome structure. Wiley, New York, pp 85–143

Nomura M (1999) Regulation of ribosome biosynthesis in Escherichia coli and Saccharomyces cerevisiae: diversity and common principles. J Bacteriol 181:6857–6864

Nomura M (2001) Ribosomal RNA genes, RNA polymerases, nucleolar structures, and synthesis of rRNA in the yeast Saccharomyces cerevisiae. Cold Spring Harb Symp Quant Biol 66:555–566. doi:10.1101/sqb.2001.66.555

Nomura M, Erdmann VA (1970) Reconstitution of 50S ribosomal subunits from dissociated molecular components. Nature 228:744–748. doi:10.1038/228744a0

Ohmayer U, Gamalinda M, Sauert M, Ossowski J, Pöll G, Linnemann J, Hierlmeier T, Perez-Fernandez J, Kumcuoglu B, Leger-Silvestre I, Faubladier M, Griesenbeck J, Woolford J, Tschochner H, Milkereit P (2013) Studies on the assembly characteristics of large subunit ribosomal proteins in S. cerevisae. PLoS One 8:e68412. doi:10.1371/journal.pone.0068412

Omer AD, Ziesche S, Decatur WA, Fournier MJ, Dennis PP (2003) RNA-modifying machines in archaea. Mol Microbiol 48:617–629. doi:10.1046/j.1365-2958.2003.03483.x

Pace NR, Sapp J, Goldenfeld N (2012) Phylogeny and beyond: scientific, historical, and conceptual significance of the first tree of life. Proc Natl Acad Sci 109:1011–1018. doi:10.1073/pnas.1109716109

Paier A, Leppik M, Soosaar A, Tenson T, Maiväli Ü (2015) The effects of disruptions in ribosomal active sites and in intersubunit contacts on ribosomal degradation in Escherichia coli. Sci Rep 5:7712. doi:10.1038/srep07712

Parker MS, Sallee FR, Park EA, Parker SL (2015) Homoiterons and expansion in ribosomal RNAs. FEBS Open Bio 5:864–876. doi:10.1016/j.fob.2015.10.005

Pederson T (2011) The nucleolus. Cold Spring Harb Perspect Biol 3:a000638. doi:10.1101/cshperspect.a000638

Pertschy B, Schneider C, Gnädig M, Schäfer T, Tollervey D, Hurt E (2009) RNA helicase Prp43 and its co-factor Pfa1 promote 20 to 18 S rRNA processing catalyzed by the endonuclease Nob1. J Biol Chem 284:35079–35091. doi:10.1074/jbc.M109.040774

Petrov AS, Gulen B, Norris AM, Kovacs NA, Bernier CR, Lanier KA, Fox GE, Harvey SC, Wartell RM, Hud NV, Williams LD (2015) History of the ribosome and the origin of translation. Proc Natl Acad Sci 112:15396–15401. doi:10.1073/pnas.1509761112

Phipps KR, Charette JM, Baserga SJ (2011) The SSU processome in ribosome biogenesis – progress and prospects. Wiley Interdiscip Rev RNA 2:1–21. doi:10.1002/wrna.57

Phung DK, Rinaldi D, Langendijk-Genevaux PS, Quentin Y, Carpousis AJ, Clouet-d'Orval B (2013) Archaeal β-CASP ribonucleases of the aCPSF1 family are orthologs of the eukaryal CPSF-73 factor. Nucleic Acids Res 41:1091–1103. doi:10.1093/nar/gks1237

Piekna-Przybylska D, Decatur WA, Fournier MJ (2008) The 3D rRNA modification maps database: with interactive tools for ribosome analysis. Nucleic Acids Res 36:D178–D183. doi:10.1093/nar/gkm855

Piir K, Paier A, Liiv A, Tenson T, Maiväli Ü (2011) Ribosome degradation in growing bacteria. EMBO Rep 12:458–462. doi:10.1038/embor.2011.47

Planta RJ (1997) Regulation of ribosome synthesis in yeast. Yeast 13:1505–1518. doi:10.1002/(SICI)1097-0061(199712)13:16<1505::AID-YEA229>3.0.CO;2-I

Polikanov YS, Melnikov SV, Söll D, Steitz TA (2015) Structural insights into the role of rRNA modifications in protein synthesis and ribosome assembly. Nat Struct Mol Biol 22:342–344. doi:10.1038/nsmb.2992

Portnoy V, Evguenieva-Hackenberg E, Klein F, Walter P, Lorentzen E, Klug G, Schuster G (2005) RNA polyadenylation in archaea: not observed in Haloferax while the exosome polynucleotidylates RNA in sulfolobus. EMBO Rep 6:1188–1193. doi:10.1038/sj.embor.7400571

Portnoy V, Schuster G (2006) RNA polyadenylation and degradation in different Archaea; roles of the exosome and RNase R. Nucleic Acids Res 34:5923–5931. doi:10.1093/nar/gkl763

Preti M, O'Donohue M-F, Montel-Lehry N, Bortolin-Cavaillé M-L, Choesmel V, Gleizes P-E (2013) Gradual processing of the ITS1 from the nucleolus to the cytoplasm during synthesis of the human 18S rRNA. Nucleic Acids Res 41:4709–4723. doi:10.1093/nar/gkt160

Raymann K, Brochier-Armanet C, Gribaldo S (2015) The two-domain tree of life is linked to a new root for the archaea. Proc Natl Acad Sci 112:6670–6675. doi:10.1073/pnas.1420858112

Redko Y, Bechhofer DH, Condon C (2008) Mini-III, an unusual member of the RNase III family of enzymes, catalyses 23S ribosomal RNA maturation in B. subtilis. Mol Microbiol 68:1096–1106. doi:10.1111/j.1365-2958.2008.06207.x

Redko Y, Condon C (2010) Maturation of 23S rRNA in Bacillus subtilis in the absence of Mini-III. J Bacteriol 192:356–359. doi:10.1128/JB.01096-09

Rivera MC, Jain R, Moore JE, Lake JA (1998) Genomic evidence for two functionally distinct gene classes. Proc Natl Acad Sci 95:6239–6244

Sanchez EM, Londei P, Amils R (1996) Total reconstitution of active small ribosomal subunits of the extreme halophilic archaeon Haloferax mediterranei. Biochim Biophys Acta 1292:140–144. doi:10.1016/0167-4838(95)00179-4

Sanchez ME, Urena D, Amils R, Londei P (1990) In vitro reassembly of active large ribosomal subunits of the halophilic archaebacterium Haloferax mediterranei. Biochemistry (Mosc.) 29:9256–9261. doi:10.1021/bi00491a021

Seistrup KH, Rose S, Birkedal U, Nielsen H, Huber H, Douthwaite S (2016) Bypassing rRNA methylation by RsmA/Dim1during ribosome maturation in the hyperthermophilic archaeon Nanoarchaeum equitans. Nucleic Acids Res. doi:10.1093/nar/gkw839

Sentenac A, Riva M (2013) Odd RNA polymerases or the A(B)C of eukaryotic transcription. Transcr Odd Pols 1829:251–257. doi:10.1016/j.bbagrm.2012.10.012

Shajani Z, Sykes MT, Williamson JR (2011) Assembly of bacterial ribosomes. Annu Rev Biochem 80:501–526. doi:10.1146/annurev-biochem-062608-160432

Sharma MR, Barat C, Wilson DN, Booth TM, Kawazoe M, Hori-Takemoto C, Shirouzu M, Yokoyama S, Fucini P, Agrawal RK (2005) Interaction of era with the 30S ribosomal subunit: implications for 30S subunit assembly. Mol Cell 18:319–329. doi:10.1016/j.molcel.2005.03.028

Sharma S, Lafontaine DLJ (2015) "View from a bridge": a new perspective on eukaryotic rRNA base modification. Trends Biochem Sci 40:560–575. doi:10.1016/j.tibs.2015.07.008

Siibak T, Remme J (2010) Subribosomal particle analysis reveals the stages of bacterial ribosome assembly at which rRNA nucleotides are modified. RNA 16:2023–2032. doi:10.1261/rna.2160010

Sirdeshmukh R, Schlessinger D (1985) Ordered processing of Escherichia coli 23S rRNA in vitro. Nucleic Acids Res 13:5041–5054

Sloan KE, Warda AS, Sharma S, Entian K-D, Lafontaine DLJ, Bohnsack MT (2016) Tuning the ribosome: the influence of rRNA modification on eukaryotic ribosome biogenesis and function. RNA Biol 2:1–16. doi:10.1080/15476286.2016.1259781

Smith JD, Dunn DB (1959) The occurrence of methylated guanines in ribonucleic acids from several sources. Biochem J 72:294–301

Spang A, Saw JH, Jorgensen SL, Zaremba-Niedzwiedzka K, Martijn J, Lind AE, van Eijk R, Schleper C, Guy L, Ettema TJG (2015) Complex archaea that bridge the gap between prokaryotes and eukaryotes. Nature 521:173–179

Srivastava AK, Schlessinger D (1990) Mechanism and regulation of bacterial ribosomal RNA processing. Annu Rev Microbiol 44:105–129. doi:10.1146/annurev.mi.44.100190.000541

Staub E, Fiziev P, Rosenthal A, Hinzmann B (2004) Insights into the evolution of the nucleolus by an analysis of its protein domain repertoire. BioEssays 26:567–581. doi:10.1002/bies.20032

Steitz JA, Tycowski KT (1995) Enhanced perspective: small RNA chaperones for ribosome biogenesis. Science 270:1626. doi:10.1126/science.270.5242.1626

Stoddard SF, Smith BJ, Hein R, Roller BRK, Schmidt TM (2015) rrnDB: improved tools for interpreting rRNA gene abundance in bacteria and archaea and a new foundation for future development. Nucleic Acids Res 43:D593–D598. doi:10.1093/nar/gku1201

Strunk BS, Loucks CR, Su M, Vashisth H, Cheng S, Schilling J, Brooks CL, Karbstein K, Skiniotis G (2011) Ribosome assembly factors prevent premature translation initiation by 40 S assembly intermediates. Science 333:1449. doi:10.1126/science.1208245

Sulthana S, Basturea GN, Deutscher MP (2016) Elucidation of pathways of ribosomal RNA degradation: an essential role for RNase E. RNA 22:1163–1171. doi:10.1261/rna.056275.116

Sulthana S, Deutscher MP (2013) Multiple exoribonucleases catalyze maturation of the 3′ terminus of 16S ribosomal RNA (rRNA). J Biol Chem 288:12574–12579. doi:10.1074/jbc.C113.459172

Talkington MWT, Siuzdak G, Williamson JR (2005) An assembly landscape for the 30S ribosomal subunit. Nature 438:628–632. doi:10.1038/nature04261

Tang TH, Rozhdestvensky TS, d'Orval BC, Bortolin M-L, Huber H, Charpentier B, Branlant C, Bachellerie J-P, Brosius J, Hüttenhofer A (2002) RNomics in archaea reveals a further link between splicing of archaeal introns and rRNA processing. Nucleic Acids Res 30:921–930

Thomas SR, Keller CA, Szyk A, Cannon JR, LaRonde-LeBlanc NA (2011) Structural insight into the functional mechanism of Nep1/Emg1 N1-specific pseudouridine methyltransferase in ribosome biogenesis. Nucleic Acids Res 39:2445–2457. doi:10.1093/nar/gkq1131

Thomson E, Ferreira-Cerca S, Hurt E (2013) Eukaryotic ribosome biogenesis at a glance. J Cell Sci 126:4815. doi:10.1242/jcs.111948

Tu C, Zhou X, Tarasov SG, Tropea JE, Austin BP, Waugh DS, Court DL, Ji X (2011) The era GTPase recognizes the GAUCACCUCC sequence and binds helix 45 near the 3′ end of 16S rRNA. Proc Natl Acad Sci U S A 108:10156–10161. doi:10.1073/pnas.1017679108

Tu C, Zhou X, Tropea JE, Austin BP, Waugh DS, Court DL, Ji X (2009) Structure of ERA in complex with the 3′ end of 16S rRNA: implications for ribosome biogenesis. Proc Natl Acad Sci U S A 106:14843–14848. doi:10.1073/pnas.0904032106

Van Buul CPJJ, Hamersma M, Visser W, Van Knippenberg PH (1984) Partial methylation of two adjacent adenosines in ribosomes from Euglena gracilis chloroplasts suggests evolutionary loss of an intermediate stage in the methyl-transfer reaction. Nucleic Acids Res 12:9205–9208. doi:10.1093/nar/12.23.9205

Van Charldorp R, Heus HA, Knippenberg PHV, Joordens J, De Bruin SH, Hilbers CW (1981) Destabilization of secondary structure in 16S ribosomal RNA by dimethylation of two adjacent adenosines. Nucleic Acids Res 9:4413–4422. doi:10.1093/nar/9.17.4413

Veith T, Martin R, Wurm JP, Weis BL, Duchardt-Ferner E, Safferthal C, Hennig R, Mirus O, Bohnsack MT, Wöhnert J, Schleiff E (2012) Structural and functional analysis of the archaeal endonuclease Nob1. Nucleic Acids Res 40:3259–3274. doi:10.1093/nar/gkr1186

Veldman GM, Klootwijk J, van Heerikhuizen H, Planta RJ (1981) The nucleotide sequence of the intergenic region between the 5.8S and 26S rRNA genes of the yeast ribosomal RNA operon. Possible implications for the interaction between 5.8S and 26S rRNA and the processing of the primary transcript. Nucleic Acids Res 9:4847–4862. doi:10.1093/nar/9.19.4847

Venema J, Tollervey D (1995) Processing of pre-ribosomal RNA in Saccharomyces cerevisiae. Yeast 11:1629–1650. doi:10.1002/yea.320111607

Vercruysse M, Köhrer C, Shen Y, Proulx S, Ghosal A, Davies BW, RajBhandary UL, Walker GC (2016) Identification of YbeY-protein interactions involved in 16S rRNA maturation and stress regulation in Escherichia coli. mBio 7:e01785-16. doi:10.1128/mBio.01785-16

Warner JR (1999) The economics of ribosome biosynthesis in yeast. Trends Biochem Sci 24:437–440. doi:10.1016/S0968-0004(99)01460-7

Williams TA, Foster PG, Nye TMW, Cox CJ, Embley TM (2012) A congruent phylogenomic signal places eukaryotes within the archaea. Proc R Soc B Biol Sci 279:4870. doi:10.1098/rspb.2012.1795

Williamson JR (2003) After the ribosome structures: how are the subunits assembled? RNA 9:165–167. doi:10.1261/rna.2164903

Woese CR, Fox GE (1977) Phylogenetic structure of the prokaryotic domain: the primary kingdoms. Proc Natl Acad Sci 74:5088–5090. doi:10.1073/pnas.74.11.5088

Woese CR, Kandler O, Wheelis ML (1990) Towards a natural system of organisms: proposal for the domains archaea, bacteria, and eucarya. Proc Natl Acad Sci 87:4576–4579. doi:10.1073/pnas.87.12.4576

Woolford JL, Baserga SJ (2013) Ribosome biogenesis in the yeast Saccharomyces cerevisiae. Genetics 195:643–681. doi:10.1534/genetics.113.153197

Woolls HA, Lamanna AC, Karbstein K (2011) Roles of Dim2 in ribosome assembly. J Biol Chem 286:2578–2586. doi:10.1074/jbc.M110.191494

Wurm JP, Meyer B, Bahr U, Held M, Frolow O, Kötter P, Engels JW, Heckel A, Karas M, Entian K-D, Wöhnert J (2010) The ribosome assembly factor Nep1 responsible for Bowen–Conradi syndrome is a pseudouridine-N1-specific methyltransferase. Nucleic Acids Res 38:2387–2398. doi:10.1093/nar/gkp1189

Xu Z, O'Farrell HC, Rife JP, Culver GM (2008) A conserved rRNA methyltransferase regulates ribosome biogenesis. Nat Struct Mol Biol 15:534–536. doi:10.1038/nsmb.1408

Yip WSV, Vincent NG, Baserga SJ (2013) Ribonucleoproteins in archaeal pre-rRNA processing and modification. Archaea 2013:614735. doi:10.1155/2013/614735

Young RA, Steitz JA (1978) Complementary sequences 1700 nucleotides apart form a ribonuclease III cleavage site in Escherichia coli ribosomal precursor RNA. Proc Natl Acad Sci U S A 75:3593–3597

Yutin N, Puigbò P, Koonin EV, Wolf YI (2012) Phylogenomics of prokaryotic ribosomal proteins. PLoS One 7:e36972. doi:10.1371/journal.pone.0036972

Zaremba-Niedzwiedzka K, Caceres EF, Saw JH, Bäckström D, Juzokaite L, Vancaester E, Seitz KW, Anantharaman K, Starnawski P, Kjeldsen KU, Stott MB, Nunoura T, Banfield JF, Schramm A, Baker BJ, Spang A, Ettema TJG (2017) Asgard archaea illuminate the origin of eukaryotic cellular complexity. Nature 19;541(7637):353–358. https://doi.org/10.1038/nature21031

Chapter 7
Structure and Function of Archaeal Ribonuclease P

Makoto Kimura, Kosuke Oshima, Xuzhu Gao, Dan Jiang, Takashi Nakashima, and Toshifumi Ueda

Abstract Ribonuclease P (RNase P) is a ribonucleoprotein complex involved in the processing of the 5′-leader sequence of precursor tRNA (pre-tRNA) and other small RNAs in all phylogenetic domains. A characteristic feature of archaeal RNase P RNAs is that they alone have, unlike bacterial counterparts, little pre-tRNA cleavage activity, but the interaction with protein components activates their catalytic activity. In addition, it has not yet been confirmed whether archaeal RNase P, like its bacterial and eukaryotic counterparts, has additional substrates in vivo. We have found that RNase P in the hyperthermophilic archaeon *Pyrococcus horikoshii* OT3 consists of RNase P RNA (*Pho*pRNA) and five protein cofactors designated *Pho*Pop5, *Pho*Rpp21, *Pho*Rpp29, *Pho*Rpp30, and *Pho*Rpp38. Biochemical and structural studies over the past 10 years have revealed that *Pho*Pop5 and *Pho*Rpp30 form a heterotetrameric complex and cooperatively activates a catalytic domain (C-domain), while *Pho*Rpp21 and *Pho*Rpp29 form a heterodimer and function together to activate a specificity domain (S-domain) in *Pho*pRNA. As for the fifth protein, *Pho*Rpp38 is involved in elevation of the optimum temperature of RNase P activity, binding to two peripheral stem-loops, including helices P12.1/12.2 and P16. Moreover, comparative analysis of the

M. Kimura (✉) • T. Nakashima
Department of Bioscience and Biotechnology, Laboratory of Biochemistry, Graduate School, Faculty of Agriculture, Kyushu University, Hakozaki 6-10-1, Higashi-ku, Fukuoka 812-8581, Japan

Laboratory of Structural Biology, Graduate School of Systems Life Sciences, Hakozaki 6-10-1, Fukuoka 812-8581, Japan
e-mail: mkimura@agr.kyushu-u.ac.jp

K. Oshima • T. Ueda
Department of Bioscience and Biotechnology, Laboratory of Biochemistry, Graduate School, Faculty of Agriculture, Kyushu University, Hakozaki 6-10-1, Higashi-ku, Fukuoka 812-8581, Japan

X. Gao • D. Jiang
Laboratory of Structural Biology, Graduate School of Systems Life Sciences, Hakozaki 6-10-1, Fukuoka 812-8581, Japan

B. Clouet-d'Orval (ed.), *RNA Metabolism and Gene Expression in Archaea*, Nucleic Acids and Molecular Biology 32, DOI 10.1007/978-3-319-65795-0_7

RNase P RNA sequences and existing crystallographic structural information of the bacterial RNase P RNAs were combined to generate a phylogenetically supported three-dimensional (3D) model of *Pho*pRNA. Recent biochemical data on the protein-*Pho*pRNA interactions localized the protein binding sites on *Pho*pRNA. Moreover, a comparative transcriptome on the hyperthermophilic archaeon *Thermococcus kodakarensis* suggested the presence of additional substrates for archaeal RNase Ps. Here, we review biochemical information on archaeal RNase Ps, mainly focusing on recent studies that allow us to generate a structural and mechanistic model for the *Pho*pRNA activation by the protein cofactors.

Keywords Hyperthermophilic archaea • Protein-RNA interaction • *Pyrococcus horikoshii* • Ribonuclease P • *Thermococcus kodakarensis*

Abbreviations

3D	Three-dimensional
C-domain	Catalytic domain
CR	Conserved region
H1 RNA	Human RNase P RNA
K-turn	Kink-turn
M1 RNA	*Escherichia coli* ribonuclease P RNA
ncRNA	Non-coding RNA
*Pho*pRNA	*Pyrococcus horikoshii* ribonuclease P RNA
pre-tRNA	Precursor tRNA
RNase P	Ribonuclease P
S-domain	Specificity domain
SL	Stem-loop
*Tko*pRNA	*Thermococcus kodakarensis* ribonuclease P RNA

7.1 Introduction

Ribonuclease P (RNase P), a ubiquitous *trans*-acting ribozyme present in all phylogenetic domains (Bacteria, Archaea, and Eukarya), catalyzes a Mg^{2+}-dependent hydrolysis to remove the 5′-leader sequence of precursor tRNA, as well as several small RNAs, such as 4.5S RNA, tmRNA, small nucleolar RNA, and riboswitches (Esakova and Krasilnikova 2010) (Fig. 7.1). Although the functionality of RNase P is almost the same in bacteria and humans, the composition is different in phylogenetic domains. Bacterial RNase P is composed of a catalytic RNA and a single protein cofactor, both of which are required for pre-tRNA processing in vivo. The RNase P from *Escherichia coli* contains a catalytic RNA

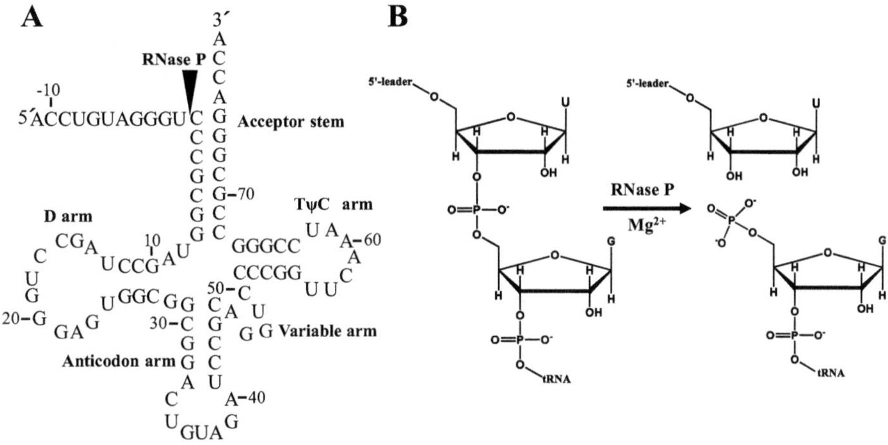

Fig. 7.1 Pre-tRNA cleavage by RNase P. (**a**) Secondary structure of *P. horikoshii* pre-tRNA^{Tyr}. An *arrow* indicates a cleavage site in pre-tRNA^{Tyr} by RNase P. (**b**) Hydrolysis of a phosphodiester bond. RNase P catalyzes a Mg^{2+}-dependent hydrolysis to remove the 5′-leader sequence of pre-tRNA

subunit termed M1 RNA and a single protein cofactor known as the C5 protein. Since Altman and co-workers discovered that M1 RNA itself can hydrolyze pre-tRNA in vitro (Guerrier-Takada et al. 1983), biochemical and structural studies on RNase P have mainly focused on bacterial enzymes (Kirsebom 2007; Smith et al. 2007). These studies showed that bacterial RNase P RNAs are composed of two domains, the substrate-binding domain (S-domain) and the catalytic domain (C-domain) (Loria and Pan 1996; Loria and Pace 2001). These domains can fold independently and the catalytic domain alone retains RNase P activity at high Mg^{2+} concentrations (Loria and Pace 2001). Furthermore, it was found that the P1-P4 multihelix junction plays a crucial role in the optimization of Mg^{2+} interactions important for catalysis. In particular, nucleotides A65 and A66 at J3/4 and helix P4 and the *pro*-Rp and *pro*-Sp non-bridging phosphate oxygen at A67 in helix P4 were assigned as binding sites for Mg^{2+} required for catalysis by M1 RNA (Christian et al. 2002). Recently, the crystal structure of bacterial RNase P in complex with tRNA has provided more insight into the structure-function relationships of bacterial RNase P (Reiter et al. 2010).

In contrast, archaeal and eukaryote RNase P consist of a single RNA and archaeal RNase P has four or five proteins, while eukaryotic RNase P consists of nine or ten proteins (Jarrous and Gopalan 2010). The RNA itself has no enzymatic activity under physiological conditions, that is, eukaryotic and archaeal RNase P RNAs cooperatively function with protein subunits in catalysis. Hence, archaeal and eukaryotic RNase Ps may serve as a model enzyme for studying how a functional RNA can be activated by protein cofactors. A highly purified nuclear

RNase P from HeLa cells has at least 10 distinct protein subunits termed Rpp14, Rpp20, Rpp21, Rpp25, Rpp29, Rpp30, Rpp38, Rpp40, hPop1, and hPop5 (Xiao et al. 2001; Jarrous and Altman 2001; Kikovska et al. 2007). Although Rpp21 and Rpp29 are known to be closely involved in the catalytic activity of human RNase P, the functional roles of the other subunits have not been established (Mann et al. 2003). Moreover, human RNase P, like bacterial RNase P, is known to be responsible for processing small non-coding RNAs (ncRNAs), including 5S rRNA and 7SL RNA, and long ncRNAs, such as metastasis associated in lung adenocarcinoma transcript 1 (MALAT1) and multiple endocrine neoplasia b (MEN-b) transcript (Wilusz et al. 2008; Sunwoo et al. 2009). Furthermore, mutation of P12 in human RNase P RNA (H1 RNA) by RNase H abolished PolIII transcription in a whole-cell extract (Reiner et al. 2006). Overexpression of the *Saccharomyces cerevisiae* nuclear RNase P RNA (RPR1) was also shown to suppress a slow-growing strain with a deletion mutation in Bdp 1, a subunit of the transcription factor TFIIB of PolIII (Ishiguro et al. 2002). These findings suggest the involvement of eukaryotic RNase P RNAs in the transcriptional regulation of PolIII.

As for archaeal RNase Ps, functional reconstitution and characterization of RNA and protein components have been reported for *Pyrococcus horikoshii* (Kouzuma et al. 2003; Fukuhara et al. 2006; Terada et al. 2006), *P. furiosus* (Tsai et al. 2006), *Methanocaldococcus jannaschii* (Chen et al. 2010), *Methanothermobacter thermautotrophicus* (Chen et al. 2010), *and Methanococcus maripaludis* RNasse Ps (Cho et al. 2010) (Table 7.1). These studies have shown that archaeal RNase P consists of RNA and five distinct proteins designated archaeal Pop5, Rpp21, Rpp29, Rpp30 and Rpp38, according to their sequence homology with the human RNase P proteins hPop5, Rpp21, Rpp29, Rpp30, and Rpp38, respectively (Jarrous and Gopalan 2010). Furthermore, they have been classified into types A and M based on the RNase P RNA's secondary structure (Harris et al.

Table 7.1 Comparison of RNase P proteins in the archaeal reconstituted RNase Ps

Pyrococcus horikoshii	*Pyrococcus furiosus*	*Methanothermobacter thermoautotrophicus*	*Methanocaldococcus jannaschii*	*Methanococcus maripaludis*
Type A			Type M	
*Pho*Pop5	Pop5	Pop5	Pop5	Pop5
*Pho*Rpp21	Rpp21	Rpp21	Rpp21	Rpp21
*Pho*Rpp29	Rpp29	Rpp29	Rpp29	Rpp29
*Pho*Rpp30	Rpp30	Rpp30	Rpp30	Rpp30
*Pho*Rpp38				Rpp38 (L7Ae)
*Pho*pRNA	RPR	RPR	RPR	RPR

P. horikoshii, *P. furiosus*, and *M. thermoautotrophicus* are classified into type A, while *M. jannaschii* and *M. maripaludis* belong to type M. The archaeal RNase P proteins were designated according to their homology to the corresponding human proteins and the prefix *Pho* is added to differentiate *P. horikoshii* proteins from homologous proteins from other organisms. The RNase P RNA in *P. horikoshii* was designated as *Pho*pRNA, while those in other archaeal RNase Ps were referred to as RPR

2001; Jarrous and Gopalan 2010) (Fig. 7.2). The type A archaeal RNase P RNAs, typified by *P. horikoshii* and *M. thermautotrophicus* RNasse Ps, resemble bacterial RNase P RNAs, although they lack helical stems P13, P14, and P18 in M1 RNA. In contrast, the type M archaeal RNase P RNAs, typified by *M. jannaschii*, are similar to the eukaryotic than to the bacterial RNase P. In our own study, we have chosen archaeal RNase Ps in the hyperthermophilic archaeon *P. horikoshii* OT3 and *Thermococcus kodakarensis* KOD1 as model enzymes to address the molecular mechanism by which RNA subunits are activated by protein cofactors. Biochemical and structural studies over the past 10 years have allowed us to gather structural and functional information about *P. horikoshii* RNase P subunits (Kimura and Kakuta 2012). Moreover, genetic analysis and comparative whole-transcriptome analysis of *T. kodakarensis* suggested that archaeal RNase Ps, as for bacterial and eukaryotic RNase Ps, have additional substrates in vivo. This article summarizes the structure-function relationships of the *P. horikoshii* and *T. kodakarensis* RNase P subunits, including the *Pho*pRNA-protein interaction, as well as a three-dimensional (3D) model of the *P. horikoshii* RNase P that is currently constructed on the basis of available biochemical data and high-resolution structures.

7.2 Essential Residues for Pre-tRNA Cleavage in *Pho*pRNA

Nucleotides A65, A66, and U69 in M1 RNA were assigned as binding sites for Mg^{2+} ions required for catalysis (Christian et al. 2002). In addition, G292 and G293, located at the loop (L15/16) between helices P15 and P16 in M1 RNA, were predicted to be involved in recognition of the acceptor end (CCA) in pre-tRNA (Kirsebom and Svard 1994). Involvement of these nucleotides in catalysis has been perfectly demonstrated by the X-ray structure of *Thermotoga maritima* RNase P in complex with tRNA (Reiter et al. 2010). Furthermore, the X-ray structure shows that A112 in CRII and G147 in CRIII are recognized by G19 in the D loop and C56 in the T loop in tRNA, respectively, by stacking interaction (Reiter et al. 2010). Although archaeal RNase Ps contain additional proteins, their RNAs retain an essential core of a conserved sequence and secondary structure (Chen and Pace 1997) (Fig. 7.2). It is thus likely that archaeal RNase P RNAs are directly involved in catalytic function. Indeed, mutational analysis of conserved nucleotide residues in *Pho*pRNA indicated that nucleotides A40, A41, and U44 at J3/4 and helix P4, and G269 and G270 located at L15/16 in *Pho*pRNA are, like the corresponding residues in M1 RNA, involved in hydrolysis by coordinating catalytic Mg^{2+} ions, and in the recognition of the acceptor end (CCA) of pre-tRNA by base-pairing, respectively (Terada et al. 2007). Recently, we further prepared several mutant RNAs, in which A105 and A205 corresponding to A112 and G147 in *T. maritima*

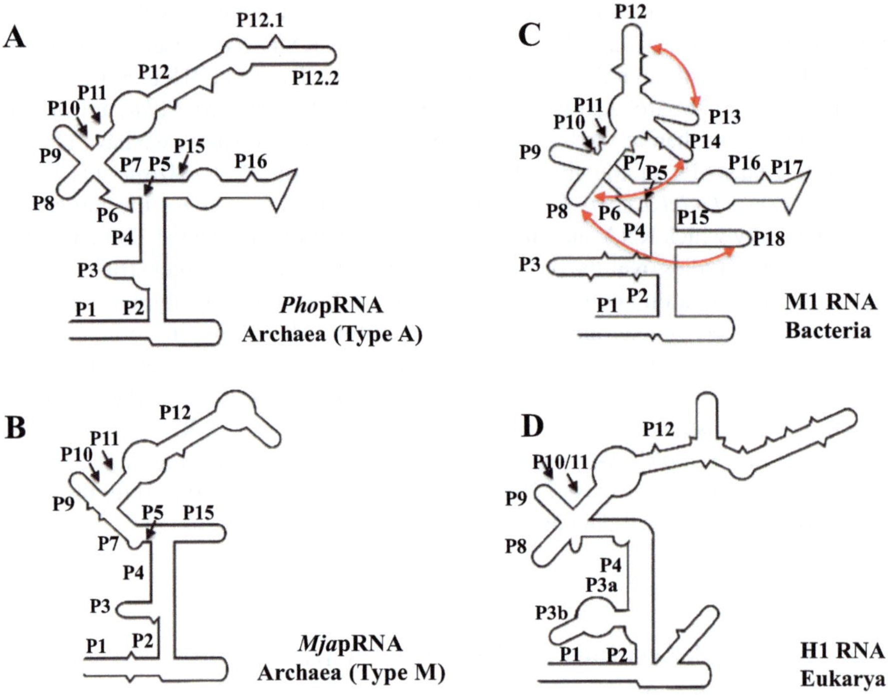

Fig. 7.2 Secondary structures of RNase P RNAs in three phylogenetic domains, Bacteria, Archaea, and Eukarya. (**a** and **b**) Schematic presentation of secondary structures of RNase P RNAs of *P. horikoshii* (*Pho*pRNA) and *M. jannaschii* (*Mja*pRNA) belonging to types A and M, respectively. (**c** and **d**) Schematic presentation of secondary structures of RNase P RNAs of *E. coli* (M1 RNA) and *H. sapiens* (H1 RNA). Helices are numbered according to the existing RNase P RNA nomenclature (Pace and Brown 1995). Interactions between secondary structures in M1 RNA are indicated by *curved red lines* with *arrows*

RNase P RNA were replaced by three other nucleotides, and reconstituted particles containing theses mutant RNAs and the five proteins were characterized with respect to pre-tRNA cleavage activity. Replacement of A105 and A205 by pyrimidine base U or C reduced the cleavage activity, and simultaneous replacement of A105 and A205 by U significantly reduced the pre-tRNA cleavage activity (Ueda et al. unpublished results). In contrast, replacements of A105 or A205 by purine base G had no influence on the cleavage activity (Ueda et al. unpublished results). This result fully supported stacking interactions observed in the X-ray structure and strongly suggests that *Pho*pRNA catalyzes the hydrolysis of pre-tRNA in approximately the same manner as bacterial RNase P RNAs, even though it has no enzymatic activity in the absence of the proteins.

7.3 Structural and Biochemical Data for the *P. horikoshii* RNase P Proteins

7.3.1 Pho*Pop5*-Pho*Rpp30*

Biochemical data suggested that *Pho*Pop5 and *Pho*Rpp30 interact with each other and synergistically activate the C-domain in *Pho*pRNA (Honda et al. 2010). The X-ray structure shows that *Pho*Pop5 and *Pho*Rpp30 fold into a heterotetramer [*Pho*Rpp30-(*Pho*Pop5)$_2$-*Pho*Rpp30], in which a homodimer of *Pho*Pop5 sits between two *Pho*Rpp30 monomers (Kawano et al. 2006) (Fig. 7.3a). *Pho*Pop5 dimerizes through a hydrogen bonding interaction from the loop between the α1 and α2 helices, and each *Pho*Pop5 interacts with two *Pho*Rpp30 molecules, where α2 and α3 in *Pho*Pop5 interact with α7 in one *Pho*Rpp30 and α8 in the other *Pho*Rpp30 molecules, respectively. The presence of two complexes in the asymmetric unit, together with gel filtration chromatography indicated that the

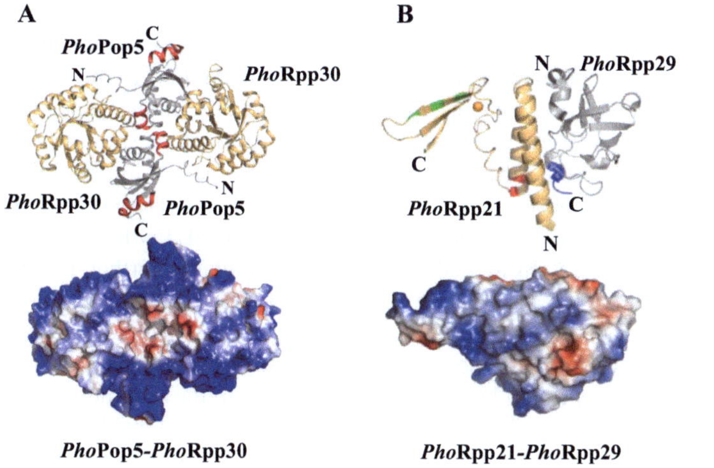

Fig. 7.3 Crystal structures of the *P. horikoshii* RNase P protein complexes. (**a**) The crystal structure of the *Pho*Pop5-*Pho*Rpp30 complex. In the *upper panel*, the crystal structure of *Pho*Pop5 (*grey*) in complex with *Pho*Rpp30 (*gold*) is presented. α-helices α2 and α4 in *Pho*Pop5, which are involved in homodimeric formation and *Pho*pRNA binding, respectively, are shown in *red*. N and C indicate the N- and C-termini of *Pho*Pop5, respectively. In the *lower panel*, the electrostatic surface potential of the *Pho*Pop5-*Pho*Rpp30 complex is shown. The surface potential is displayed as a color gradient from *red* (negative) to *blue* (positive), showing a unique charge distribution on the molecular surface. (**b**) The crystal structure of the *Pho*Rpp21-*Pho*Rpp29 complex. In the *upper panel*, the crystal structure of *Pho*Rpp21 (*gold*) in complex with *Pho*Rpp29 (*grey*) is presented. Lys53, Lys54, and Lys56 at the N-terminal helix (α2), Arg82, Arg84, and Arg86 at the C-terminal β-strand (β1) in *Pho*Rpp21, and the C-terminal residues in *Pho*Rpp29, which play a crucial role in *Pho*pRNA activation, are shown in *red*, *green*, and *blue*, respectively. In the *lower panel*, electrostatic surface potential of the *Pho*Rpp21-*Pho*Rpp29 complex is shown in the same manner as described above (For interpretation of the references to colors in this Figure Legend, the reader is referred to the web version of this paper)

heterotetramer is stable in solution and represents a functional state in the cell (Kawano et al. 2006; Hamasaki et al. 2016). A gel filtration study further indicated that *Pho*Rpp30 exists as a monomer, whereas *Pho*Pop5 is an oligomer in solution (Hamasaki et al. 2016). Examining the *Pho*Pop5-*Pho*Rpp30 interaction, the N-terminal residues Leu5 to Pro9 in *Pho*Pop5 interact with residues Pro110 to Ile112 in *Pho*Rpp30 by hydrophobic interactions (Kawano et al. 2006). These findings suggested that *Pho*Rpp30 assists *Pho*Pop5 in attaining a functionally active conformation by shielding hydrophobic surfaces of *Pho*Pop5.

As described above, *Pho*Pop5 dimerizes through a hydrogen bonding interaction from the loop between the α1 and α2 helices. The reconstituted particle containing the *Pho*Pop5 mutant termed ΔL43-48, in which the α1-α2 loop in *Pho*Pop5 was deleted, had significantly reduced pre-tRNA cleavage activity (Hazeyama et al. 2013). Furthermore, reconstitution experiments indicated that deletion of the C-terminal helices α4 and α5 (3_{10}-helix) significantly influenced the pre-tRNA cleavage activity, while that of α5 had little effect on this activity (Hazeyama et al. 2013). These results indicate that the heterotetrameric structure is essential for activation of *Pho*pRNA, and that C-terminal helix α4 in *Pho*Pop5 plays a crucial role in the activation of *Pho*pRNA. Recently, we found that *Pho*pRNA mutants ΔP3 and ΔP16, in which stem-loops SL3 and SL16 containing P3 and P16 in *Pho*pRNA were deleted, respectively, had little ability to bind *Pho*Pop5 and *Pho*Rpp30, suggesting that the *Pho*Pop5-*Pho*Rpp30 complex specifically recognizes P3 and P16 (Ueda et al. 2014). Moreover, surface plasmon resonance (SPR) analysis revealed that the tetramer strongly interacts with an oligonucleotide including the nucleotide sequence of SL3 in *Pho*pRNA (Hamasaki et al. 2016). In contrast, *Pho*Pop5 had markedly reduced affinity to SL3, whereas *Pho*Rpp30 had little affinity to SL3. SPR studies of *Pho*Pop5 mutants further revealed that the C-terminal helix (α4) in *Pho*Pop5 function as a molecular recognition element for SL3. These results, together with available data, allow us to generate a structural and mechanistic model for the *Pho*pRNA activation by *Pho*Pop5 and *Pho*Rpp30, in which the two C-terminal helices (α4) of *Pho*Pop5 in the tetramer whose formation is assisted by *Pho*Rpp30 act as binding elements and bridge P3 and P16 in *Pho*pRNA, thereby stabilizing a double stranded RNA structure (P4) containing catalytic Mg^{2+} ions (Fig. 7.4a).

7.3.2 Pho*Rpp21*-Pho*Rpp29*

*Pho*Rpp21 and *Pho*Rpp29 were suggested to activate the S-domain in *Pho*pRNA cooperatively (Honda et al. 2010). X-ray crystallographic analysis revealed that *Pho*Rpp21 and *Pho*Rpp29 fold into a heterodimeric structure, where the N-terminal two helices (α1 and α2) in *Pho*Rpp21 predominantly interact with the N-terminal extended structure, the β-strand (β2), and the C-terminal helix (α3) in *Pho*Rpp29 (Honda et al. 2008) (Fig. 7.3b). The truncation of the 31 N-terminal residues in *Pho*Rpp29 abolished the ability to interact with *Pho*Rpp21 and also reduced RNase

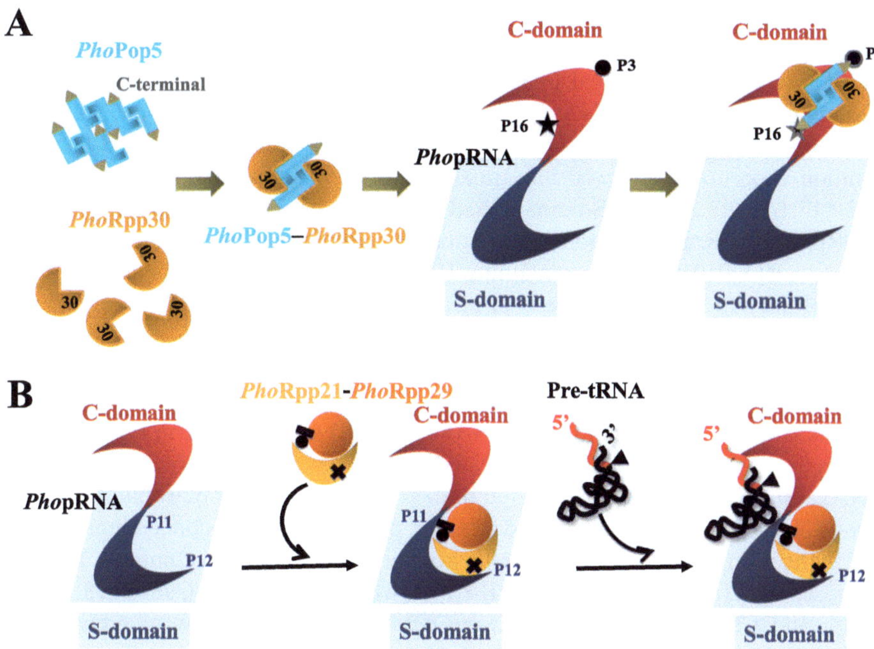

Fig. 7.4 The proposed molecular mechanism by which *Pho*Pop5-*Pho*Rpp30 and *Pho*Rpp21-*Pho*Rpp29 activate *Pho*pRNA. (**a**), The proposed mechanism by which the *Pho*Pop5-*Pho*Rpp30 complex activates *Pho*pRNA. *Pho*Rpp30 (*brown*) exists as a monomer, whereas *Pho*Pop5 (*cyan*) is an oligomer in solution. Hydrophobic interactions of *Pho*Rpp30 with *Pho*Pop5 avoid the self-oligomerization of *Pho*Pop5, which results in attaining a functional dimeric conformation of *Pho*Pop5 in the heterotetramer. Then, the two *Pho*Pop5 C-terminal helices α4 shown in triangles in the tetramer bind the stem-loop structures containing the P3 (*filled circle*) and P16 (*filled astrisk*) helices, and thereby stabilize an appropriate conformation of *Pho*pRNA. (**b**), The proposed mechanism by which the *Pho*Rpp21-*Pho*Rpp29 complex activates *Pho*pRNA. The *Pho*Rpp21-*Pho*Rpp29 complex binds the loop between P11 and P12 helices through Arg residues (*filled cross*) at the C-terminal β-strand (β1) and serves as a scaffold for *Pho*Rpp29 so as to optimize structural conformation of a positively charged edge composed of Lys residues (*filled circle*) in *Pho*Rpp21 and C-terminal residues (*filled bar*) in *Pho*Rpp29, which stabilizes stacking interactions between *Pho*pRNA and pre-tRNA. *filled triangle* indicates the processing site in pre-tRNA (For interpretation of the references to colors in this Figure Legend, the reader is referred to the web version of this paper)

P activity. These results indicated that the heterodimerization of *Pho*Rpp21 and *Pho*Rpp29 plays an important role in the function of *P. horikoshii* RNase P.

To elucidate the molecular basis for their cooperativity, we first analyzed binding ability to *Pho*pRNA using a pull-down assay. The result showed that *Pho*Rpp21 is able to bind to *Pho*pRNA in the absence of *Pho*Rpp29, whereas *Pho*Rpp29 alone has reduced affinity to *Pho*pRNA, suggesting that *Pho*Rpp21 primarily functions as a binding element for *Pho*pRNA in the *Pho*Rpp21-*Pho*Rpp29 complex (Jiang et al. 2017). Mutational analyses further suggested that although Lys53, Lys54, and Lys56 at the N-terminal helix (α2) and

10 C-terminal residues in *Pho*Rpp29 contribute little to the *Pho*pRNA binding, they are essential for *Pho*pRNA activation (Jiang et al. 2017). In contrast, Arg82, Arg84, and Arg86 at the C-terminal β-strand (β1) in *Pho*Rpp21 are suggested to act as binding residues to *Pho*pRNA.

As for binding sites of the *Pho*Rpp21-*Pho*Rpp29 complex on *Pho*pRNA, reconstitution experiments showed that truncation of stem-loops containing P8, P9, or P12/P12.1/P12.2 in the S-domain had little influence on the binding of the *Pho*Rpp21-*Pho*Rpp29 complex. Additionally, *Pho*Rpp21 and *Pho*Rpp29 could bind to a P11/P12 fragment containing P11 and P12/P12.1/P12.2 helices with the same affinity to *Pho*pRNA (Ueda et al. unpublished results). Collectively, it was suggested that either of two single-stranded loops, A105-A116 or G202-G209, connecting P11 and P12/P12.1/P12.2 helices are involved in binding to the *Pho*Rpp21-*Pho*Rpp29 complex (Jiang et al. 2017). There are universally conserved regions CRII (A107-A111 in *Pho*pRNA) and CRIII (U203-A207 in *Pho*pRNA) in the single strand loops connecting P11 and P12 helices in the S-domains in RNase P RNAs. To localize the binding region of the *Pho*Rpp21-*Pho*Rpp29 complex on the *Pho*pRNA S-domain, we examined whether deletion of loops connecting P11 and P12 helices in *Pho*pRNA could influence the binding of the complex. For this purpose, ΔCRII, ΔCRIII, and ΔCRII/III, in which nucleotides C98-G118, A198-U214, and C98-G118/A198-U214 were deleted, respectively, were prepared by in vitro transcription, and these mutant RNAs were incubated with five *P. horikoshii* RNase P proteins. The resulting reconstituted particles were subjected to glycerol density-gradient ultracentrifugation, and the proteins that bound ΔCRII, ΔCRIII, or ΔCRII/III were analyzed by SDS-PAGE analysis. The result showed that the *Pho*Rpp21 and *Pho*Rpp29 protein bands from the reconstituted particles containing ΔCRIII or ΔCRII/III appeared to become weaker than those from the particles containing the wild type *Pho*pRNA or ΔCRII (Jiang et al. 2017). This result suggested that nucleotides A198 to U214 containing CRIII form a main binding site for the *Pho*Rpp21-*Pho*Rpp29 complex. Collectively, the present results suggest that *Pho*Rpp21 binds the loop between P11 and P12 helices through Arg residues at the C-terminal β-strand (β1) and serves as a scaffold for *Pho*Rpp29 so as to optimize conformation of its N-terminal helix (α2), as well as C-terminal residues in *Pho*Rpp29 for RNase P activity, which presumably stabilize the interaction of *Pho*pRNA with pre-tRNA (Fig. 7.4b).

7.3.3 Pho*Rpp38*

*Pho*Rpp38 belongs to the ribosomal protein L7Ae family that specifically recognizes a kink-turn (K-turn) motif (Fukuhara et al. 2006). The K-turn motif identified in the *Haloarcula marismortui* large ribosomal subunit has a kink in the phosphodiester backbone that causes a sharp turn in the RNA helix. The standard K-turn comprises an asymmetric internal loop (a typically 3-nucleotide bulge) flanked by C-G rich pairs on one side and two consecutive *trans*-Hoogsteen-sugar

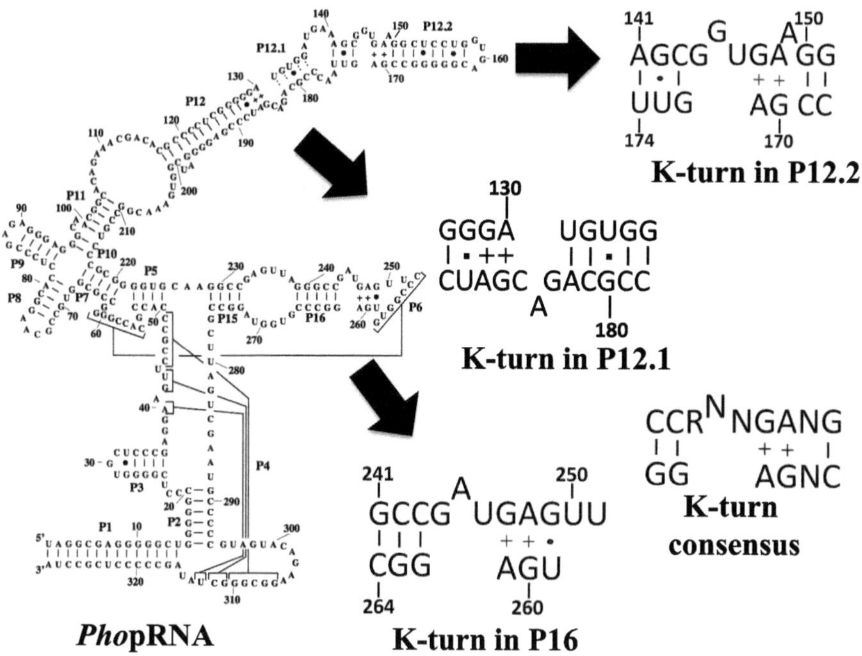

Fig. 7.5 Possible K-turn motifs in *Pho*pRNA. Secondary structures of the three putative K-turn motifs in P12.1, P12.2, and P16 helices, and the consensus K-turn motif are presented

edge G•A and A•G base pairs on the other (Klein et al. 2001) (Fig. 7.5). The adenine nucleobase in the G•A and A•G base pairs make key cross-strand hydrogen bonds that stabilize the kinked conformation (McPhee et al. 2014). A previous biochemical study showed that *Pho*Rpp38 specifically binds to two stem-loops, SL12 and SL16, containing helices P12.1/12.2 and P15/16 respectively, in *Pho*pRNA (Fukuhara et al. 2006). Sequence comparison of a consensus sequence of the K-turn with nucleotides in SL12 and SL16 showed that three regions, G127-G135/C178-C189 in P12.1, A141-G151/C168-U174 in P12.2, and G241-U251/U259-C264 in P16, are predicted to be folded into the K-turn (Oshima et al. 2016) (Fig. 7.5). In order to gain insight into the *Pho*Rpp38 binding mode to *Pho*pRNA, we determined the crystal structure of *Pho*Rpp38 in complex with the RNA fragment (SL12M), including A141-G151/C168-U174 in P12.2, at a resolution of 3.4 Å (Oshima et al. 2016) (Fig. 7.6). The structure revealed that Lys35 on the β-strand (β1) and Asn38, Glu39, and Lys42 on the α-helix (α2) in *Pho*Rpp38 interact with characteristic G•A and A•G pairs in SL12M, where Ile93, Glu94, and Val95, on a loop between α4 and β4 in *Pho*Rpp38, interact with the 3-nucleotide bulge (G-G-U) in SL12M. Structure-based mutational analysis indicated that amino acid residues involved in the binding to SL12 are also responsible for the binding to K-turns in P12.1 and P16 (Oshima et al. 2016). A pull-down assay further suggested the presence of the three K-turns in *Pho*pRNA (Oshima et al. 2016). These results suggested that each *Pho*Rpp38 binds to the three K-turns in SL12 and SL16 in *Pho*pRNA.

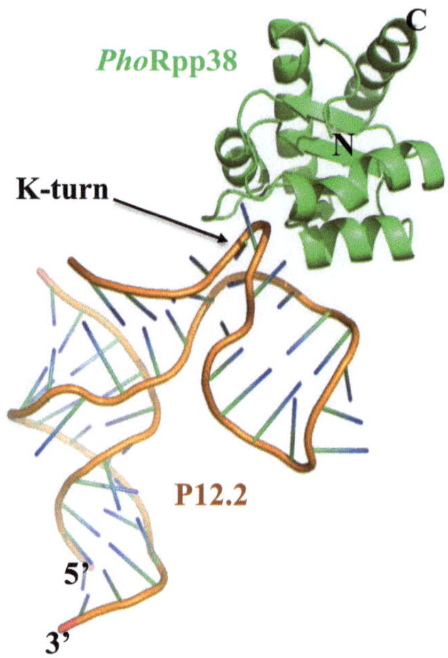

Fig. 7.6 Structures of *Pho*Rpp38 bound SL12M. The crystal structure of *Pho*Rpp38 in complex with SL12M (PDB ID. 5DCV). The protein is represented by its *ribbon diagram*, while SL12M is drawn in *stick form* (For interpretation of the references to colors in this Figure Legend, the reader is referred to the web version of this paper)

7.4 Do Archaeal RNase Ps Have Additional Substrates In Vivo?

Bacterial RNase P is involved in processing for a variety of ncRNAs, such as precursors to 4.5S RNA and riboswitches (Esakova and Krasilnikov 2010). In addition, human RNase P is known to be responsible for processing several ncRNAs (Wilusz et al. 2008; Sunwoo et al. 2009) and suggested to be involved in the transcriptional regulation of PolIII (Reiner et al. 2006; Ishiguro et al. 2002). Despite functional information having become available for bacterial and eukaryotic RNase Ps, the biological function of archaeal RNase P has not yet been clarified.

To gain insight into the biological function of archaeal RNase P in vivo, we chose *T. kodakarensis*, because genetic manipulation systems have been established (Sato et al. 2003; Takemasa et al. 2011). For this purpose, we first examined the biochemical and structural correspondence between *P. horikoshii* and *T. kodakarensis* RNase P subunits. The ribonuclease P (RNase P) proteins in *T. kodakarensis* were prepared and characterized with respect to pre-tRNA cleavage activity using the reconstitution system of the well-studied *P. horikoshii* RNase P. The reconstituted particle containing the *T. kodakarensis* subunit in place of the *P. horikoshii* counterpart retained pre-tRNA cleavage activity comparable to that of the reconstituted *P. horikoshii* RNase P (Suematsu et al. 2015). Moreover, we determined crystal structures of *Tko*Rpp30 (*Pho*Rpp30 homolog) alone and in complex with *Tko*Pop5 (*Pho*Pop5 homolog) (Suematsu et al. 2015). Like their

P. horikoshii counterparts, *Tko*Rpp30 and *Tko*Pop5 fold into a TIM barrel and RRM-like fold, respectively. This finding demonstrates that RNase P proteins in *T. kodakarensis* and *P. horikoshii* are interchangeable, and that their 3D structures are highly conserved.

We previously reported that the reconstituted particles containing ΔP3 or ΔP16, in which the stem-loops including helices P3 or P16 in *Pho*pRNA were individually deleted, had reduced activity (20–65%), although that containing ΔP8, in which the stem-loop including helix P8 in *Pho*pRNA was deleted, retained considerable levels of activity (80–100%) (Ueda et al. 2014). To obtain archaeal RNase Ps with impaired activity, we prepared archaeal mutants KUWΔP3, KUWΔP8, and KUWΔP16, in which the gene segments encoding stem-loops containing helices, respectively, P3, P8 and P16 in RNase P RNA (*Tko*pRNA) of the hyperthermophilic archaeon *T. kodakarensis* were deleted. Phenotypic analysis showed that KUWΔP3 and KUWΔP16 grew slowly compared with wild-type *T. kodakarensis* KUW1, while KUWΔP8 displayed no difference from *T. kodakarensis* KUW1 (Ueda et al. 2015). RNase P isolated using an affinity-tag from KUWΔP3 had reduced pre-tRNA cleavage activity compared with that from *T. kodakarensis* KUW1. Moreover, quantitative RT-PCR (qRT-PCR) and Northern blots analyses of KUWΔP3 showed greater accumulation of unprocessed transcripts for pre-tRNAs than that of *T. kodakarensis* KUW1 (Ueda et al. 2015). Subsequent whole-transcriptome analysis of *T. kodakarensis* KUW1 and KUWΔP3 using deep sequencing showed three operons containing genes, Tk0179-Tk0181, Tk1103-Tk1106, and Tk2018-Tk2020, were accumulated more abundantly in KUWΔP3 than in *T. kodakarensis* KUW1 (Suematsu et al. unpublished results). This information suggests that archaeal RNase P, like bacterial RNase P, has RNA substrates addition to pre-tRNA. Further study should allow for the comprehensive identification of RNA substrates of archaeal RNase Ps.

7.5 Future Directions

Despite significant progress in determining the crystal structure of proteins, the structure of *Pho*pRNA remains elusive. Hence, comparative analysis of the RNase P RNA sequences and existing crystallographic structural information of the bacterial RNase P RNAs were combined to generate a phylogenetically supported 3D model of *Pho*pRNA (Zwieb et al. 2011). Based on the all results, together with available biochemical data, we constructed a 3D model of *Pho*pRNA in complex with all five proteins as well as tRNA (Fig. 7.7). Details of the construction of the 3D model will be described elsewhere. The bacterial RNase P RNAs have helical stems P13, P14, and P18, which are absent from the archaeal RNase P RNAs (Fig. 7.2) (Torres-Larios et al. 2005; Reiter et al. 2010). It is known that tetraloop-helix interactions between P8 and P14, P12 and P13, and P8 and P18 in the bacterial RNase P RNA position the two domains (C- and S-domains) correctly to permit catalysis. A shortened RNA and an increase in the number of proteins in

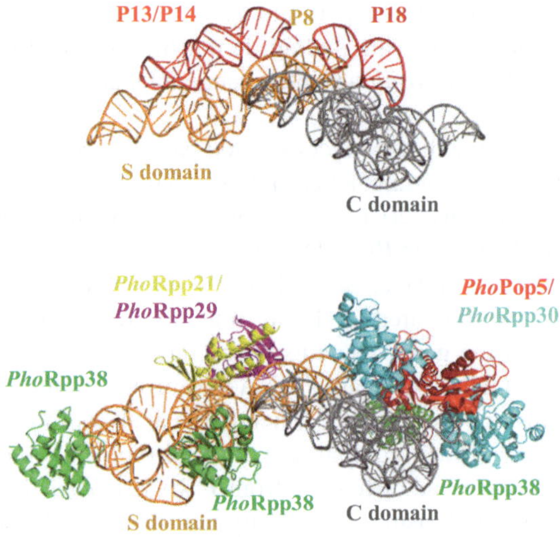

Fig. 7.7 Structural comparison of the bacterial RNase P RNA with the 3D model of *P. horikoshii* RNase P. The bacterial A-type RNase P RNA in *T. maritima* (*top*) is made up of two layers; the large layer of the structure contains most of the universally conserved regions, whereas the second layer comprises helical stems P13, P14, and P18 shown in *red*, which are absent in the archaeal RNase P RNAs. It is assumed that the tetraloop-helix interactions between P8 and P14 and P8 and P18 in the bacterial RNase P RNA are replaced by the interaction of the loop connecting P11 and P12 with the *Pho*Rpp21-*Pho*Rpp29 complex and the cross-linking of P3 and P16 by the *Pho*Pop5-*Pho*Rpp30 complex in *P. horikoshii* RNase P (*bottom*), respectively. The C- and S-domains in the bacterial RNase P RNA and *Pho*pRNA are in *gray* and *gold*, respectively. *Pho*Pop5, *Pho*Rpp21, *Pho*Rpp29, *Pho*Rpp30, and *Pho*Rpp38 are shown in *red*, *violet*, *yellow*, *cyan*, and *green*, respectively (For interpretation of the references to colors in this Figure Legend, the reader is referred to the web version of this paper)

archaeal RNase Ps suggest that some structural roles of eubacterial RNase P RNA may be delegated to the proteins in archaeal RNase P. On the basis of the 3D model of *P. horikoshii* RNase P, it is likely that the tetraloop-helix interactions between P8 and P14 and P8 and P18 in the bacterial RNase P RNA are replaced by the interaction of the loop connecting P11 and P12 with the *Pho*Rpp21-*Pho*Rpp29 complex and the cross-linking of P3 and P16 by the *Pho*Pop5-*Pho*Rpp30 complex in archaeal RNase P RNAs, respectively. Demonstration of the presented 3D model by crystallographic analysis or electron microscopic analysis will provide the structural basis for the mechanism of protein-mediated activation of *Pho*pRNA, and also deliver insight into the molecular evolution of RNase Ps in the three phylogenetic domains of life.

Acknowledgements We thank the past and present members of our group, particularly, M. Hamasaki and K. Hazeyama for their contribution to the study on *Pho*Pop5 and *Pho*Rpp30. We are also grateful to Drs. Y. Ishino (Kyushu University), S. Ishino (Kyushu University), and Y. Kawarabayasi (National Institute of Advanced Industrial Science and Technology, Japan) for their help with gene manipulation of *T. kodakarensis* and deep sequencing. This work was supported in part by a grant-in-aid for scientific research from the Ministry of Education, Culture, Sports, Science, and Technology of Japan (no. 15H04487 to M.K.).

References

Chen JL, Pace NR (1997) Identification of the universally conserved core of ribonuclease P RNA. RNA 3:557–560

Chen W-K, Pulukkunat DK, Cho I-M, Tsai H-Y, Gopalan V (2010) Dissecting functional cooperation among protein subunits in archaeal RNase P, a catalytic ribonucleoprotein complex. Nucleic Acids Res 38:8316–8327

Cho I-M, Lai LB, Susanti D, Mukhopadhyay B, Gopalan V (2010) Ribosomal protein L7Ae is a subunit of atchaeal RNase P. Proc Natl Acad Sci USA 107:14573–14578

Christian EL, Kaye NM, Harris ME (2002) Evidence for a polynuclear metal ion binding site in the catalytic domain of ribonuclease P RNA. EMBO J 21:2253–2262

Esakova O, Krasilnikov AS (2010) Of proteins and RNA: the RNase P/MRP family. RNA 16:1725–1747

Fukuhara H, Kifusa M, Watanabe M, Terada A, Honda T, Numata T, Kakuta Y, Kimura M (2006) A fifth protein subunit Ph1496p elevates the optimum temperature for the ribonuclease P activity from *Pyrococcus horikoshii* OT3. Biochem Biophys Res Commun 343:956–964

Guerrier-Takada C, Gardiner K, Marsh T, Pace N, Altman S (1983) The RNA moiety of ribonuclease P is the catalytic subunit of the enzyme. Cell 35:849–857

Hamasaki M, Hazeyama K, Iwasaki F, Ueda T, Nakashima T, Kakuta Y, Kimura M (2016) Functional implication of the heteroterameric formation of archaeal homologues of human RNase P protein pairs Pop5 with Rpp30. J Biochem 159:31–40

Harris JK, Haas ES, Williams D, Frank DN, Brown JW (2001) New insight into RNase P RNA structure from comparative analysis of the archaeal RNA. RNA 7:220–232

Hazeyama K, Ishihara M, Ueda T, Nishimoto E, Nakashima T, Kakuta Y, Kimura M (2013) Extra-structural elements in the RNA recognition motif in archaeal Pop5 play a crucial role in the activation of RNase P RNA from *Pyrococcus horikoshii* OT3. Biochem Biophys Res Commun 440:594–598

Honda T, Kakuta Y, Kimura K, Saho J, Kimura M (2008) Structure of an archaeal homolog of the human protein complex Rpp21-Rpp29 that is a key core component for the assembly of active ribonuclease P. J Mol Biol 384:652–662

Honda T, Hara T, Nan J, Zhang X, Kimura M (2010) Archaeal homologs of human RNase P protein pairs Pop5 with Rpp30 and Rpp21 with Rpp29 work on distinct functional domains of the RNA subunit. Biosci Biotechnol Biochem 74:266–273

Ishiguro A, Kassavetis GA, Geiduschek EP (2002) Essential roles of Bdp 1, a subunit of RNA polymerase III initiation factor TFIIIB, in transcription and tRNA processing. Mol Cell Biol 22:3264–3275

Jarrous N, Altman S (2001) Human ribonuclease P. Methods Enzymol 342:93–100

Jarrous N, Gopalan V (2010) Archaeal/eukaryal RNase P: subunits, functions and RNA diversification. Nucleic Acids Res 38:7885–7894

Jiang D, Izumi K, Ueda T, Oshima K, Nakashima T, Kimura M (2017) Functional characterization of archaeal homologues of a human nuclear RNase P proteins Rpp21 and Rpp29 provides insights into the molecular basis of their cooperativity in catalysis. Biochem Biophys Res Commun 482:68–74

Kawano S, Nakashima T, Kakuta Y, Tanaka I, Kimura M (2006) Crystal structure of protein Ph1481p in complex with protein Ph1877p of archaeal RNase P from *Pyrococcus horikoshii* OT3: implication of dimer formation of the holoenzyme. J Mol Biol 357:583–591

Kikovska E, Svard SG, Kirsebom LA (2007) Eukaryotic RNase P RNA mediates cleavage in the absence of protein. Proc Natl Acad Sci USA 104:2062–2067

Kimura M, Kakuta Y (2012) In: Satyanarayana T et al (eds) Microorganisms in sustainable agriculture and biotechnology, Chap 23. Springer Science+Business Media B.V., Amsterdam, pp 487–508

Kirsebom LA (2007) RNase P RNA mediated cleavage: substrate recognition and catalysis. Biochimie 89:1183–1194

Kirsebom LA, Svard SG (1994) Base pairing between Escherichia coli RNase P RNA and its substrate. EMBO J 13:4870–4876

Klein DJ, Schmeing TM, Moore PB, Steitz TA (2001) The kink-turn: a new RNA secondary structure motif. EMBO J 20:4214–4221

Kouzuma Y, Mizoguchi M, Takagi H, Fukuhara H, Tsukamoto M, Numata T, Kimura M (2003) Reconstitution of archaeal ribonuclease P from RNA and four protein components. Biochem Biophys Res Commun 306:666–673

Loria A, Pace NR (2001) Modular construction for function of a ribonucleoprotein enzyme: the catalytic domain of *Bacillus subtilis* RNase P complexed with *B. subtilis* RNase P protein. Nucleic Acids Res 29:1892–1897

Loria A, Pan T (1996) Domain structure of the ribozyme from eubacterial ribonuclease P. RNA 2:551–563

Mann H, Ben-Asouli Y, Schein A, Moussa S, Jarrous N (2003) Eukaryotic RNase P: role of RNA and protein subunits of a primordial catalytic ribonucleoprotein in RNA-based catalysis. Mol Cell 12:925–935

McPhee SA, Huang L, Lilley DMJ (2014) A critical base pair in k-turns that confers folding characteristics and correlates with biological function. Nat Commun 5:5127

Oshima K, Kakiuchi Y, Tanaka Y, Ueda T, Nakashima T, Kimura M, Yao M (2016) Structural basis for recognition of a kink-turn motif by an archaeal homologue of human RNase P protein Rpp38. Biochem Biophys Res Commun 474:541–546

Pace NR, Brown JW (1995) Evolutionary perspective on the structure and function of ribonuclease P, a ribozyme. J Bacteriol 177:1919–1928

Reiner R, Ben-Asouli Y, Krilovetzky I, Jarrous N (2006) A role for the catalytic ribonucleoprotein RNase P in RNA polymerase III transcription. Gene Dev 20:1621–1635

Reiter NJ, Osterman A, Torres-Larios A, Swinger KK, Pan T, Mondragon A (2010) Structure of a bacterial ribonuclease P holoenzyme in complex with tRNA. Nature 468:784–789

Sato T, Fukui T, Atomi H, Imanaka T (2003) Targeted gene disruption by homologous recombination in the hyperthermophilic archaeon *Thermococcus kodakaraensis* KOD1. J Bacteriol 185:210–220

Smith JK, Hsieh J, Fierke CA (2007) Importance of RNA-protein interactions in bacterial ribonuclease P structure and catalysis. Biopolymers 87:329–338

Suematsu K, Ueda T, Nakashima T, Kakuta Y, Kimura K (2015) On archaeal homologs of the human RNase P proteins Pop5 and Rpp30 in the hyperthermophilic archaeon *Thermococcus kodakarensis*. Biosci Biotechnol Biochem 79:952–959

Sunwoo H, Dinger ME, Wilusz JE, Amaral PP, Mattick JS, Spector DL (2009) MEN epsilon/beta nuclear-retained non-coding RNAs are up-regulated upon muscle differentiation and are essential components of paraspeckles. Genome Res 19:347–359

Takemasa R, Yokooji Y, Yamatsu A, Atomi H, Imanaka T (2011) *Thermococcus kodakarensis* as a host for gene expression and protein secretion. Appl Environ Microbiol 77:2392–2398

Terada A, Honda T, Fukuhara H, Hada K, Kimura M (2006) Characterization of the archaeal ribonuclease P proteins from *Pyrococcus horikoshii* OT3. J Biochem 140:293–298

Terada A, Yoshida T, Kimura M (2007) Identification of nucleotide residues essential for RNase P activity from the hyperthermophilic archaeaon *Pyrococcus horikoshii* OT3. Biosci Biotechnol Biochem 71:1940–1945

Torres-Larios A, Swinger KK, Krasilnikov AS, Pan T, Mondragon A (2005) Crystal structure of the RNA component of bacterial ribonuclease P. Nature 437:584–587

Tsai HY, Pulukkunat DK, Woznick WK, Gopalan V (2006) Functional reconstitution and characterization of *Pyrococcus furiosus* RNase P. Proc Natl Acad Sci USA 103:16147–16152

Ueda T, Yamaguchi H, Miyanoshita M, Nakashima T, Kakuta Y, Kimura M (2014) Characterization of the peripheral structures of archaeal RNase P RNA from *Pyrococcus horikoshii* OT3. J Biochem 155:25–33

Ueda T, Ishino S, Suematsu K, Nakashima T, Kakuta Y, Ishino Y, Kimura M (2015) Mutation of the gene encoding the ribonuclease P RNA in the hyperthermophilic archaeon *Thermococcus kodakarensis* causes decreased growth rate and impaired processing of tRNA precursors. Biochem Biophys Res Commun 468:660–665

Wilusz JE, Freier SM, Spector DL (2008) 3′ end processing of a long nuclear-retained noncoding RNA yields a tRNA-like cytoplasmic RNA. Cell 135:919–932

Xiao S, Houser-Scott F, Engelke DR (2001) Eukaryotic ribonuclease P: increased complexity to cope with the nuclear pre-tRNA pathway. J Cell Physiol 187:11–20

Zwieb C, Nakao Y, Nakashima T, Takagi H, Goda S, Andersen ES, Kakuta Y, Kimura M (2011) Structural modeling of RNase P RNA from the hyperthermophilic archaeon *Pyrococcus horikoshii* OT3. Biochem Biophys Res Commun 414:517–522

Chapter 8
Function and Biosynthesis of the Universal tRNA Modification N^6-Threonylcarbamoyl-Adenosine

Adeline Pichard-Kostuch, Marie-Claire Daugeron, Patrick Forterre, and Tamara Basta

Abstract Transfer RNAs (tRNAs) are essential components of the translation machinery that reads the genetic message and translates it into polypeptides. Apart from the four canonical nucleotides A, C, U and G, tRNAs contain a variety of modified nucleosides which are formed enzymatically during maturation process. Nucleosides 34 (wobble base) and 37 (dangling base) in the anticodon loop of tRNAs are frequently modified. These modifications were shown to be important for the efficiency and fidelity of translation. This chapter focuses on one of these modified nucleosides, N6-threonylcarbamoyl adenosine (t^6A_{37}), which occurs exclusively at the position 37 of all tRNAs decoding ANN codons (N = A, C, U or G). Initial biochemical and structural studies established the function of this universal modification in facilitating the binding of tRNA on ribosome and preventing frameshifting in course of translation. Recently, the genes encoding the synthetic machinery for the biosynthesis of t^6A_{37} were discovered in the three domains of life providing the possibility to study the reaction mechanism and measure the effect of t^6A_{37} on cellular physiology. We describe in this chapter these experimental characterizations with the focus on the archaeal genes and proteins and the homologous eukaryotic system.

Keywords tRNA modification • N6-threonylcarbamoyl-adenosine • Sua5/TsaC protein family • Kae1/TsaD/Qri7 protein family

A. Pichard-Kostuch • M.-C. Daugeron • T. Basta (✉)
Institute for Integrative Biology of the Cell (I2BC), CEA, CNRS, Univ. Paris-Sud, Université Paris-Saclay, 91198 Gif-sur-Yvette cedex, France
e-mail: tamara.basta@i2bc.paris-saclay.fr

P. Forterre
Institute for Integrative Biology of the Cell (I2BC), CEA, CNRS, Univ. Paris-Sud, Université Paris-Saclay, 91198 Gif-sur-Yvette cedex, France

Unité Biologie Moléculaire du Gène chez les Extrêmophiles, Institut Pasteur, 25 rue du Docteur Roux, 75015 Paris, France

B. Clouet-d'Orval (ed.), *RNA Metabolism and Gene Expression in Archaea*, Nucleic Acids and Molecular Biology 32, DOI 10.1007/978-3-319-65795-0_8

8.1 Introduction

Translation of the genetic message into proteins is a complex multistep process involving a multitude of actors such as ribosomes, translation factors, tRNA synthetases and tRNAs. Despite its complexity, translation is remarkably accurate, in average only one error occurs for 10^3–10^4 amino acids incorporated (Kurland 1992). While single misreading errors generally yield functional polypeptides, the cases where translational reading frame is shifted lead invariably to production of aberrant peptides and/or short peptides in cases where the frameshift introduces a premature stop codon (Yourno et al. 1970; Jackman and Alfonzo 2013). It was early on recognized that the canonical Watson-Crick pairing between the codon and anticodon was not sufficient to explain the observed error rate hinting that modified nucleosides in the tRNA molecules could be important (Agris 2004).

The striking feature of tRNAs from all organisms is the large number (approximately 90), and the variety of chemical structures of modified nucleosides found within these molecules (Juhling et al. 2009; Cantara et al. 2011; Machnicka et al. 2013, 2014). The survey of several different organisms from the three domains of life showed that modifications are found on about 12% of nucleosides with a median value of eight modifications per tRNA species (Phizicky and Alfonzo 2010). In archaea, at least 47 different modifications have been identified with some of them, such as archaeosine or agmatidine, being specific to archaea (Phillips and de Crecy-Lagard 2011; Grosjean et al. 2008). Chemical modifications can occur at the base of the nucleosides and/or at the 2′ OH group of the ribose. They are introduced during the maturation process of the tRNA molecules by various enzymes and can be simple additions of a methyl group or complex multistep reactions such as the formation of tricyclic wyosine derivatives (Grosjean 2009; Urbonavicius et al. 2014).

Large body of physical evidence supports the claim that modifications in the body of tRNA molecules are important for their stability and the correct folding, whereas many of the modifications in the anticodon loop have significant effects on efficiency and accuracy of translation (El Yacoubi et al. 2012; Phizicky and Alfonzo 2010; Agris et al. 2007; Agris 2008). The most effective for the fidelity of translation are the modified nucleosides at the wobble position 34 of the anticodon and at the purine bases at the position 37 3′ adjacent to the anticodon, also called the dangling base. These nucleosides have the capacity to restrict the dynamics of the anticodon stem loop and shape its structure such that it fits the decoding site in the ribosome (Agris 2008). More than 70% of tRNA species are modified at the position 37 (Juhling et al. 2009; Machnicka et al. 2013). Among those, of special interest is N^6-threonylcarbamoyladenosine (t^6A_{37}) which, together with N^1-methylguanosine (m^1G), is the only modified nucleoside that exists in tRNAs of all domains of life and also in mitochondria and chloroplasts (Grosjean et al. 1995).

t^6A_{37} was first isolated in the late 1960s but the enzymes responsible for its synthesis were identified 40 years later. This breakthrough discovery provided the necessary knowledge to start investigating the reaction mechanism leading to the biosynthesis of t^6A_{37} and the impact of the t^6A_{37} on the cellular physiology and the accuracy of translation. In this chapter, we provide an overview on these experimental characterizations with the focus on the archaeal genes and proteins and on the homologous eukaryotic system. For an in-depth review of t^6A_{37} synthesis in all domains of life the reader is referred to the paper by Thiaville et al. (2014b).

8.2 N^6-Threonylcarbamoyladenosine: The "Anti-slip" Nucleoside

N^6-threonylcarbamoyladenosine is found exclusively at the position 37, next to the anticodon, of almost all tRNAs decoding ANN codons (where N = A, U, G, C) (Fig. 8.1). The only mysterious exception to this rule is the absence of this modification in the initiator $tRNA_{Met}$ (anticodon CAU) of archaea, bacteria and organels, whereas the cytoplasmic initiator $tRNA_{Met}$ (CAU) of eukaryotes always contain t^6A_{37} (Machnicka et al. 2013). In a few tRNAs t^6A_{37} can be additionally methylated on the threonyl moiety (hnt^6A_{37}) or on the N6 position of the adenine base ($m^6t^6A_{37}$) and it may also contain a methylthio group in position 2 of the purine ring ($ms^2t^6A_{37}$). In some bacteria, fungi, protists and plants, t^6A_{37} is enzymatically converted to a circularized derivative ct^6A_{37} (Fig. 8.1) (Miyauchi et al. 2013). How widely distributed is ct^6A_{37} among the living organisms and whether it occurs in Archaea remains to be investigated.

It has been proposed, initially by Dube and colleagues, that one of the roles of t^6A_{37} is to prevent mispairing between the first nucleotide of the codon and the third nucleoside of the anticodon, thus ensuring correct reading of the genetic code on the

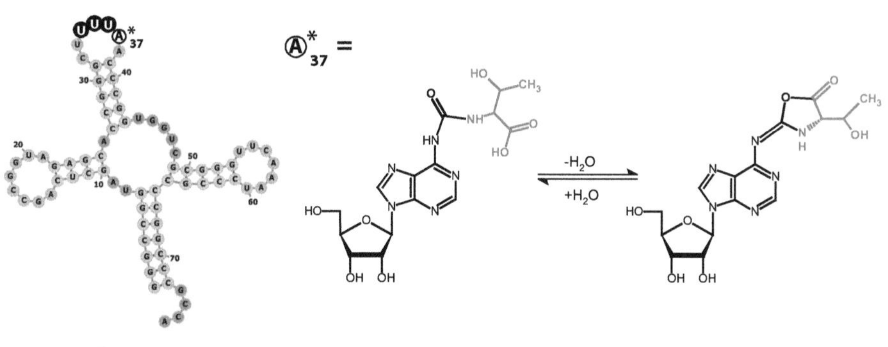

Fig. 8.1 t^6A_{37} modification and its cyclic derivative. On the *left* is depicted the 2D model of the tRNALys (UUU) from *Pyrococcus furiosus* DSM3638. The anticodon UUU is indicated with *black circles*. Adenosine at position 37 is indicated in *white circle*. On the *right* is the chemical structure of t^6A_{37}-modified nucleoside and its cyclic derivative ct^6A_{37}. Chemical groups originating from threonine or bicarbonate are indicated in *grey* and *bold*, respectively. The conversion of t^6A_{37} into ct^6A_{37} is catalyzed in *E. coli* by CsdL, an ATP-dependent dehydratase. The 2D model of the tRNALys (UUU) was generated using Forna software (Kerpedjiev et al. 2015)

ribosome (Dube et al. 1968). In agreement with this prediction, early biochemical studies showed that the presence of t^6A_{37} on tRNA had a small but significant stabilization effect on the ribosome-mediated codon binding for several tRNA species from *E. coli* and yeast (Miller et al. 1976; Weissenbach and Grosjean 1981). The stabilization effect of about 1.7-fold could be measured both for the canonical U–A pairing (where U is the third nucleotide of the anticodon and A is the first nucleotide of the codon) and for U–G mismatched base pairing. This indicated that prevention of mispairing between the first nucleoside of the codon and the third nucleoside of the anticodon is probably not due to t^6A_{37} alone (Weissenbach and Grosjean 1981). It was therefore suggested that the main role of t^6A_{37} is to prevent the "slipping" of the codon against the anticodon and thus to maintain the correct translational reading frame. Structural analysis of $tRNA_{Lys}$ (UUU) decoding at the ribosome A site revealed that the t^6A_{37} enhanced the stability of the anticodon-codon base pairing by creating cross-strand base stacking interactions with the first position of the codon (Murphy et al. 2004; Vendeix et al. 2012). Similar observations were made using molecular dynamics simulations of anticodon stem loop (ASL) of $tRNA_{Ile}$ (CAU) which contains t^6A_{37} (Sonawane and Sambhare 2015). The t^6A_{37} has a planar structure that acts as a third heterocycle to stack with neighboring bases and it also engages into van der Waals contacts that influence the stacking interactions. As a result of the planar heterocyclic structure and the conformational restrictions imposed by the bulky threonyl group, the t^6A_{37} does not stack with U36 of the anticodon but instead forms stacking interactions with the A1 of the codon. These structural data explain the increased stability of the codon–anticodon interactions for t^6A_{37} modified tRNA versus non-modified tRNA observed previously (Grosjean et al. 1976; Konevega et al. 2004; Yarian et al. 2000; Yarian et al. 2002). The structural data further revealed that the presence of the t^6A_{37} in the anticodon loop impairs the intra-loop base pairing with the invariant U_{33} nucleoside which is critical for adopting the canonical U-turn backbone structure of tRNA. This creates an open conformation of the anticodon-loop which is beneficial for the binding of the tRNA on the ribosomal A site and on the mRNA (Murphy et al. 2004; Yarian et al. 2000; Vendeix et al. 2012; Stuart et al. 2000; Sundaram et al. 2000). Taken together, the biochemical and structural data indicated that t^6A_{37} has a fundamental role in translation by enhancing the binding of the tRNA on the ribosome and preventing translational frameshifting.

8.3 Biosynthesis of N^6-Threonylcarbamoyladenosine in the Three Domains of Life

The t^6A_{37} nucleoside was first isolated in 1969 from several species of tRNA decoding ANN codons (Chheda et al. 1969; Ishikura et al. 1969; Schweizer et al. 1969). The initial *in vivo* studies of the enzymatic pathway leading to the formation of t^6A_{37} established that L-threonine is the precursor for the side chain of the t^6A_{37}

(Chheda et al. 1972; Powers and Peterkofsky 1972). *In vitro* studies that followed, demonstrated using partially purified *E. coli* cell extracts that reaction additionally required bicarbonate, ATP and Mg^{2+} suggesting the formation of a yet unidentified adenylate intermediate (Elkins and Keller 1974; Körner and Söll 1974). However, the possibility that carbamoyl-phosphate is one of the required intermediates has been ruled out (Powers and Peterkofsky 1972). Microinjection of tRNA transcripts into *Xenopus laevis* ovocytes demonstrated that, in addition to the targeted A_{37}, only U_{36} is absolutely required, however, the A_{38} significantly facilitated the quantitative transformation of A_{37} to t^6A_{37}. The integrity of the L-shaped tRNA architecture was shown to be another strict requirement for the action of the *X. laevis* enzymes, although some local perturbations of the 3D structure of the tRNA were allowed (Morin et al. 1998). Although these early studies made much progress in the understanding of the reaction pathway, they failed to identify the t^6A_{37} synthetic enzymes.

The breakthrough came four decades after the discovery of t^6A_{37}, in 2009, when de Crecy-Lagard laboratory predicted, using comparative genomic approach and literature mining, that the enzymes of the universal COG0009 family TsaC*/Sua5 (*previously called YrdC, see also Table 8.1 for nomenclature) were involved in the synthesis of t^6A_{37}. This prediction was validated by showing that the *SUA5* null strain of *S. cerevisiae* lacked the t^6A_{37} modified tRNA (El Yacoubi et al. 2009). The

Table 8.1 Names for t^6A_{37} synthesis proteins

	Bacteria		Eukarya		Archaea
	Escherichia coli	*Bacillus subtilis*	*Saccharomyces cerevisiae*	*Homo sapiens*	*Pyrococcus abyssi*
Universal	**TsaC** (YrdC/RimN)	TsaC2 (YwlC)	**Sua5** (Tcs2)	IRIP (Tcs1)	**Sua5** (Tcs2)
	TsaD (YgjD)	TsaD (YdiE)	**Kae1** (Tcs3)	**OSGEP** (Tcs3)	**Kae1** (Tcs3)
			[**Qri7**]mt (Tcs4)	[OSGEPL1]mt (Tcs4)	
Eukarya/ Archaea specific			**Bud32** (Tcs5)	**PRPK** (Tcs5)	**Bud32** (Tcs5)
			Pcc1 (Tcs6)	**LAGE3** (Tcs6)	**Pcc1** (Tcs6)
			Cgi121 (Tcs7)	TPRKB (Tcs7)	**Cgi121** (Tcs7)
			Gon7 (Tcs8)		
				C14ORF142	
Bacteria specific	**TsaB** (YeaZ)	TsaB (YdiC)			
	TsaE (YjeE)	TsaE (YdiB)			

Names used in the main text are in bold. For further details concerning the nomenclature, the reader is referred to the paper by Thiaville and colleagues (Thiaville et al. 2014b)

same group reported 2 years later the functional link between a second ubiquitous enzyme family COG0533 TsaD*/Kae1 (*previously called YgjD) and t^6A_{37} formation (El Yacoubi et al. 2011; Srinivasan et al. 2011). However, using these two universal enzymes it was not possible to reconstitute the t^6A_{37} reaction *in vitro* indicating that other proteins were required.

8.3.1 The Diversity of t^6A_{37} Synthetases

In bacteria, the clue for the remaining t^6A_{37} biosynthetic proteins came from the investigation of the interaction network of the TsaD protein (Fig. 8.2) (Handford et al. 2009). Two more proteins involved in t^6A_{37} synthesis were identified: TsaB (previously called YeaZ), a paralog of TsaD protein, and TsaE (previously called YjeE), a highly conserved bacterial P-loop ATPase (Teplyakov et al. 2002; Allali-Hassani et al. 2004). TsaC, TsaD, TsaB and TsaE of *E. coli* were shown to be required and sufficient for the formation of t^6A_{37} *in vitro* (Deutsch et al. 2012). The three proteins TsaD, TsaB and TsaE form a ternary complex required for the t^6A_{37} synthesis *in vitro* (Zhang et al. 2015b; Nichols et al. 2013). Based on the crystal structure of the stable heterodimer TsaD-TsaB, the small angle X-ray scattering (SAXS) experiments and isothermal titration calorimetry experiments (ITC) a model structure of the ternary complex was proposed where TsaE binds at the TsaD-TsaB interface only in presence of ATP (Zhang et al. 2015b; Nichols et al. 2013).

In archaea and eukaryotes, Kae1, the ortholog of TsaD, was first identified as a component of the conserved EKC/KEOPS (Endopeptidase-like Kinase Chromatin-associated/Kinase Endopeptidase and Other Proteins of Small size) complex involved in telomere maintenance and transcription in *S. cerevisiae* cells (Downey et al. 2006; Kisseleva-Romanova et al. 2006). The partner proteins of Kae1 in this complex were the natural candidates for the missing t^6A_{37} synthetic proteins: Bud32 is an ancient RIO-like atypical protein serine/threonine kinase conserved in all archaeal and eukaryotic species, while Pcc1 and Cgi121 possess novel folds with no inferable function (Facchin et al. 2003; Mao et al. 2008). A fifth member of the complex, Gon7, is found in Fungi. Recently, a global search for EKC/KEOPS interactants in human cell line identified a fifth member of the human complex called C14ORF142 (Wan et al. 2016b). Despite low sequence conservation between C14ORF142 and Gon7 these proteins share several biophysical and biochemical properties suggesting that they are orthologous. The structure of the Kae1 from *Pyrococcus abyssi* with a bound ATP analog revealed a bilobal architecture belonging to the ASKHA/HALF (Acetate and Sugar Kinases/Hsc70/Actin) superfamily and an iron ion in between the two lobes (Hecker et al. 2007; Hecker et al. 2008). Bud32, also called PRPK (p53-related protein kinase) received much attention in the past because it was shown to interact with and phosphorylate the human tumor suppressor p53 protein (Abe et al. 2001; Facchin et al. 2003). Intriguingly, Bud32 lacks the conventional structural elements necessary for the

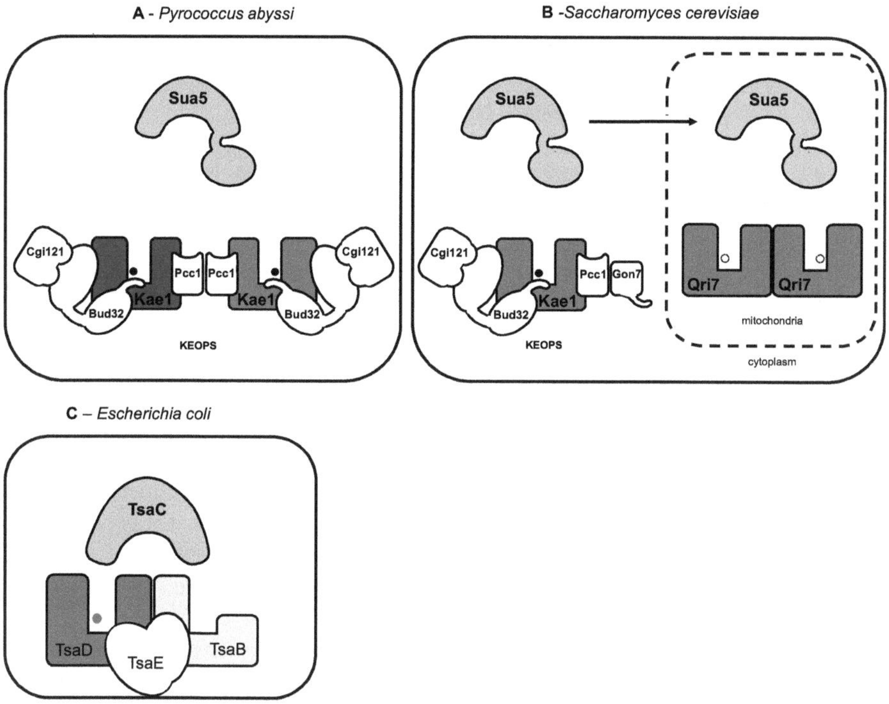

Fig. 8.2 Diversity of the t^6A_{37} synthetases in the three domains of life. In all organisms, the core t^6A_{37} synthetase is made of the two universal enzyme families: Sua5/TsaC and Kae1/TsaD/Qri7 (in *grey*). Sua5 proteins contain an additional C-terminal domain of unknown function which is missing in TsaC homologs. In archaea and eukaryotes (**a** and **b**, respectively), Kae1 forms together with Bud32, Pcc1 and Cgi121 the EKC/KEOPS complex (Endopeptidase-like Kinase Chromatin-associated/Kinase Endopeptidase and Other Proteins of Small size). In *S. cerevisiae* the additional protein Gon7, found only in fungi, interacts with Pcc1. In *P. abyssi* the EKC/KEOPS complex dimerizes via Pcc1 homo-dimerization. The iron ion (indicated by a *black dot*) in the active site of Kae1 is essential for its activity. The mitochondrial ortholog of TsaD, Qri7, requires no accessory proteins and forms homodimers. Metal ion is found in the active site of Qri7 but its chemical nature remains unknown. Because mitochondria lacks the gene for Sua5/TsaC, the Sua5 protein has to be imported (*black arrow*) from the cytoplasm into mitochondria to complete the mitochondrial t^6A_{37} synthetase. (**c**) In *Escherichia coli* TsaD forms a stable dimer with TsaB, an inactive paralog of TsaD. The binding of the bacterial protein TsaE, at the interface of the TsaD-TsaB dimer, is ATP dependent and required for the bacterial complex to be active. A zinc ion (indicated by a *grey dot*), bound to the active site of TsaD, is essential for its activity

substrate recognition by the canonical Ser/Thr kinases as well as a lysyl residue that participates in the catalysis (Mao et al. 2008; Facchin et al. 2002).

The model structure of the EKC/KEOPS complex (Kae1/Bud32/Pcc1/Cgi121) was derived from a series of overlapping structures and showed a linear arrangement where Pcc1 interacts with Kae1 which interacts with Bud32 which in turn interacts with Cgi121 (Mao et al. 2008). The deletion mutants of *bud32* and *pcc1* showed decreased levels of t^6A_{37} in *S. cerevisiae* thus providing the functional link between EKC/KEOPS and t^6A_{37} synthesis (Daugeron et al. 2011; El Yacoubi et al. 2011; Srinivasan et al. 2011). In line with these results, t^6A_{37} synthetic reaction was reconstituted *in vitro* using purified Sua5 and EKC/KEOPS proteins from *Pyrococcus abyssi* and *S. cerevisiae* (Perrochia et al. 2013a).

In addition to Kae1, eukaryotes carry a second version of this enzyme, named Qri7 in *S. cerevisiae*, which operates in mitochondria. The phylogenetic analysis of the TsaD/Kae1/Qri7 family of proteins showed that Qri7 orthologs cluster together with the TsaD bacterial counterparts indicating that they originate from ancestral bacterial-type of organism (Hecker et al. 2007). It seems however that the ancestral bacterial gene encoding TsaC/Sua5 proteins has been lost in eukaryotic cells since only the nuclear copy of *SUA5* gene is found in sequenced genomes of eukaryotes. This gene was found to have two alternative translation starting sites yielding two different polypeptides, one that is cytoplasmic and a second one which is exported to mitochondria (Thiaville et al. 2014a). Interestingly, global subcellular localization proteomic studies in *S. cerevisiae* failed to detect any of the KEOPS complex proteins in the mitochondria suggesting that Qri7 may act alone (Ghaemmaghami et al. 2003; Huh et al. 2003). This hypothesis was confirmed since it was possible to reconstitute *in vitro* the t^6A_{37} synthetic reaction by combining only Sua5 and Qri7 indicating that mitochondria possess a minimalist t^6A_{37} synthetase (Thiaville et al. 2014a; Wan et al. 2013).

The ubiquitous distribution of t^6A_{37} in the modern organisms strongly suggests that this modification was already present in the Last Universal Common Ancestor (LUCA) of the three domains of life indicating that this ancestral organism already possessed rather sophisticated translational apparatus. The minimal t^6A_{37} synthetase of mitochondria could mimic the ancestral synthetase that was used by the LUCA to synthesize t^6A_{37}. In this scenario, different set of accessory proteins was acquired in the course of evolution in the branches leading to bacteria or archaea/eukaryotes to yield contemporary DEZ or EKC/KEOPS complexes, respectively. Interestingly, this scenario implies that the root of the tree of life, the position of which is a matter of a hot debate, lies between Bacteria and a clade grouping Archaea and Eukarya (Arkarya *sensu* Forterre 2015) as initially proposed by Carl Woese. Alternatively, one could imagine that LUCA already possessed a set of accessory proteins, say DEZ complex, and that this set was replaced by a set of non-homologous proteins in the branch leading to archaea/eukaryotes.

8.3.2 Reaction Mechanism for the Synthesis of N^6-Threonylcarbamoyladenosine

Despite the differences in their protein composition, the three types of the known t^6A_{37} synthetases (bacterial, archaeal/eukaryotic and mitochondrial) use the same global reaction mechanism for the formation of t^6A_{37}. This reaction proceeds in two consecutive steps: first, TsaC/Sua5 enzymes catalyze the condensation of threonine, ATP and CO_2/bicarbonate into an activated adenylate intermediate threonylcarbamoyl-AMP (TC-AMP); and second, the TC-AMP is taken up by TsaD/Kae1/Qri7 enzymes which transfer the threonylcarbamoyl moiety to the tRNA acceptor molecule (Fig. 8.3) (Deutsch et al. 2012; Perrochia et al. 2013a; Lauhon 2012; Wan et al. 2013; Thiaville et al. 2014a).

The reaction intermediate TC-AMP was shown to be unstable under physiological conditions (pH 7.5, 37 °C, 2 mM $MgCl_2$), with a half-life of 3.5 min, suggesting that in the cell TC-AMP needs to be protected from hydrolysis. Direct channeling of TC-AMP from the active site of Sua5 to the active site of Kae1 was proposed as transfer mechanism, based on the observation that the rate of the first step of the reaction is slightly higher than the rate of the overall reaction (Deutsch et al. 2012; Lauhon 2012). Consistent with this hypothesis, the t^6A_{37} formation is much less efficient if Sua5 and Qri7 are separated physically by a semipermeable membrane that allows the diffusion of the small substrate molecules but not that of proteins or tRNA (Wan et al. 2013). Stable interaction between Sua5 and Kae1/Qri7 proteins could so far not be detected despite several attempts suggesting that the interaction (if it exists) is transient (Perrochia et al. 2013a).

Remarkably, the Sua5/TsaC family of enzymes consists of two distinct enzyme versions: TsaC proteins are single domain proteins whereas Sua5 proteins are

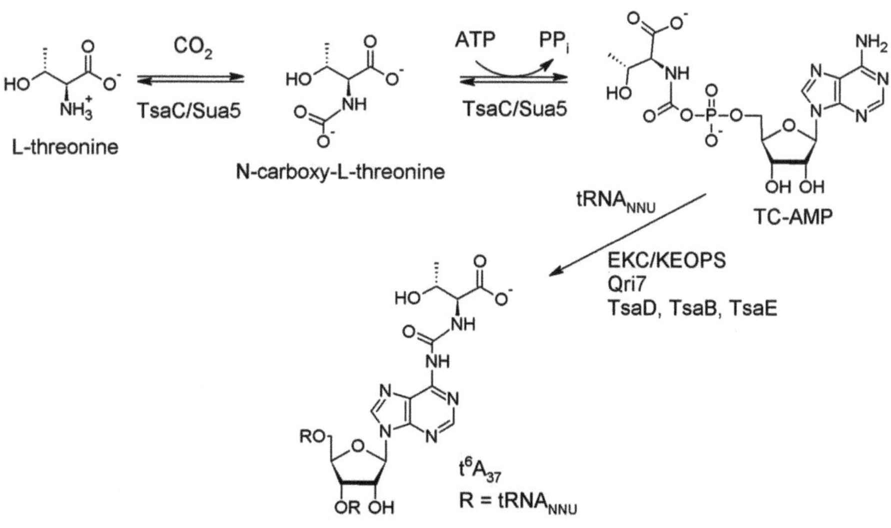

Fig. 8.3 Proposed reaction mechanism for the biosynthesis of t^6A_{37}. The initial step in the reaction is catalyzed by TsaC/Sua5 protein family. It is proposed that these enzymes directly catalyze the formation of N-carboxy-L-threonine by shifting the equilibrium in favor of this product. This intermediate is subsequently positioned to form the adenylate by attack on the alpha phosphate of ATP to give the product TC-AMP. This unstable intermediate is subsequently transferred by an unknown mechanism to EKC/KEOPS in archaea/eukaryotes or Qri7 in mitochondria or ternary bacterial complex composed of TsaD, TsaB and TsaE in bacteria. These proteins bind tRNA (NNU) and transfer the threonyl-carbamoyl moiety in ATP independent fashion to the N6 of adenosine 37. Figure adapted from Lauhon (2012)

longer and contain a TsaC-like catalytic domain linked to a C-terminal extension of about 100 amino acids (Thiaville et al. 2014b). It is currently unknown whether the C-terminal domain of Sua5 proteins functions in t^6A_{37} synthesis. No clear phylogenetic or lifestyle pattern could be established that could explain the existence of two different enzymes for the formation of TC-AMP (Thiaville et al. 2014b). The overall structure of the N-terminal TsaC-like domain of Sua5 protein from *Sulfolobus tokodaii* is very similar to that of TsaC of *E. coli* and it resembles a baseball glove with a central concave cavity lined with a positive electrostatic potential (Agari et al. 2008; Kuratani et al. 2011; Teplova et al. 2000). The reexamination of the structure of Sua5 from *S. tokodaii* revealed the presence of TC-AMP in the central cavity (Parthier et al. 2012). The strictly conserved motif $K_xR_x(\sim50)S_xN$ was shown to be essential for the activity and involved in the ATP binding (El Yacoubi et al. 2009; Kuratani et al. 2011). The studies of the TsaC protein of *E. coli* in solution using nuclear magnetic resonance (NMR) indicated that the binding of L-threonine may be required for the productive binding of ATP (Harris et al. 2015). This hypothesis is supported by the observation that the low ATPase activity of Sua5 protein of *P. abyssi* is significantly boosted up by the addition of L-threonine in the reaction mixture (Perrochia et al. 2013a). Several studies have shown that *E. coli* TsaC selectively binds hypomodified tRNA (El Yacoubi et al. 2009; Harris et al. 2011; Teplova et al. 2000), but this seems not to be the case for the Sua5 proteins of *P. abyssi* and *S. cerevisiae* (Perrochia et al. 2013a) indicating that functional differences may exist between TsaC and Sua5 proteins.

The reaction leading to the formation of TC-AMP requires the formation of two covalent bonds, one between the threonine and CO_2/HCO_3- and one between the CO_2/HCO_3- and ATP molecule. The results of kinetic studies of *in vitro* TC-AMP production by Sua5 from *Bacillus subtilis* suggested that this enzyme directly catalyzes N-carboxy-L-threonine formation by shifting the equilibrium towards this product (Lauhon 2012). Lauhon proposed that the role of the Sua5 in the reaction is to position the two substrates next to each other in the active site and provide favorable environment for the reaction to occur. Once formed, the N-carboxy-L-threonine could be correctly positioned to form the adenylate by attack on the alpha phosphate of the ATP molecule to give the final product TC-AMP as hypothesized in a structural work on HypF protein (Petkun et al. 2011).

A substantial amount of comparative structural and biochemical evidence strongly indicates that the Kae1/TsaD/Qri7 proteins bind the TC-AMP molecule and transfer the TC moiety to tRNA. The comparison of the Kae1 structure of *P. abyssi* complexed to AMP-PNP with the structures of two enzymes, HypF and TobZ, which catalyze the transfer of carbamoyl moiety from carbamoyl-adenylate to the final acceptor molecule, revealed that the nucleotide binding residues in the active sites of these proteins are conserved (Hecker et al. 2007; Petkun et al. 2011; Parthier et al. 2012). The mutation of these residues in Kae1, TsaD or Qri7 proteins results in a total abolishment of t^6A_{37} synthetic activity which is consistent with their role in binding of TC-AMP and catalysis of the TC transfer to tRNA (Wan et al. 2013; Mao et al. 2008; Perrochia et al. 2013b). Common feature of all Kae1/

TsaD/Qri7 family members is the conservation of two histidine residues which participate in the coordination of an iron atom (Kae1) or zinc atom (TsaD) in the active site. The mutation of these residues results in a severe growth phenotype, comparable to that of a *kae1* null mutant, in *S. cerevisiae* cells and leads to 90% of loss of activity *in vitro* for the *P. abyssi* t^6A_{37} synthetase (Mao et al. 2008; Perrochia et al. 2013b). These results indicated that the metal ion is essential for the biosynthesis of t^6A_{37} but its exact role in the reaction remains to be established.

8.3.3 The Role of the Partner Proteins in the Synthesis of t^6A_{37}

The catalysis of the t^6A_{37} synthetic reaction by the different synthetases relies *stricto sensu* on the action of the two universal protein families, TsaC/Sua5 and TsaD/Kae1/Qri7. This is exemplified by the minimal mitochondrial t^6A_{37} synthetase composed of only Sua5 and Qri7 (Wan et al. 2013; Thiaville et al. 2014a). This observation leads to one of the most intriguing questions concerning the t^6A_{37} synthetic pathway: are the accessory proteins in the KEOPS complex required for the t^6A_{37} synthesis and if so what is their function in this process?

The role of the partner proteins in the EKC/KEOPS complex of *P. abyssi* was studied in some detail (Perrochia et al. 2013b). It was demonstrated *in vitro* that Pcc1, Kae1 and Bud32 form a minimal functional unit while the Cgi121 stimulates the reaction probably by stabilizing the complex. The latter hypothesis is based on the observation that a stable trimeric complex Pcc1-Kae1-Bud32 could not be isolated suggesting that the binding of Cgi121 to Bud32 provokes a chain reaction of conformational changes necessary for the formation of the KEOPS complex. It appears therefore that two different strategies were developed to stabilize the complexes that catalyze the last step of the t^6A_{37} synthesis, i.e. the bacterial synthetases use the binding of ATP, whereas the archaeal/eukaryal systems use a protein, Cgi121.

The occurrence of a Ser/Thr protein kinase in the KEOPS complex was puzzling since the t^6A_{37} synthetic reaction in principle does not require such enzymatic activity. Consistent with this notion, Bud32 was shown *in vitro* to hydrolyze ATP to ADP and inorganic phosphate which were released in the reaction medium (Perrochia et al. 2013b). Thus, Bud32 acts as a P-loop ATPase and not as a kinase when bound to Kae1 in the EKC/KEOPS complex. Interestingly, another atypical RIO-type kinase called Rio2 exhibits robust ATPase activity, which is required for ribosome biogenesis in yeast (Ferreira-Cerca et al. 2012). It remains to be investigated if other RIO-type orthologs with non-canonical kinase signatures follow the same trend or if Bud32 and Rio2 are exceptional.

The EKC/KEOPS complex containing a catalytic mutant of Bud32 lost completely the t^6A_{37} synthetic activity indicating that the ATPase activity of Bud32 is essential for the t^6A_{37} synthesis. It was demonstrated that the transfer of the threonylcarbamoyl moiety to tRNA does not require ATP hydrolysis, therefore the need for the ATPase activity of the EKC/KEOPS complex remains currently unexplained (Lauhon 2012). Intriguingly, the bacterial ternary complex TsaD-TsaB-TsaE also possess the ATPase activity via the TsaE protein, however this activity is not required for the t^6A_{37} synthesis (Deutsch et al. 2012; Zhang et al. 2015b).

The last step of the t^6A_{37} synthesis requires the binding of the tRNA to the EKC/KEOPS complex or the bacterial ternary complex. Kae1 of *P. abyssi* binds tightly to tRNA in EMSA essays, however, the binding profile suggest the formation of aggregates. The addition of Pcc1 results in a binding pattern almost identical to the one determined for the whole EKC/KEOPS complex of *P. abyssi* suggesting that Pcc1-Kae1 complex forms the tRNA binding core. The binding of the tRNA seems to proceed in cooperative manner whereby the binding of one molecule of tRNA facilitates the binding of a second one (Perrochia et al. 2013a). This feature may be related to the fact that EKC/KEOPS complex of *P. abyssi* forms dimers in solution whereby Pcc1 functions as a dimerization module (see Sect. 8.3.4).

Based on these biochemical data and the previous structural work a tentative model was proposed for the function of the KEOPS complex in the synthesis of t^6A_{37} (Perrochia et al. 2013b) (Fig. 8.4).

8.3.4 Formation of Dimeric Complexes Is a Shared Characteristic of Kae1/TsaD/Qri7 Family of Proteins

Dimerization of the Kae1/TsaD/Qri7 proteins is a functionally important leitmotif found in all t^6A_{37} synthetases. However, the way the dimerization is achieved is not conserved which raises the critical question of what purpose the divergent modes of dimerization serve in the t^6A_{37} biosynthetic mechanism.

Qri7 proteins form homodimers in solution, TsaD proteins form heterodimers with their inactive paralog TsaB and Kae1 proteins dimerize indirectly via Pcc1 dimerization module in archaea (but not in yeast) and this mode of dimerisation is conserved in humans (Wan et al. 2013; Handford et al. 2009; Mao et al. 2008; Costessi et al. 2012). Remarkably, despite different modes of dimerisation all Kae1/TsaD/Qri7 proteins engage the same surface for the monomer-monomer interaction suggesting that this interaction is functionally important (Wan et al. 2013). In line with this hypothesis, the preservation of the dimerization interface of the Qri7 protein was found to be critical for *in vitro* synthesis of t^6A_{37} and to support yeast growth *in vivo* (Wan et al. 2013). However, in the case of TsaD-TsaB dimer *in vitro* and *in vivo* data are less congruent: *tsaD* null mutant could be functionally complemented only by a coexpression of heterologous TsaD-TsaB pairs indicating that the capacity to form the heterodimer is essential *in vivo* (Wan et al. 2013; El Yacoubi et al. 2011). On the other hand, Zhang and colleagues reported that the mutated ternary complex with disrupted interaction between TsaD and TsaB still

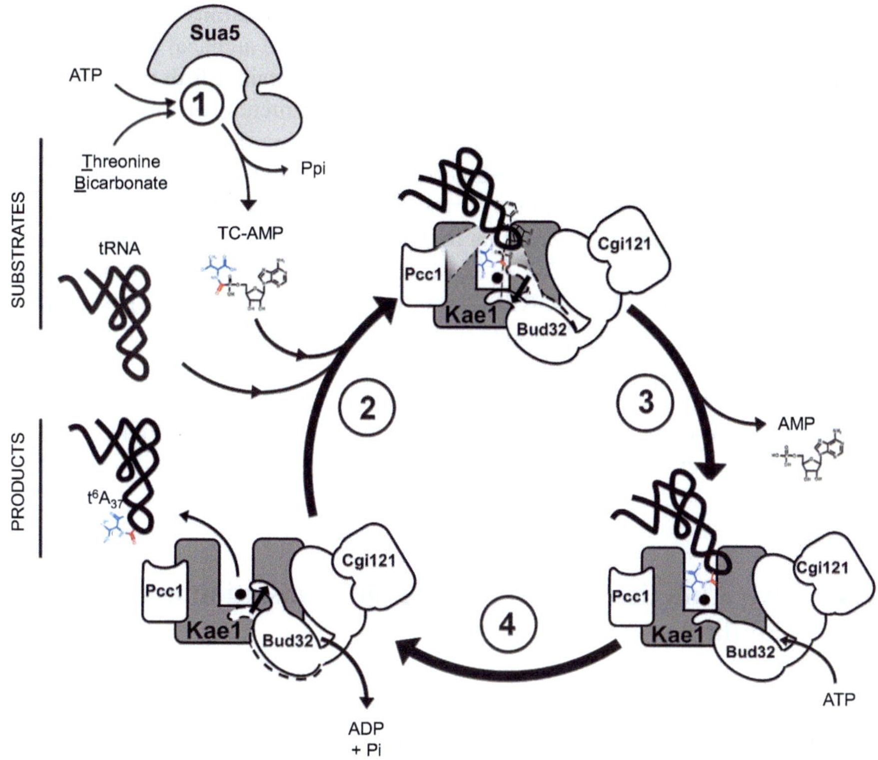

Fig. 8.4 Putative mechanism for the catalysis of the last step in the biosynthesis of t^6A_{37} modification by the EKC/KEOPS complex. (1) Formation of TC-AMP: Sua5 catalyzes the condensation of threonine, bicarbonate and ATP leading to the formation of an unstable TC-AMP intermediate and release of inorganic pyrophosphate. (2) Binding of tRNA and TC-AMP to EKC/KEOPS: TC-AMP binds into the active site of Kae1 and interacts directly with the iron atom via threonyl part of the molecule. Binding of tRNA to the complex induces conformational changes in the complex, including the movement of the C-terminal tail of Bud32 (indicated with an *arrow*). Pcc1 and Kae1 are involved in the major part of contacts (*gray triangles*) with tRNA, whereas Bud32 participates in binding of tRNA via C-terminal tail. Anticodon loop carrying the target nucleotide A37 is positioned at the entrance to the active site cavity of Kae1 next to the TC-AMP intermediate and iron atom. (3) Transfer of L-threonyl-carbamoyl to tRNA: Threonyl-carbamoyl moiety is transferred to A37 of substrate tRNA in an ATP-independent fashion and AMP is released. (4) Release of t6A37-modified tRNA: ATP hydrolysis catalyzed by Bud32 powers the conformational changes, in particular motion of the C-terminal tail of Bud32 (indicated by *arrow*), which leads to the dissociation of modified tRNA from the EKC/KEOPS complex. The resulting KEOPS complex is competent for another catalytic cycle. Figure from Perrochia et al. (2013b)

retained 60% of the t^6A_{37} synthetic activity *in vitro* (Zhang et al. 2015b). Even more confusing data were reported for the EKC/KEOPS complex: the mutations at the Pcc1 dimerization interface provoke a severe growth defect in yeast which is compatible with a total loss of t^6A_{37} synthetic activity (Mao et al. 2008). However, corresponding mutations are totally neutral for the *in vitro* activity of the archaeal

EKC/KEOPS (Wan et al. 2016a)! These conflicting observations may point to the existence of intrinsic mechanistic differences between the archaeal and eukaryotic complexes. Indeed, the yeast and human complexes contain a fifth member Gon7 and C14ORF142, respectively. Notably, the crystal structure of the yeast Gon7-Pcc1 heterodimer revealed that Pcc1 engaged Gon7 using the same surface previously shown to mediate Pcc1 homodimerization (Zhang et al. 2015a). Indeed, the stoichiometry of 1:1 was determined for the heterodimer Pcc1-Gon7 indicating that the binding of Gon7 to Pcc1 prevents the dimerization of the yeast EKC/KEOPS complex. Analysis of human EKC/KEOPS complex revealed that it formed monomers in solution only if C14ORF142 was part of the complex suggesting that, similar to Gon7, C14ORF142 binds to LAGE3 (Pcc1 ortholog) and impairs its capacity to dimerize (Wan et al. 2016b).

Despite some paradoxical data, the dimerization capacity of the Qri7/Kae1/TsaD seems to be invariably essential for the t^6A_{37} synthesis *in vivo*. The challenging task remains to uncover the function of this peculiar feature for the biosynthesis of t^6A_{37}.

8.4 t^6A_{37} Function *In Vivo*: Effect on Translation and Cell Viability

Once the function of the TsaC/Sua5 and Kae1/TsaD/Qri7 protein families and their partners in the synthesis of t^6A_{37} was established, it became possible to study the effect of this modification on translation *in vivo* and test how the loss of t^6A_{37} affects cell viability.

The early studies of the function of Sua5 identified this protein as a suppressor gene in a screen for aberrant upstream AUG codon start of *CYC1* gene (Na et al. 1992). *SUA5* gene (suppressor of upstream ATG) could partially restore the production of Cyc1 protein without altering the transcription start of the mRNA indicating that Sua5 was implicated in fidelity of start site selection during translation initiation. In the light of the new findings, it can be now proposed that the loss of *SUA5* and thereby the absence of t^6A_{37} on $tRNA_{Meti}$ (CAU) would have adverse effect on the efficiency of base pairing between the start codon and the anticodon. Failure to bind the first AUG codon (the upstream one) would result in the continued scanning for the following start codon thus accounting for the partial rescue of the Cyc1 synthesis (Lin et al. 2010). In agreement with this proposal several studies showed that the *KAE1* or *SUA5* deletion strains of *S. cerevisiae* exhibited increased leaky scanning through upstream start codons. In addition to this defect, the loss of these genes increased the frequency of +1 frameshifting and read-through of stop codons by two- to threefold as compared to controls (Lin et al. 2010; El Yacoubi et al. 2011; Daugeron et al. 2011).

Recently, ribosome profiling approach was used to measure the genome scale impact of the loss of t^6A_{37} modification on the translation in *S. cerevisiae* (Thiaville et al. 2016). This study showed that the translation ambiguities (increased translation initiation at upstream AUG codons, increased frameshifting) were significantly more frequent in the *SUA5* null mutant versus the WT strain, however, these were not global catastrophic alterations of the reading frame but rather specific events at discreet sequences or codons. One of the major roles of t^6A_{37} was found to be in homogenizing the elongation process by slowing down the elongation rate for codons decoded by high abundance tRNAs and I_{34}:C_3 pairs while accelerating the elongation rate of rare tRNAs and G_{34}:U_3 pairs (Thiaville et al. 2016). The global impact of the loss of t^6A_{37} was recently investigated in a multicellular organism, the fly *Drosophila melanogaster* (Ibar et al. 2013; Rojas-Benitez et al. 2015; Lin et al. 2015). These studies found that the t^6A_{37} levels in this organism determine the potential for cell growth. They further exposed a regulatory relationship between the availability of t^6A_{37} modified tRNAs and target of rapamycin (TOR) kinase activity which acts as the central regulator of cell growth in eukaryotes. Strongly proliferating tissues were greatly affected by the loss of t^6A_{37} while non-proliferating tissues were less affected highlighting tissue-specific requirements in metazoal context. In *Plasmodium falciparum* $Kae1^{api}$ protein functioning in the apicoplast, a chloroplast derived organelle, was reported to be essential for blood stage parasite development (Mallari et al. 2014).

Although serious translational defects could be identified *in vivo*, the absence of t^6A_{37} is not lethal for the cells of *S. cerevisiae* in laboratory conditions. The mutant cells containing no detectable levels of t^6A_{37} could be obtained, however these are seriously affected and exhibit very slow growth phenotype with four- to fivefold decrease in doubling time (MC Daugeron, personal communication). This phenotype could be assigned to the knockout of the *SUA5*, *PCC1*, *KAE1*, *BUD32* genes and fungi specific *GON7* gene (Mao et al. 2008; Kisseleva-Romanova et al. 2006; Downey et al. 2006; El Yacoubi et al. 2009; Srinivasan et al. 2011). The *CGI121* deletion was reported to have little effect on t^6A_{37} levels in yeast and these mutant cells showed WT growth rate (Downey et al. 2006; Srinivasan et al. 2011). In view of these data it seems reasonable to assume that the cells missing the t^6A_{37} modification would be most likely counter-selected in authentic environmental growth conditions.

Only two studies investigated the essentiality of t^6A_{37} synthetic genes in archaea. All t^6A_{37} synthesis genes were found to be essential in the methanogenic archaeon *Methanococcus maripaludis* using genome wide transposon mutagenesis (Sarmiento et al. 2013). In halophile *Haloferax volcanii* Kae1-Bud32 fusion protein and Cgi121 were found to be essential while Pcc1 deletion mutants displayed reduced growth and slightly lower levels of t^6A_{37} modification (Naor et al. 2012). Why is the t^6A_{37} modification essential in these archaea but not in yeast? One explanation, proposed by Naor and colleagues, is that C_{34} (wobbling base in the anticodon) in $tRNA_{Ile}$ (CAU), must be modified to agmatidine (agm^2C) in Archaea (and not in Eukaryotes). Agmatidine allows the modified $tRNA_{Ile}$ (CAU) to decode AUA isoleucine codons and not methionine AUG codons (Mandal et al. 2010; Ikeuchi et al. 2010). This modification is synthetized by the action of TiaS proteins which were shown to be essential in *H. volcanii* (Blaby et al. 2010). If t^6A_{37}

modified tRNA is required for the synthesis of agmatidine then the lack of t^6A_{37} would indirectly become essential. This appealing hypothesis remains to be tested.

8.5 Are t^6A_{37} Archaeal/Eukaryotic Synthetic Machineries Multifunctional?

Deletion of non-essential t^6A_{37} synthetic genes in archaea and eukaryotes yielded mutant strains which grew slowly and, intriguingly, exhibited a plethora of phenotypes unrelated to direct translational defects. *H. volcanii pcc1* deletion mutant displayed higher DNA content per cell and abnormally high levels of Advanced Glycated End Products (AGEs) which are highly stable toxic glycated proteins and lipids involved in various human degenerative pathologies and aging (Naor et al. 2012). An impressive number of different phenotypes were recorded in *S. cerevisiae* mutant cells: shorter telomeres (Wan et al. 2013; Downey et al. 2006), modified transcription patterns (Kisseleva-Romanova et al. 2006), increased chromosome instability (Ben-Aroya et al. 2008), respiration deficiency (Hecker et al. 2009; Oberto et al. 2009; Lin et al. 2010) protein folding defects, sensitivity to stress (heat, ethanol, salt), increased AGEs levels and sensitivity to TOR pathway inhibitors (Thiaville et al. 2016). These observations raised an intriguing question of whether the t^6A_{37} synthetic genes are directly involved in these processes in the cell or are all these pleiotropic phenotypes merely indirect consequences of translational defects.

Similar pleiotropic phenotypes (respiration defect, shortening of telomeres) were linked to the loss of another tRNA modification 5-methoxycarbonylmethyliouridine (mcm^5U) and its thiolated derivative (mcm^5s^2U) found at position 34 (wobble nucleoside) of several tRNAs (Bjork et al. 2007; Chen et al. 2011). Interestingly, the overexpression of a single $tRNA_{Lys}$ (UUU) (which is also modified by t^6A_{37}) suppressed all the mcm^5s^2U phenotypes. Thiaville and colleagues recently showed that the overexpression of individual ANN-tRNAs (including $tRNA_{Lys}$ (UUU) did not restore the slow growth of *SUA5* deletion strain (Thiaville et al. 2016). They also excluded the possibility that t^6A_{37} modification is required for the synthesis of mcm^5s^2U or vice-versa. The loss of t^6A_{37} thus appears to have more complex consequences than those of the loss of mcm^5s^2U. Even if this study identified specific pathways (such as arginine biosynthesis) or proteins that seem to be more affected by the loss of t^6A_{37}, it remains difficult to assign specific molecular processes to the observed phenotypes.

Presently we therefore cannot unambiguously exclude the possibility that the proteins required for t^6A_{37} synthesis have an expanded role outside of its known tRNA-modifying activity. There is a growing body of evidence that this may indeed be the case, at least in eukaryotes. In yeast, the complementation of *KAE1* null mutant by the mitochondrial ortholog *QRI7* (that functions without partner proteins) yields wild type levels of t^6A_{37} but the strain exhibits short telomeres phenotype suggesting that KEOPS complex (or some of its components) may be involved in telomere homeostasis (Wan et al. 2013). In human cells, it was shown that the tumor antigen PRAME (PReferentially expressed Antigen in MElanoma) specifically interacts with OSGEP (=KAE1) and LAGE3 (=PCC1) and that it can recruit a Cul2 ubiquitin ligase to KEOPS. It was further demonstrated that KEOPS subunits associate with PRAME target sites on chromatin (Costessi et al. 2012). In the parasite *P. falciparum* a high number of protein interactants was identified for $Kae1^{api}$ linking it to a ribosome-associated complex (Mallari et al. 2014). Finally, it was recently shown in rats that PRPK (=BUD32) is important for the axonal elongation of neurons which is promoted by Rab35 protein. PRPK-dependent Rab35 phosphorylation induces the degradation of this protein via ubiquitine/proteasome pathway. Interestingly, the PRPK activity is regulated by microtubule-associated protein 1B (MAP 1B) which interacts with PRPK and prevents it from phosphorylating Rab35 (Villarroel-Campos et al. 2016).

8.6 Conclusion and Outlook

In the late 1960s when the t^6A_{37} modification was discovered in tRNAs decoding ANN codons its importance for the decoding accuracy was immediately predicted based on its strategic position next to the anticodon. The early biochemical studies and later structural work reinforced this prediction by showing that t^6A_{37} enhanced the codon-anticodon binding and maintained the anticodon loop in an open conformation. Based on this data it was deduced that t^6A_{37} would be important for maintaining the translational reading frame and facilitating the binding of the tRNA on the ribosome. The recent discovery of the t^6A_{37} synthetic genes in the three domains of life has finally allowed to extend this work and investigate the function of t^6A_{37} *in vivo*. In eukaryotic model organism *S. cerevisiae* the predictions based on *in vitro* data were confirmed but it was also revealed that the loss of t^6A_{37} has no catastrophic global effect on the proteome and that the major role of t^6A_{37} is in fact in regulating elongation rate at specific codons. In multicellular organism *Drosophila melanogaster* it was revealed that strongly proliferating tissues have higher requirements for t^6A_{37} and that t^6A_{37} levels determine the cell growth potential. From wider perspective, this indicates that t^6A_{37}-modified tRNAs have a relevant role in modulating protein expression and in consequence, the study of the relationship of t^6A_{37} modification with protein synthesis regulation in physiological and stress conditions as well as in disease, may emerge as a fertile area of research.

The genes encoding the EKC/KEOPS complex and its function in t^6A_{37} biosynthesis are highly conserved in Archaea and Eukaryotes. Even so, several significant differences can be observed between archaeal and eukaryotic EKC/KEOPS complexes. One of those is the occurrence in fungi of the fifth member Gon7 which was found to be required for the t^6A_{37} synthesis but its role is unknown. Sequence similarity searches have, so far, identified no homologs in Archaea or other Eukaryotes suggesting this to be fungi-specific feature. This indicates that additional domain-specific functionalities may have arisen in course of evolution. How much variability and "invention" occurred in course of evolution remains to be investigated by isolating and characterizing EKC/KEOPS complexes from different archaeal and eukaryotic lineages.

Even if significant progress has been made, the reaction mechanism leading to the synthesis of the t^6A_{37}, is far from being entirely understood and will require further biochemical and structural characterization. The complex reaction performed by the TsaC/Sua5 enzymes which need to catalyze the formation of two covalent bonds and combine three substrates into an activated TC-AMP intermediate presently remains hypothetical. Another open question is the functional difference between the two distinct variants of TC-AMP producing enzymes, a longer Sua5 and a shorter TsaC where only the TsaC-like domain is homologous. Whether the C-terminal part of Sua5 proteins which is missing in TsaC counterparts, is functionally linked to the t^6A_{37} synthesis or involved in some other processes in cells or both remains to be elucidated. It is presently unclear how the transfer of the unstable TC-AMP intermediate from TsaC/Sua5 proteins to the EKC/KEOPS complex is achieved. A direct contact between the two catalytic subunits TsaC/Sua5 and TsaD/Kae/Qri7 and channeling of the TC-AMP between the two active sites is suspected but not yet proven. Finally, almost nothing is known about the mode of tRNA binding by the KEOPS complex or Qri7 proteins: how do accessory proteins participate, how is the substrate tRNA recognized and distinguished from the non-substrate tRNA, how does Qri7 achieves tRNA binding and recognition in absence of accessory proteins, is Bud32 ATPase activity related to tRNA binding?

With only one exception all *in vivo* work concerning the archaeal/eukaryotic t^6A_{37} synthetic machinery was done in eukaryotic model organisms. This led to a discovery of several interactions with proteins unrelated to t^6A_{37} synthesis. This strongly indicates that EKC/KEOPS and/or some of its subunits have expanded role outside of its known tRNA-modifying activity. Bud32 protein is particularly interesting in this context because it seems to act as a *bona fide* kinase in human and rat cells, but it functions as ATPase in the context of the EKC/KEOPS complex. These data raise the interesting possibility, that EKC/KEOPS complex contributes yet another layer of regulation in eukaryotes by controlling the amount of free Bud32 and thus its kinase activity. And how about archaea? Does archaeal Bud32 orthologs act as kinases and if so what are their substrate proteins? Although the exploration of the *in vivo* functions of archaeal t^6A_{37} machinery is still in its infancy, it is our feeling that this avenue of research reserves many surprises to come.

References

Abe Y, Matsumoto S, Wei S, Nezu K, Miyoshi A, Kito K, Ueda N, Shigemoto K, Hitsumoto Y, Nikawa J, Enomoto Y (2001) Cloning and characterization of a p53-related protein kinase expressed in interleukin-2-activated cytotoxic T-cells, epithelial tumor cell lines, and the testes. J Biol Chem 276(47):44003–44011

Agari Y, Sato S, Wakamatsu T, Bessho Y, Ebihara A, Yokoyama S, Kuramitsu S, Shinkai A (2008) X-ray crystal structure of a hypothetical Sua5 protein from Sulfolobus tokodaii strain 7. Proteins 70(3):1108–1111

Agris PF (2004) Decoding the genome: a modified view. Nucleic Acids Res 32(1):223–238. Review

Agris PF (2008) Bringing order to translation: the contributions of transfer RNA anticodon-domain modifications. EMBO Rep 9(7):629–635. https://doi.org/10.1038/embor.2008.104

Agris PF, Vendeix FA, Graham WD (2007) TRNA's wobble decoding of the genome: 40 years of modification. J Mol Biol 366:1–13

Allali-Hassani A, Campbell TL, Ho A, Schertzer JW, Brown ED (2004) Probing the active site of YjeE: a vital Escherichia coli protein of unknown function. Biochem J 384:577–584

Ben-Aroya S, Coombes C, Kwok T, O'Donnell KA, Boeke JD, Hieter P (2008) Toward a comprehensive temperature-sensitive mutant repository of the essential genes of Saccharomyces cerevisiae. Mol Cell 30:248–258

Bjork GR, Huang B, Persson OP, Bystrom AS (2007) A conserved modified wobble nucleoside (mcm5s2U) in lysyl-tRNA is required for viability in yeast. RNA 13(8):1245–1255

Blaby IK, Phillips G, Blaby-Haas CE, Gulig KS, El Yacoubi B et al (2010) Towards a systems approach in the genetic analysis of archaea: accelerating mutant construction and phenotypic analysis in Haloferax volcanii. Archaea 2010:426239

Cantara WA, Crain PF, Rozenski J, McCloskey JA, Harris KA, Zhang X, Vendeix FA, Fabris D, Agris PF (2011) The RNA Modification Database, RNAMDB: 2011 update. Nucleic Acids Res 39(Database issue):D195–D201. https://doi.org/10.1093/nar/gkq1028

Chen C, Huang B, Eliasson M, Ryden P, Bystrom AS (2011) Elongator complex influences telomeric gene silencing and DNA damage response by its role in wobble uridine tRNA modification. PLoS Genet 7(9):e1002258

Chheda GB, Hall RH, Mozejko J, Magrath DI, Schweizer MP, Stasiuk L, Taylor PR (1969) Aminoacyl nucleosides. VI. Isolation and preliminary characterization of threonyladenine derivatives from transfer ribonucleic acid. Biochemistry 8(8):3278–3282. No abstract available

Chheda GB, Hong CI, Piskorz CF, Harmon GA (1972) Biosynthesis of N-(purin-6-ylcarbamoyl)-L-threonine riboside. Incorporation of L-threonine in vivo into modified nucleoside of transfer ribonucleic acid. Biochem J 127(3):515–519

Costessi A, Mahrour N, Sharma V, Stunnenberg R, Stoel MA, Tijchon E, Conaway JW, Conaway RC, Stunnenberg HG (2012) The human EKC/KEOPS complex is recruited to Cullin2ubiquitin ligases by the human tumour antigen PRAME. PLoS One 7:e42822

Daugeron MC, Lenstra TL, Frizzarin M, El Yacoubi B, Liu X, Baudin-Baillieu A, Lijnzaad P, Decourty L, Saveanu C, Jacquier A, Holstege FC, de Crécy-Lagard V, van Tilbeurgh H, Libri D (2011) Gcn4 misregulation reveals a direct role for the evolutionary conserved EKC/KEOPS in the t6A modification of tRNAs. Nucleic Acids Res 39:6148–6160

Deutsch C, El Yacoubi B, de Crécy-Lagard V, Iwata-Reuyl D (2012) Biosynthesis of threonylcarbamoyl adenosine (t6A), a universal tRNA nucleoside. J Biol Chem 287 (17):13666–13673. https://doi.org/10.1074/jbc.M112.344028

Downey M, Houlsworth R, Maringele L, Rollie A, Brehme M, Galicia S, Guillard S, Partington M, Zubko MK, Krogan NJ, Emili A, Greenblatt JF, Harrington L, Lydall D, Durocher D (2006) A genome-wide screen identifies the evolutionarily conserved KEOPS complex as a telomere regulator. Cell 124(6):1155–1168

Dube SK, Marcker KA, Clark BF, Cory S (1968) Nucleotide sequence of N-formyl-methionyl-transfer RNA. Nature 218(5138):232–233

El Yacoubi B, Lyons B, Cruz Y, Reddy R, Nordin B, Agnelli F, Williamson JR, Schimmel P, Swairjo MA, de Crécy-Lagard V (2009) The universal YrdC/Sua5 family is required for the formation of threonylcarbamoyladenosine in tRNA. Nucleic Acids Res 37(9):2894–2909. https://doi.org/10.1093/nar/gkp152

El Yacoubi B, Hatin I, Deutsch C, Kahveci T, Rousset J-P, Iwata-Reuyl D, Murzin AG, de Crécy-Lagard V (2011) A role for the universal Kae1/Qri7/YgjD (COG0533) family in tRNA modification. EMBO J 30:882–893

El Yacoubi B, Bailly M, de Crécy-Lagard V (2012) Biosynthesis and function of posttranscriptional modifications of transfer RNAs. Annu Rev Genet 46:69–95. https://doi.org/10.1146/annurev-genet-110711-155641. Review

Elkins BN, Keller EB (1974) The enzymatic synthesis of N-(purin-6-ylcarbamoyl)threonine, an anticodon-adjacent base in transfer ribonucleic acid. Biochemistry 13(22):4622–4628

Facchin S, Lopreiato R, Stocchetto S, Arrigoni G, Cesaro L, Marin O, Carignani G, Pinna LA (2002) Structure-function analysis of yeast piD261/Bud32, an atypical protein kinase essential for normal cell life. Biochem J 364(Pt 2):457–463

Facchin S, Lopreiato R, Ruzzene M, Marin O, Sartori G, Götz C, Montenarh M, Carignani G, Pinna LA (2003) Functional homology between yeast piD261/Bud32 and human PRPK: both phosphorylate p53 and PRPK partially complements piD261/Bud32 deficiency. FEBS Lett 549 (1–3):63–66

Ferreira-Cerca S, Sagar V, Schäfer T, Diop M, Wesseling AM, Lu H, Chai E, Hurt E, LaRonde-LeBlanc N (2012) ATPase-dependent role of the atypical kinase Rio2 on the evolving pre-40S ribosomal subunit. Nat Struct Mol Biol 19(12):1316–1323. https://doi.org/10.1038/nsmb.2403

Forterre P (2015) The universal tree of life: an update. Front Microbiol 6:717. https://doi.org/10.3389/fmicb.2015.00717

Ghaemmaghami S, Huh W-K, Bower K, Howson RW, Belle A, Dephoure N, O'Shea EK, Weissman JS (2003) Global analysis of protein expression in yeast. Nature 425:737–741. https://doi.org/10.1038/nature02046

Grosjean H (2009) Nucleic acids are not boring long polymers of only four types of nucleotides: a guided tour. In: Grosjean H (ed) DNA and RNA modification enzymes: structure, mechanism, function and evolution. Landes Bioscience, Austin, TX, pp 1–18

Grosjean H, Söll DG, Crothers DM (1976) Studies of the complex between transfer RNAs with complementary anticodons. I. Origins of enhanced affinity between complementary triplets. J Mol Biol 103(3):499–519

Grosjean H, Sprinzl M, Steinberg S (1995) Posttranscriptionally modified nucleosides in tRNA: their locations and frequencies. Biochimie 77:139–141

Grosjean H, Gupta R, Maxwell ES (2008) Modified nucleotides in archaeal RNAs. In: Blum P (ed) Archaea: new models for prokaryotic biology. Horizon Scientific Press/Caister Academic Press, Norwich/Norfolk, pp 171–196

Handford JI, Ize B, Buchanan G, Butland GP, Greenblatt J, Emili A, Palmer T (2009) Conserved network of proteins essential for bacterial viability. J Bacteriol 191(15):4732–4749. https://doi.org/10.1128/JB.00136-09

Harris KA, Jones V, Bilbille Y, Swairjo MA, Agris PF (2011) YrdC exhibits properties expected of a subunit for a tRNA threonylcarbamoyl transferase. RNA 17(9):1678–1687. https://doi.org/10.1261/rna.2592411

Harris KA, Bobay BG, Sarachan KL, Sims AF, Bilbille Y, Deutsch C, Iwata-Reuyl D, Agris PF (2015) NMR-based structural analysis of threonylcarbamoyl-AMP synthase and its substrate interactions. J Biol Chem 290(33):20032–20043. https://doi.org/10.1074/jbc.M114.631242

Hecker A, Leulliot N, Gadelle D, Graille M, Justome A, Dorlet P, Brochier C, Quevillon-Cheruel-S, Le Cam E, van Tilbeurgh H, Forterre P (2007) An archaeal orthologue of the universal protein Kae1 is an iron metalloprotein which exhibits atypical DNA-binding properties and apurinic-endonuclease activity in vitro. Nucleic Acids Res 35(18):6042–6051

Hecker A, Lopreiato R, Graille M, Collinet B, Forterre P, Libri D, van Tilbeurgh H (2008) Structure of the archaeal Kae1/Bud32 fusion protein MJ1130: a model for the eukaryotic EKC/KEOPS subcomplex. EMBO J 27(17):2340–2351. https://doi.org/10.1038/emboj.2008.157

Hecker A, Graille M, Madec E, Gadelle D, Le Cam E, van Tilbergh H, Forterre P (2009) The universal Kae1 protein and the associated Bud32 kinase (PRPK), a mysterious protein couple probably essential for genome maintenance in Archaea and Eukarya. Biochem Soc Trans 37 (Pt 1):29–35. https://doi.org/10.1042/BST0370029

Huh WK, Falvo JV, Gerke LC, Carroll AS, Howson RW, Weissman JS, O'Shea EK (2003) Global analysis of protein localization in budding yeast. Nature 425(6959):686–691

Ibar C, Cataldo VF, Vásquez-Doorman C, Olguín P, Glavic A (2013) Drosophila p53-related protein kinase is required for PI3K/TOR pathway-dependent growth. Development 140 (6):1282–1291. https://doi.org/10.1242/dev.086918

Ikeuchi Y, Kimura S, Numata T, Nakamura D, Yokogawa T et al (2010) Agmatine-conjugated cytidine in a tRNA anticodon is essential for AUA decoding in Archaea. Nat Chem Biol 6:277–282

Ishikura H, Yamada Y, Murao K, Saneyoshi M, Nishimura S (1969) The presence of N-[9-(beta-D-ribofuranosyl)purin-6-ylcarbamoyl]threonine in serine, methionine and lysine transfer RNA's from Escherichia coli. Biochem Biophys Res Commun 37(6):990–995. No abstract available

Jackman JE, Alfonzo JD (2013) Transfer RNA modifications: nature's combinatorial chemistry playground. WIREs RNA 4:35–48. https://doi.org/10.1002/wrna.1144

Jühling F, Mörl M, Hartmann RK, Sprinzl M, Stadler PF, Pütz J (2009) tRNAdb 2009: compilation of tRNA sequences and tRNA genes. Nucleic Acids Res 37:D159–D162

Kerpedjiev P, Hammer S, Hofacker IL (2015) Forna (force-directed RNA): simple and effective online RNA secondary structure diagrams. Bioinformatics 31(20):3377–3379

Kisseleva-Romanova E, Lopreiato R, Baudin-Baillieu A, Rousselle JC, Ilan L, Hofmann K, Namane A, Mann C, Libri D (2006) Yeast homolog of a cancer-testis antigen defines a new transcription complex. EMBO J 25(15):3576–3585

Konevega AL, Soboleva NG, Makhno VI, Semenkov YP, Wintermeyer W, Rodnina MV, Katunin VI (2004) Purine bases at position 37 of tRNA stabilize codon-anticodon interaction in the ribosomal A site by stacking and Mg2+-dependent interactions. RNA 10(1):90–101

Körner A, Söll D (1974) N-(purin-6-ylcarbamoyl)threonine: biosynthesis in vitro in transfer RNA by an enzyme purified from Escherichia coli. FEBS Lett 39(3):301–306. No abstract available

Kuratani M, Kasai T, Akasaka R, Higashijima K, Terada T, Kigawa T, Shinkai A, Bessho Y, Yokoyama S (2011) Crystal structure of Sulfolobus tokodaii Sua5 complexed with L-threonine and AMPPNP. Proteins 79(7):2065–2075. https://doi.org/10.1002/prot.23026

Kurland CG (1992) Translational accuracy and the fitness of bacteria. Annu Rev Genet 26:29–50

Lauhon CT (2012) Mechanism of N6-threonylcarbamoyladenonsine (t(6)A) biosynthesis: isolation and characterization of the intermediate threonylcarbamoyl-AMP. Biochemistry 51 (44):8950–8963. https://doi.org/10.1021/bi301233d

Lin CA, Ellis SR, True HL (2010) The Sua5 protein is essential for normal translational regulation in yeast. Mol Cell Biol 30:354–363

Lin CJ, Smibert P, Zhao X, Hu JF, Ramroop J, Kellner SM, Benton MA, Govind S, Dedon PC, Sternglanz R, Lai EC (2015) An extensive allelic series of Drosophila kae1 mutants reveals diverse and tissue-specific requirements for t6A biogenesis. RNA 21(12):2103–2118. https://doi.org/10.1261/rna.053934.115

Machnicka MA, Milanowska K, Osman Oglou O, Purta E, Kurkowska M, Olchowik A, Januszewski W, Kalinowski S, Dunin-Horkawicz S, Rother KM, Helm M, Bujnicki JM, Grosjean H (2013) MODOMICS: a database of RNA modification pathways–2013 update. Nucleic Acids Res 41(Database issue):D262–D267. https://doi.org/10.1093/nar/gks1007

Machnicka MA, Olchowik A, Grosjean H, Bujnicki JM (2014) Distribution and frequencies of post-transcriptional modifications in tRNAs. RNA Biol 11(12):1619–1629. https://doi.org/10.4161/15476286.2014.992273. Review

Mallari JP, Oksman A, Vaupel B, Goldberg DE (2014) Kinase-associated endopeptidase 1 (Kae1) participates in an atypical ribosome-associated complex in the apicoplast of Plasmodium falciparum. J Biol Chem 289(43):30025–30039. https://doi.org/10.1074/jbc.M114.586735

Mandal D, Kohrer C, Su D, Russell SP, Krivos K et al (2010) Agmatidine, a modified cytidine in the anticodon of archaeal tRNA(Ile), base pairs with adenosine but not with guanosine. Proc Natl Acad Sci U S A 107:2872–2877

Mao DY, Neculai D, Downey M, Orlicky S, Haffani YZ, Ceccarelli DF, Ho JS, Szilard RK, Zhang W, Ho CS, Wan L, Fares C, Rumpel S, Kurinov I, Arrowsmith CH, Durocher D, Sicheri F (2008) Atomic structure of the KEOPS complex: an ancient protein kinase-containing molecular machine. Mol Cell 32(2):259–275. https://doi.org/10.1016/j.molcel.2008.10.002

Miller JP, Hussain Z, Schweizer MP (1976) The involvement of the anticodon adjacent modified nucleoside N-(9-(-D-ribofuranosyl)-purine-6-ylcarbamoyl)-threonine in the biological function of E. coli tRNAIle. Nucleic Acids Res 3:1185–1201

Miyauchi K, Kimura S, Suzuki T (2013) A cyclic form of N6-threonylcarbamoyladenosine as a widely distributed tRNA hypermodification. Nat Chem Biol 9(2):105–111. https://doi.org/10.1038/nchembio.1137

Morin A, Auxilien S, Senger B, Tewari R, Grosjean H (1998) Structural requirements for enzymatic formation of threonylcarbamoyladenosine (t6A) in tRNA: an in vivo study with Xenopus laevis oocytes. RNA 4(1):24–37

Murphy FV IV, Ramakrishnan V, Malkiewicz A, Agris PF (2004) The role of modifications in codon discrimination by tRNA-(Lys)UUU. Nat Struct Mol Biol 11:1186–1191

Na JG, Pinto I, Hampsey M (1992) Isolation and characterization of SUA5, a novel gene required for normal growth in Saccharomyces cerevisiae. Genetics 131(4):791–801

Naor A, Thiaville PC, Altman-Price N, Cohen-Or I, Allers T, de Crécy-Lagard V, Gophna U (2012) A genetic investigation of the KEOPS complex in halophilic Archaea. PLoS One 7(8): e43013. https://doi.org/10.1371/journal.pone.0043013

Nichols CE, Lamb HK, Thompson P, El Omari K, Lockyer M, Charles I, Hawkins AR, Stammers DK (2013) Crystal structure of the dimer of two essential Salmonella typhimurium proteins, YgjD & YeaZ and calorimetric evidence for the formation of a ternary YgjD-YeaZ-YjeE complex. Protein Sci 22(5):628–640. https://doi.org/10.1002/pro.2247

Oberto J, Breuil N, Hecker A, Farina F, Brochier-Armanet C, Culetto E, Forterre P (2009) Qri7/OSGEPL, the mitochondrial version of the universal Kae1/YgjD protein, is essential for mitochondrial genome maintenance. Nucleic Acids Res 37(16):5343–5352. https://doi.org/10.1093/nar/gkp557

Parthier C, Görlich S, Jaenecke F, Breithaupt C, Bräuer U, Fandrich U, Clausnitzer D, Wehmeier UF, Böttcher C, Scheel D, Stubbs MT (2012) The O-carbamoyltransferase TobZ catalyzes an ancient enzymatic reaction. Angew Chem Int Ed Eng 51(17):4046–4052. https://doi.org/10.1002/anie.201108896

Perrochia L, Crozat E, Hecker A, Zhang W, Bareille J, Collinet B, van Tilbeurgh H, Forterre P, Basta T (2013a) In vitro biosynthesis of a universal t6A tRNA modification in Archaea and Eukarya. Nucleic Acids Res 41(3):1953–1964. https://doi.org/10.1093/nar/gks1287

Perrochia L, Guetta D, Hecker A, Forterre P, Basta T (2013b) Functional assignment of KEOPS/EKC complex subunits in the biosynthesis of the universal t6A tRNA modification. Nucleic Acids Res 41(20):9484–9499. https://doi.org/10.1093/nar/gkt720

Petkun S, Shi R, Li Y, Asinas A, Munger C, Zhang L, Waclawek M, Soboh B, Sawers RG, Cygler M (2011) Structure of hydrogenase maturation protein HypF with reaction intermediates shows two active sites. Structure 19(12):1773–1783

Phillips G, de Crécy-Lagard V (2011) Biosynthesis and function of tRNA modifications in Archaea. Curr Opin Microbiol 14(3):335–341. https://doi.org/10.1016/j.mib.2011.03.001. Review

Phizicky EM, Alfonzo JD (2010) Do all modifications benefit all tRNAs? FEBS Lett 584:265–271

Powers DM, Peterkofsky A (1972) Biosynthesis and specific labeling of N-(purin-6-ylcarbamoyl) threonine of Escherichia coli transfer RNA. Biochem Biophys Res Commun 46(2):831–838. No abstract available

Rojas-Benitez D, Thiaville PC, de Crécy-Lagard V, Glavic A (2015) The levels of a universally conserved tRNA modification regulate cell growth. J Biol Chem 290(30):18699–18707. https://doi.org/10.1074/jbc.M115.665406

Sarmiento F, Mrázek J, Whitman WB (2013) Genome-scale analysis of gene function in the hydrogenotrophic methanogenic archaeon Methanococcus maripaludis. Proc Natl Acad Sci U S A 110(12):4726–4731. https://doi.org/10.1073/pnas.1220225110

Schweizer MP, Chheda GB, Baczynskyj L, Hall RH (1969) Aminoacyl nucleosides: VII. N-(Purin-6-ylcarbamoyl)threonine. A new component of transfer ribonucleic acid. Biochemistry 8:3283–3289

Sonawane KD, Sambhare SB (2015) The influence of hypermodified nucleosides lysidine and t(6) A to recognize the AUA codon instead of AUG: a molecular dynamics simulation study. Integr Biol (Camb) 7(11):1387–1395. https://doi.org/10.1039/c5ib00058k

Srinivasan M, Mehta P, Yu Y, Prugar E, Koonin EV, Karzai AW, Sternglanz R (2011) The highly conserved KEOPS/EKC complex is essential for a universal tRNA modification, t6A. EMBO J 30(5):873–881. https://doi.org/10.1038/emboj.2010.343

Stuart JW, Gdaniec Z, Guenther R, Marszalek M, Sochacka E, Malkiewicz A, Agris PF (2000) Functional anticodon architecture of human tRNALys3 includes disruption of intraloop hydrogen bonding by the naturally occurring amino acid modification, t6A. Biochemistry 39:13396–13404

Sundaram M, Durant PC, Davis DR (2000) Hypermodified nucleosides in the anticodon of tRNALys stabilize a canonical U turn structure. Biochemistry 39:12575–12584

Teplova M, Tereshko V, Sanishvili R, Joachimiak A, Bushueva T, Anderson WF, Egli M (2000) The structure of the yrdC gene product from Escherichia coli reveals a new fold and suggests a role in RNA binding. Protein Sci 9(12):2557–2566

Teplyakov A, Obmolova G, Tordova M, Thanki N, Bonander N, Eisenstein E, Howard AJ, Gilliland GL (2002) Crystal structure of the YjeE protein from Haemophilus influenzae: a putative ATPase involved in cell wall synthesis. Proteins 48:220–226

Thiaville PC, El Yacoubi B, Perrochia L, Hecker A, Prigent M, Thiaville JJ, Forterre P, Namy O, Basta T, de Crécy-Lagard V (2014a) Cross kingdom functional conservation of the core universally conserved threonylcarbamoyladenosine tRNA synthesis enzymes. Eukaryot Cell 13(9):1222–1231. https://doi.org/10.1128/EC.00147-14

Thiaville PC, Iwata-Reuyl D, de Crécy-Lagard V (2014b) Diversity of the biosynthesis pathway for threonylcarbamoyladenosine (t(6)A), a universal modification of tRNA. RNA Biol 11 (12):1529–1539. https://doi.org/10.4161/15476286.2014.992277. Review

Thiaville PC, Legendre R, Rojas-Benítez D, Baudin-Baillieu A, Hatin I, Chalancon G, Glavic A, Namy O, de Crécy-Lagard V (2016) Global translational impacts of the loss of the tRNA modification t6A in yeast. Microb Cell 3(1):29–45

Urbonavičius J, Meškys R, Grosjean H (2014) Biosynthesis of wyosine derivatives in tRNA(Phe) of Archaea: role of a remarkable bifunctional tRNA(Phe):m1G/imG2 methyltransferase. RNA 20(6):747–753. https://doi.org/10.1261/rna.043315.113. Review

Vendeix FA, Murphy FV IV, Cantara WA, Leszczyńska G, Gustilo EM, Sproat B, Malkiewicz A, Agris PF (2012) Human tRNA(Lys3)(UUU) is pre-structured by natural modifications for cognate and wobble codon binding through keto-enol tautomerism. J Mol Biol 416 (4):467–485. https://doi.org/10.1016/j.jmb.2011.12.048

Villarroel-Campos D, Henríquez DR, Bodaleo FJ, Oguchi ME, Bronfman FC, Fukuda M, Gonzalez-Billault C (2016) Rab35 functions in axon elongation are regulated by P53-related protein kinase in a mechanism that involves Rab35 protein degradation and the microtubule-associated protein 1B. J Neurosci 36(27):7298–7313. https://doi.org/10.1523/JNEUROSCI.4064-15.2016

Wan LC, Mao DY, Neculai D, Strecker J, Chiovitti D, Kurinov I, Poda G, Thevakumaran N, Yuan F, Szilard RK, Lissina E, Nislow C, Caudy AA, Durocher D, Sicheri F (2013) Reconstitution and characterization of eukaryotic N6-threonylcarbamoylation of tRNA using a minimal enzyme system. Nucleic Acids Res 41(12):6332–6346. https://doi.org/10.1093/nar/gkt322

Wan LC, Pillon MC, Thevakumaran N, Sun Y, Chakrabartty A, Guarné A, Kurinov I, Durocher D, Sicheri F (2016a) Structural and functional characterization of KEOPS dimerization by Pcc1 and its role in t6A biosynthesis. Nucleic Acids Res 44(14):6971–6980

Wan LC, Maisonneuve P, Szilard RK, Lambert JP, Ng TF, Manczyk N, Huang H, Laister R, Caudy AA, Gingras AC, Durocher D, Sicheri F (2016b) Proteomic analysis of the human KEOPS complex identifies C14ORF142 as a core subunit homologous to yeast Gon7. Nucleic Acids Res. https://doi.org/10.1093/nar/gkw1181

Weissenbach J, Grosjean H (1981) Effect of threonylcarbamoyl modification (t6A) in yeast tRNA Arg III on codon-anticodon and anticodon-anticodon interactions. A thermodynamic and kinetic evaluation. Eur J Biochem 116(1):207–213

Yarian C, Marszalek M, Sochacka E, Malkiewicz A, Guenther R, Miskiewicz A, Agris PF (2000) Modified nucleoside dependent Watson-Crick and wobble codon binding by tRNALysUUU species. Biochemistry 39(44):13390–13395

Yarian C, Townsend H, Czestkowski W, Sochacka E, Malkiewicz AJ, Guenther R, Miskiewicz A, Agris PF (2002) Accurate translation of the genetic code depends on tRNA modified nucleosides. J Biol Chem 277:16391–16395

Yourno J, Kohno T, Roth JR (1970) Enzyme evolution: generation of a bifunctional enzyme by fusion of adjacent genes. Nature 228(5274):820–824. No abstract available

Zhang W, Collinet B, Graille M, Daugeron MC, Lazar N, Libri D, Durand D, van Tilbeurgh H (2015a) Crystal structures of the Gon7/Pcc1 and Bud32/Cgi121 complexes provide a model for the complete yeast KEOPS complex. Nucleic Acids Res 43(6):3358–3372. https://doi.org/10.1093/nar/gkv155

Zhang W, Collinet B, Perrochia L, Durand D, van Tilbeurgh H (2015b) The ATP-mediated formation of the YgjD-YeaZ-YjeE complex is required for the biosynthesis of tRNA t6A in Escherichia coli. Nucleic Acids Res 43(3):1804–1817. https://doi.org/10.1093/nar/gku1397

Chapter 9
Evolution of C/D Box sRNAs

Vanessa Tripp and Lennart Randau

Abstract Nucleotide modifications are important for the proper folding and the stability of most structured RNA molecules. The ribosomal RNAs of Archaea and Eukaryotes contain large numbers of 2′-*O*-methyl and pseudouridine modifications. A family of small RNAs, the C/D box s(no)RNAs, utilizes base complementarity with target RNAs to guide methyltransferase activity. Thus, the guide sequences of C/D box sRNAs determine the scope of 2′-O-methylations within archaeal and eukaryotic organisms. In this chapter, we describe the general architecture and functionality of C/D box sRNAs with a focus on the accelerated evolution of their guide sequences. The plasticity of C/D box sRNA biogenesis allows for the evolution of novel guide RNA sequence pairs and permits adjustments of the cellular modification landscape.

9.1 Introduction

Ribosome biogenesis requires carefully coordinated processes of transcription, processing and folding of ribosomal RNA (rRNA) and their assembly with ribosomal proteins. Post-transcriptional modifications play important roles for the structural stability of the ribosome and support proper RNA folding (Helm 2006; Herschlag et al. 1993; Watkins and Bohnsack 2012; Gaspin et al. 2000; Steitz and Tycowski 1995; Polikanov et al. 2015). In archaea and eukaryotes, the most common RNA nucleotide modifications are 2′-*O*-methylations of the ribose moiety or pseudouridylations. The presence of these modifications is not restricted to rRNA nucleotides as they can also be found in small nuclear RNA (snRNA) in eukaryotes and transfer RNA (tRNA) in archaea (Omer et al. 2000; Maden et al. 1995; Kiss-Laszlo et al. 1996; Tycowski et al. 1998). Pseudouridylation increases the hydrogen bonding potential of RNA nucleotides. In contrast, the introduction of

V. Tripp • L. Randau (✉)
Max Planck Institute for Terrestrial Microbiology, Karl-von-Frisch Strasse 10, 35043 Marburg, Germany
e-mail: lennart.randau@mpi-marburg.mpg.de

B. Clouet-d'Orval (ed.), *RNA Metabolism and Gene Expression in Archaea*, Nucleic Acids and Molecular Biology 32, DOI 10.1007/978-3-319-65795-0_9

2′-*O*-methylation modifications increases RNA stability against hydrolysis and ribonucleolytic cleavage by masking the hydrogen bonding potential. Additionally, due to steric effects, the 2′-*O*-methylated ribose favors the C3′-endo conformation which increases the melting temperature of RNA sequences (Yokoyama et al. 1987; Ge and Yu 2013). RNA modifications are introduced by a huge variety of enzymes that can be divided into stand-alone proteins and members of ribonucleoprotein complexes. Large numbers of specific methyltransferases and pseudouridine synthases exist that are capable of generating methylated nucleotides and pseudouridines in rRNA, tRNA and other target RNA molecules. These stand-alone variants are predominant in Bacteria. In this chapter, we will focus on ribonucleoprotein complexes that utilize small RNAs to guide the introduction of RNA modifications in Archaea and Eukaryotes.

9.2 C/D Box sRNA Features

Pseudouridylations and 2′-*O*-methyl modifications are introduced into the target RNA molecules by site-specific RNA-guided mechanisms. Small RNA molecules serve as guides in ribonucleoprotein complexes and direct their modification activity using base complementarity with the target RNA molecules (Omer et al. 2000; Maxwell and Fournier 1995; Balakin et al. 1996). Pseudouridylations are guided by H/ACA box small nucleolar (sno)RNAs in eukaryotes and their archaeal homologues are termed H/ACA box sno-like RNAs (H/ACA box sRNA). Similarly, C/D box small nucleolar (sno)RNAs guide the introduction of 2′-*O*-methylations in eukaryotes and their homologues are termed C/D box sno-like RNAs (C/D box sRNA) in archaea (Kiss-Laszlo et al. 1996; Omer et al. 2000; Ganot et al. 1997; Ni et al. 1997). These RNA families are characterized by conserved sequence elements and a signature secondary structure. C/D box s(no)RNAs possess two conserved sequence elements, the boxC (consensus sequence: RUGAUGA) and boxD (consensus sequence: CUGA) motifs, which are situated close to the 5′ and 3′ terminus, respectively. Both elements can be duplicated in the central region (termed boxC′, boxD′), but these two boxes are less conserved in eukaryotes (Kiss-Laszlo et al. 1998; Szewczak et al. 2002). Upon C/D box s(no)RNA folding, the boxC and boxD sequences partially base-pair, resulting in a helix-internal loop-helix structure termed kink-turn (k-turn) (Watkins et al. 2000; Klein et al. 2001) (Fig. 9.1).

A terminal helix of the k-turn motif that is formed by the sequences upstream of boxC and downstream of boxD (helixI), is common for eukaryotic C/D box snoRNAs, but is absent in most archaeal C/D box sRNAs (Omer et al. 2000; Su et al. 2013; Randau 2012; Gaspin et al. 2000). The sequences of the boxC and boxD are part of the asymmetric internal loop with three unpaired nucleotides at one side and an internal helix (helixII). The internal helix typically consists of two sheared GA base pairs, a mismatch pair and a Watson-Crick base pair (Klein et al. 2001).

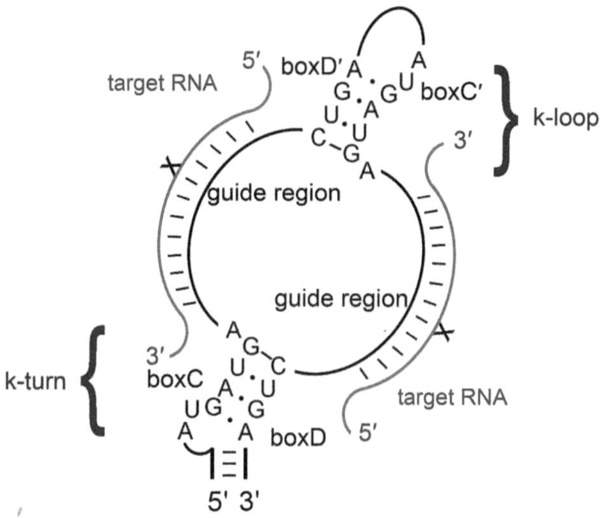

Fig. 9.1 Schematic representation of a C/D box s(no)RNA. The conserved boxC/C′ and boxD/D′ sequences base-pair and k-turn and k-loop structures are formed. The guide regions show complementarity to the target RNA and the nucleotide that is complementary to the fifth nucleotide upstream of the boxD/D′ motif becomes 2′-*O*-methylated

The k-turn motif can be also found in different other RNAs e.g. rRNAs, H/ACA box s(no)RNAs, human ribonuclease P and untranslated regions of mRNA including several riboswitches (Klein et al. 2001; Rozhdestvensky et al. 2003; Reiter et al. 2010; Zago et al. 2005; Blouin and Lafontaine 2007; Heppell and Lafontaine 2008; Lilley 2014). Typically, a second version of the k-turn motif can be found in C/D box s(no)RNAs due to base pairing between the boxC′ and boxD′ sequences. In this case, the terminal helix is replaced by a loop which coined the term k-loop (Nolivos et al. 2005).

The sequences that are located between the boxC and boxD′ motif, as well as between the boxC′ and boxD motif, respectively, usually show complementarity to the sequences of the target RNA and serve as guide sequences that determine the sites for the 2′-*O*-methylation reaction. The guide sequences usually have a length of 10–12 nt in archaea and 10–21 nt in eukaryotes. The length of the guide sequences is crucial for the 2′-*O*-methylation reaction and the modification is introduced site-specifically at the nucleotide of the target RNA that is complementary to the fifth nucleotide upstream of the boxD/D′ motif (Kiss-Laszlo et al. 1996; Tran et al. 2005). A Watson-Crick base pair is required at this position and extended Watson-Crick base pairs along the guide RNA-target RNA duplex are crucial for efficient methylation (Omer et al. 2002; Appel and Maxwell 2007; Cavaille et al. 1996; Ziesche et al. 2004). In yeast, the region between the boxC′ and boxD′ sequences has additional complementarity with the target RNA, which stimulates methylation efficiency (van Nues et al. 2011).

9.3 Organization of C/D Box sRNPs

C/D box s(no)RNAs form ribonucleoprotein complexes (RNPs) with three highly conserved core proteins in archaea and four core proteins in eukaryotes. For archaea, crystal structures of individual heterologously produced C/D box sRNP proteins, reconstituted subcomplexes and the complete C/D box sRNP provide insights into the composition, assembly and function of these complexes. L7Ae is the primary RNA binding protein that binds and stabilizes the k-turn and k-loop motifs (Kuhn et al. 2002; Zago et al. 2005). It has an additional function as ribosomal protein subunit (Ban et al. 2000). Crucial features for L7Ae binding to the C/D box sRNA are (i) the terminal stem at the C/D box sRNA 5′ and 3′ ends, which juxtaposes the boxC and boxD sequences, (ii) the two sheared GA base pairs formed by pairing of the boxC and boxD sequences and (iii) the uridine of boxC that is part of the internal loop (Kuhn et al. 2002).

Nop5 possesses several important domains that are required for the assembly of the complex. The carboxy-terminal domain exhibits RNA-binding sites and interacts with the L7Ae-C/D box sRNA complex (Aittaleb et al. 2003). Additionally, electrostatic interactions exist between Nop5 and the C/D box sRNA guide sequences (Lapinaite et al. 2013; Ghalei et al. 2010). The coiled-coil domains of two Nop5 molecules interact, leading to the dimerization of the proteins. The amino-terminal domain of Nop5 interacts with fibrillarin, the third core protein of archaeal C/D box sRNPs (Aittaleb et al. 2003; Gagnon et al. 2012). Fibrillarin uses S-adenosyl-L-methionine as methyl group donor for the 2′-*O*-methylation reaction. Independent methyl transfer activity of fibrillarin could not be observed indicating that the activity is dependent on C/D box sRNP formation (Omer et al. 2002; Wang et al. 2000). In the absence of L7Ae and the C/D box sRNA, fibrillarin and Nop5 build stable complexes with two subunits of both proteins (Aittaleb et al. 2003; Zhang et al. 2006). L7Ae can only bind to this subcomplex in the presence of a C/D box sRNA. It is not clear whether a step-wise assembly of all proteins occurs *in vivo* or whether a Nop5-fibrillarin subcomplex is preformed that binds to the L7Ae-C/D box sRNA subcomplex (Bower-Phipps et al. 2012).

L7Ae is also a member of H/ACA RNPs and recognizes k-turn structures in H/ACA RNAs. H/ACA RNPs contain four proteins whose core structure is conserved between Archaea and Eukaryotes. Apart from L7Ae, these complexes contain Cbf5 (a pseudouridine synthase), Nop10 (an elongated protein that stabilizes active site structure of Cbf5) and finally Gar1 (a small basic protein that interacts with Cbf5) (Hamma and Ferre-D'Amare 2010). H/ACA box sRNAs usually contain one or two stem-loop structures that each contain a single L7Ae binding site (Rozhdestvensky et al. 2003) and a larger bulge. These bulge nucleotides exhibit complementary to the target RNA molecules and are termed pseudouridylation pockets.

9.4 The Structure of C/D Box sRNPs

Slightly conflicting biochemical and structural studies of the archaeal C/D box sRNP described the native complex either as a monomer with one C/D box sRNA and two copies of each protein or as a dimer of two C/D box sRNAs and four copies of each protein. The first electron microscopy (EM) structure of a C/D box sRNP was obtained for an *in vitro* reconstituted complex with recombinant *Methanocaldococcus jannaschii* proteins. This complex revealed a di-sRNP architecture (Bleichert et al. 2009; Bower-Phipps et al. 2012). The crystal structure of a reconstituted complex with *Pyrococcus furiosus* proteins supported the existence of this di-sRNP architecture (Xue et al. 2010). However, a crystal structure of an *in vitro* reconstituted C/D box sRNP with recombinant *Sulfolobus solfataricus* proteins indicated a mono-sRNP structure (Lin et al. 2011). The difference of the assembly approaches resided in the structure of the utilized C/D box sRNA. The reconstitution of the *M. jannaschii* and *P. furiosus* complexes was performed with *in vitro* transcribed natural C/D box sRNA (Bleichert et al. 2009). The reconstitution of the *S. solfataricus* complex was achieved with an artificial two-stranded RNA without a k-loop motif (Lin et al. 2011). Therefore, it is plausible that differences in the C/D box sRNP reconstitution procedures occurred due to the usage of different RNAs as it was shown that C/D box sRNP assembly strategies with non-natural C/D box sRNAs can lead to monomeric complex assemblies (Bower-Phipps et al. 2012; Bleichert and Baserga 2010).

Subsequent NMR spectroscopy (nuclear magnetic resonance spectroscopy) and small-angle X-ray and neutron scattering (SAXS and SANS) experiments of reconstituted *P. furiosus* C/D box sRNPs indicated the presence of dimeric structures in which the two C/D box sRNAs exhibit an antiparallel orientation. Furthermore, the Nop5 subunits were shown to interact with each other via their coiled-coil domains but the proteins appear not to be strictly assigned to one guide sequence (Lapinaite et al. 2013). This structure also suggested that target RNAs can likely be 2′-*O*-methylated in a sequential manner. Sequential 2′-*O*-methylation was indeed observed previously and a dimeric structure allows for the proposal of a mechanism (Singh et al. 2004). It is proposed that all guide regions form base pairs with the target RNA, but only the two diagonally opposing fibrillarin subunits are in contact with the target RNA at the same time and therefore only two modifications can be introduced at the same time. The modifications that are guided by the C/D box sRNA regions located between the boxC and boxD′ motifs are targeted first and more efficiently (Lapinaite et al. 2013; Graziadei et al. 2016). It is intriguing to speculate that this sequential methylation can be used to regulate the spatiotemporal dynamics of rRNA folding.

Recent cryo-EM studies of reconstituted *M. jannaschii* complexes suggest a different orientation of the two C/D box sRNAs in the dimeric complex. Both C/D box sRNAs might have a parallel orientation and interactions between the stem ends of both C/D box sRNAs might stabilize the interaction and the dimeric C/D box sRNP structure as disruption of C/D box sRNA stems leads to less efficient dimeric assemblies (Yip et al. 2016). The different C/D box sRNA orientation, in comparison to the *P. furiosus* NMR spectroscopy structure, might result from the usage of a non-natural occurring C/D box sRNA in the *P. furiosus* complex in which the k-loop is substituted by a k-turn (Yip et al. 2016; Lapinaite et al. 2013). It remains to be determined how the complex is stabilized in C/D box sRNAs without external stems, a phenomenon that is often found in archaeal C/D box sRNAs.

In eukaryotes, orthologues of the three archaeal C/D box sRNP forming proteins exist. The protein that stabilizes the k-turn motif is the 15.5 kDa protein (Snu13 in yeast). K-loop motif binding could not be shown *in vitro* for the eukaryotic protein. This indicates that C/D box s(no)RNP assembly differs between archaea and eukaryotes (Gagnon et al. 2010; Szewczak et al. 2002; Watkins et al. 2000; Granneman et al. 2009; Watkins et al. 2002). The structural subunits of the eukaryotic C/D box snoRNP are Nop56 and Nop58, which exhibit highly similar sequences. They are predicted to form a heterodimer and replace the archaeal Nop5 as the coiled-coil domains of the two proteins show a similar sequence to the coiled-coil domain in Nop5 (Aittaleb et al. 2003; Newman et al. 2000; Caffarelli et al. 1998; Granneman et al. 2009). Nop56 interacts with the guide regions of the snoRNA suggesting its involvement in substrate recognition. Nop58 is required for snoRNA stability (Lafontaine and Tollervey 1999, 2000; van Nues et al. 2011; Granneman et al. 2009). The catalytic subunit of the eukaryotic C/D box snoRNPs is fibrillarin (Nop1 in yeast) and S-adenosyl-L-methionine is used as methyl group donor for the 2′-*O*-methylation reaction (Tyc and Steitz 1989; Ochs et al. 1985; Schimmang et al. 1989; Galardi et al. 2002). Moreover, several snoRNP-specific proteins exist in eukaryotes that have functions in snoRNP assembly and snoRNA protection during the maturation process (Watkins and Bohnsack 2012).

9.5 C/D Box sRNA Methylation Targets in Ribosomal RNAs

In archaea, the organisms' growth temperatures were found to correlate with the number of C/D box sRNA genes. For example, only seven C/D box sRNA genes were identified in the mesophilic archaeon *Methanococcus maripaludis*, while RNA-Seq analyses of the small RNome of the thermophilic archaea *Ignicoccus hospitalis* and *Methanopyrus kandleri* revealed 128 and 127 C/D box sRNA genes, respectively (Noon et al. 1998; Dennis et al. 2001; Su et al. 2013; Randau 2012; Omer et al. 2000; Grosjean et al. 2008). It is plausible that higher growth temperatures result in an increased demand for RNA-modifications to aid in rRNA structure stabilization and RNA folding coordination (Su et al. 2013). The sequences of the C/D box sRNA guide regions allow for the analysis of potential modified nucleotides in rRNAs and for the identification of conserved modification sites in target RNA molecules of multiple organisms.

Maps with computationally predicted 2′-*O*-methylation target sites exist for several model archaea (e.g. *Methanococcus maripaludis*, *Nanoarchaeum equitans*,

Sulfolobus acidocaldarius, *Thermoproteus tenax*, *Methanopyrus kandleri*, *Ignicoccus hospitalis* and *Pyrobaculum calidifontis*). In total, these maps include target predictions of 489 C/D box sRNAs from which 719 modification target sites were predicted in 16S and 23S rRNA molecules (Dennis et al. 2015) (Table 9.1, Fig. 9.2).

Pseudouridylations and base modifications are rare in archaea. Consequently, the predicted 2′-*O*-methylation sites comprise the majority of rRNA modifications in the analyzed archaea (Noon et al. 1998; Su et al. 2013). Mass-spectrometric analyses determined 67 2′-*O*-methylated nucleotides, 9 pseudouridylations and at least 11 base modifications in the 16S and 23S rRNA of *S. solfataricus*. Of those, the presence of two 5-methylcytidines could be verified via RNA-Seq of samples after bisulfate treatment which converts C residues but not 5-methylated C residues into U residues (Noon et al. 1998; Edelheit et al. 2013). For *Haloferax* volcanii, 4 of 13 modifications are pseudouridylations (Grosjean et al. 2008). The 5S rRNA of *S. acidocaldarius*, *S. solfataricus*, *Pyrodictium occultum*, *Halobacterium halobium* and *Haloarcula marismortui* was analyzed for modifications with mass-spectrometric approaches. One 2′-*O*-methylation could be identified for the *Sulfolobus* species and in *P. occultum*, a 2′-*O*-methylation and a base modification were detected (Bruenger et al. 1993; Kirpekar et al. 2000).

Comparison of predicted C/D box guide RNA targets in the 16S and 23S rRNAs of seven archaea unraveled that the distribution of the modifications is not random and identified several methylation hotspots. In the 23S rRNA, the modification hotspots are located in conserved regions of the domains II, IV and V that are

Table 9.1 2′-*O*-methylation sites in archaea

Organism	Growth temperature	No. of C/D box sRNA genes	Predicted rRNA 2′-*O*-methylation sites
Nanoarchaeum equitans	80–100 °C	26	28
Ignicoccus hospitalis	80–100 °C	128	198
Methanococcus maripaludis C5	35–40 °C	7	9
Methanopyrus kandleri	84–110 °C	127	273
Pyrobaculum calidifontis	90–100 °C	88	118
Sulfolobus acidocaldarius	67–80 °C	61	55
Thermoproteus tenax	70–97 °C	52	69
		No. of specific methyltransferases	**rRNA methylation sites**
Escherichia coli	~37 °C	23	24

C/D box sRNAs were identified for a number of archaeal model organisms and 2′-*O*-methylation targets were predicted for the respective 16S and 23S rRNA molecules (Dennis et al. 2015). Note, that guides can lack targets or have multiple targets. For comparison, *E. coli* harbors 24 methylated rRNA nucleotides (and 10 pseudouridines) that are introduced by specific methyltransferases without RNA guidance

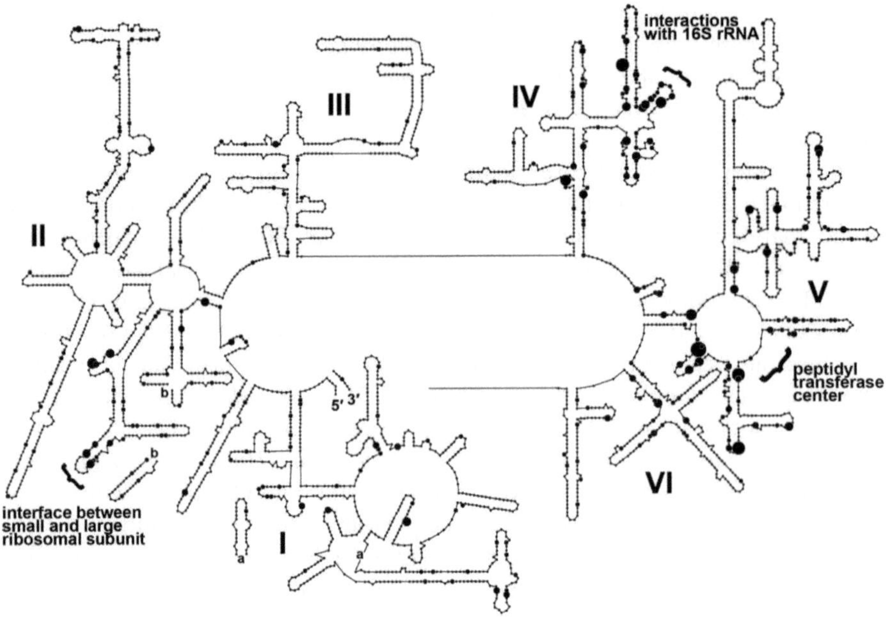

Fig. 9.2 Distribution of 2′-*O*-ribose methylation sites in archaeal 23S rRNA. Predicted 2′-*O*-ribose methylation sites of seven archaeal species are mapped onto the consensus secondary structures of the rRNAs (Cannone et al. 2002)

important for the functionality of the ribosome (Fig. 9.2). These regions comprise the catalytic peptidyl transferase center responsible for peptide bond formation and release, the interface between the large and the small ribosomal subunit and direct interaction sites with the 16S rRNA. A comparative genomics approach identified nine modifications in the *Haloferax volcanii* 23S rRNA (three 2′-*O*-methylation, four pseudouridylations, two base modifications) and all of these modifications are also located in these conserved regions (Petrov et al. 2013; Cate et al. 1999; Hury et al. 2006; Dennis et al. 2015; Grosjean et al. 2008). For *H. marismortui*, the 23S rRNA domains II, IV and V were assayed for modifications and in total eight modifications were identified (three 2′-*O*-methylation, three pseudouridylations, two base modifications) whose positions correspond with identified modifications of *H. volcanii* (Kirpekar et al. 2005; Grosjean et al. 2008). Hotspots for methylated nucleotides are usually found in regions that are not protected by ribosomal proteins in the assembled ribosome (Ban et al. 2000).

In the 16S rRNA, the locations of guide-sequence-deduced modifications are more varied, but most predicted 2′-*O*-methylation target sites are located in the central core which connects the four 16S rRNA domains (Fig. 9.3). This core is formed in the early stage of the rRNA assembly and is located close to the decoding center which monitors codon-anticodon base pairing (Wimberly et al. 2000; Dennis et al. 2015). In the 16S rRNA of *H. volcanii*, three of the four modifications are

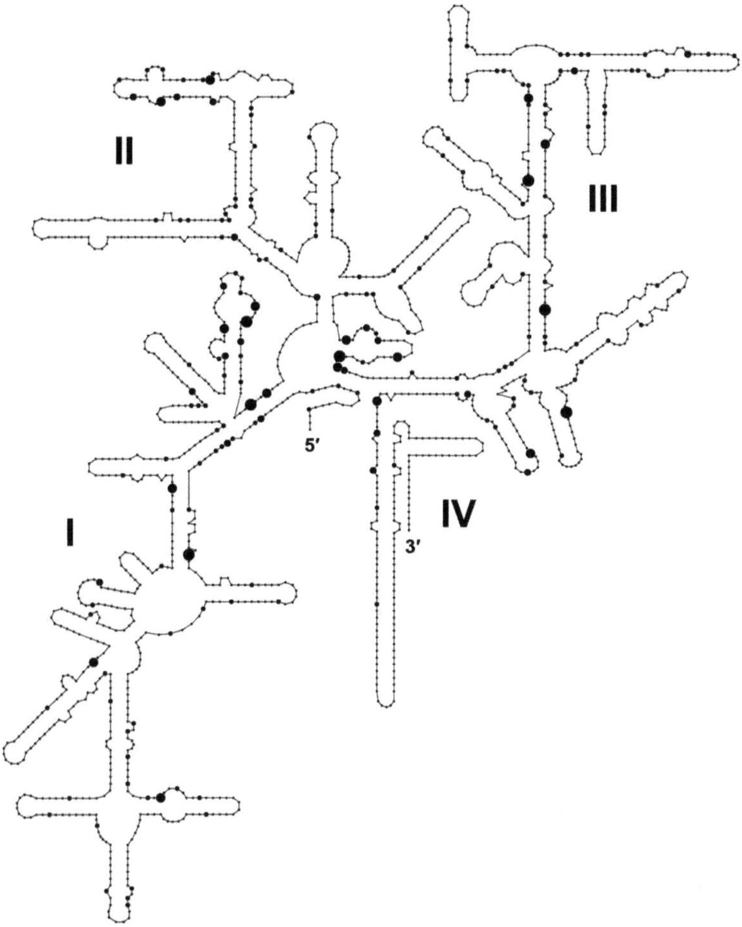

Fig. 9.3 Distribution of 2′-*O*-ribose methylation sites in archaeal 16S. Predicted 2′-*O*-ribose methylation sites of seven archaeal species are mapped onto the consensus secondary structures of the rRNAs (Cannone et al. 2002)

found in the terminal helix 44 that is part of the 30S ribosomal subunit interface and its upper part is also important for decoding activity (Grosjean et al. 2008).

Experimentally verified modifications maps of eukaryotes and bacteria revealed a conservation of modification hotspots in all three domains of life (Decatur and Fournier 2002; Mengel-Jorgensen et al. 2006). It should be noted that 2′-*O*-methylations and pseudouridylations are rare in bacteria. Instead, base modifications are more common which are introduced in an RNA-independent fashion by site or region specific enzymes (Decatur and Fournier 2002). 2′-*O*-methylations are predominant in archaea, while pseudouridylations are also frequently found in eukaryotes. Of the 109 modified yeast nucleotides, 55 are 2′-*O*-methylated, 44 are pseudouridylated and 10 possess base methylations (Decatur and Fournier 2002).

The 24 modified nucleotides in the 23S rRNA of *Escherichia coli* and the 74 modified nucleotides in the 23S rRNA of yeast occur in the center and front of the large ribosomal subunit (LSU) and comprise sites of the peptidyl transferase center, the polypeptide exit channel and the subunit bridges. In the 16S rRNA, the modifications cluster in the region that connects the four domains (Decatur and Fournier 2002). A crystal structure of the *Thermus thermophilus* ribosome revealed the presence of modifications. Most modified nucleotides are oriented towards other rRNA helices and form additional hydrophobic contacts with neighboring nucleotides. Additionally, some modifications are located at the interface between the large and the small ribosomal subunit or form contacts with the ribosome ligands in the P site. 18 of the 23 modifications are in common with *E. coli* (Polikanov et al. 2015). The high conservation of modifications at individual nucleotides in bacteria might be due to the modification mechanism via specific enzymes. Additionally, no connection between growth temperature of bacterial organisms and the number of modifications exists, which contrasts observations in archaea.

Modified nucleotides are often found at hotspots in specific rRNA regions or structural elements in archaea and eukaryotes, but the exact position of the modified nucleotide is often not conserved. Some conservation is observed for haloarchaea. The *H. volcanii* 23S rRNA exhibits nine modifications and *H. marismortui* eight modifications which are all at identical positions in *H. volcanii* (Grosjean et al. 2008; Kirpekar et al. 2005). However, a broader view of C/D box guide RNA targets in seven model archaea did not reveal a single nucleotide in 16S or 23S rRNA that is targeted in all organisms. Only one nucleotide of the 23S rRNA is targeted in six of the analyzed archaea and a total of three nucleotides are targeted by five archaea. These hotspots for conserved modifications were found in 23S rRNA helix 68 and helices 90–93 (Dennis et al. 2015). The low conservation of modifications at specific nucleotides was also observed in bacteria and eukaryotes. In agreement, mutational studies revealed that the modification of most specific individual nucleotides can often be removed without adverse effects. In contrast, the deletion of clusters of modifications was shown to cause cell growth defects suggesting that the impact of modifications is increased by the accumulation of modifications in specific regions (King et al. 2003; Liang et al. 2009). The individual removal of 23S rRNA 2′-*O*-modifications in *E. coli* revealed that only the loss of the single modification at position U2552, close to the peptidyl transferase center, leads to growth and translational defects (Caldas et al. 2000; Purta et al. 2009; Lovgren and Wikstrom 2001). The deletion of all 2′-*O*-methylations is not possible in yeast (Schimmang et al. 1989). Interestingly, *Geobacillus stearothermophilus* and *Thermus aquaticus* ribosomes can be assembled using *in vitro* transcribed RNA that lacks all modifications and the ribosomal activity was shown to be reduced but not abolished. Seven modifications close to the peptidyl transferase center in the 23S rRNA were shown to be important for *in vitro* reconstitution of active *E. coli* ribosomes (Green and Noller 1996, 1999; Khaitovich et al. 1999).

In summary, a conserved modification pattern exists in all three domains of life although different kinds of machineries are responsible for the formation of the modifications. Most modifications are located in ancient core regions of the rRNAs that are important for ribosome integrity and functionality. The presence of modifications contributes to the folding, structural stabilization and function of the rRNAs. As these regions include RNA-RNA interaction sites between the 23S rRNA and the 16S rRNA, as well as between the 23S rRNA and the tRNAs in the peptidyl transferase center, it is very likely that these interactions are stabilized by these modifications (Hansen et al. 2002a; Cate et al. 1999). Additionally, the regions in which methylation hotspots can be observed are not protected by ribosomal proteins and the modifications may help in the stabilization of the rRNA structure in these unprotected regions (Ban et al. 2000). Nevertheless, partial modification of individual rRNA nucleotides could be observed in organisms of all three domains of life (Buchhaupt et al. 2014; Andersen et al. 2004; Hansen et al. 2002b; Noon et al. 1998). For *S. solfataricus*, mass-spectrometric analyses showed variations in the $2'$-*O*-methylation pattern at different growth temperatures that might result from a differential expression of the C/D box sRNAs (Noon et al. 1998). Thus, hypomodified nucleotides add to the heterogeneity of the rRNAs and the ribosome.

9.6 C/D Box sRNAs as RNA Chaperones

The observed low conservation of modifications of individual nucleotides suggests that a low selection pressure for modifications at specific nucleotides allows for the presence of dynamic guide regions. Consequently, evolution permits changes of guide regions to occur faster than changes in the target rRNA sequences. Nucleotide substitutions can occur in the guide regions of C/D box sRNAs which subsequently lead to mismatches between the guide and the target RNA. The accumulation of mutations leads to orphan guides that are free to evolve guides for new methylation targets (Dennis et al. 2001). This scenario would also explain the large amount of orphan guides that do not show complementarity to known target RNAs (Huttenhofer et al. 2001; Chen et al. 2008; Omer et al. 2000). For example, one third of the guide regions of 489 C/D box sRNAs from seven model archaea did not show clear rRNA targets (Dennis et al. 2015). Some of the most conserved guide sequences were found to be the result of convergent evolution. Here, guide sequences that are shared between archaeal organisms can e.g. be found upstream of the D′ box in one organism and upstream of the D box in a different organism. Thus, different pairings with the second guide sequence of these individual C/D box sRNAs are created.

The analysis of these two linked guide regions that are present in every archaeal C/D box sRNAs suggested an additional consequence of this architectural feature. It was hypothesized that the presence of two guide sequences in each C/D box sRNA (and four in a C/D box sRNP) has evolved to provide a regulatory mechanism for rRNA folding. C/D box sRNAs were identified that target positions that are widely separated in the primary sequence but located in close proximity in the secondary structure (Dennis et al. 2015; Gaspin et al. 2000). Thus, guide-target interactions could play a role in bridging distant rRNA sequences and might facilitate substructure formation which can be stabilized by subsequent methylation activity. In this model, C/D box sRNAs should be considered to represent RNA chaperones. The accelerated evolution of C/D box sRNA guides creates novel pairs of guide sequences and beneficial RNA chaperone activity can be selected for.

9.7 C/D Box sRNA Biogenesis

The genomic organization of C/D box s(no)RNAs and their transcription is highly variable in archaea and eukaryotes. In eukaryotes, a trend exists from lower to higher eukaryotes towards the reduction of the number of independent promoters by arranging C/D box snoRNA genes in polycistrons or within introns (Dieci et al. 2009). In yeast, as well as in plants, most snoRNA genes are transcribed from independent RNA polymerase II (or less frequently RNA polymerase III) promoters either as mono- or polycistronic transcripts. In plants, C/D box snoRNA genes exist almost exclusively as polycistronic clusters (Li et al. 2005; Liang et al. 2002; Brown et al. 2003; Dieci et al. 2009; Leader et al. 1997). Additionally, dicistronic tRNA-C/D box snoRNA genes are reported for plants (Kruszka et al. 2003; Barbezier et al. 2009). In vertebrates, only few snoRNA genes are transcribed from independent promoters. In these organisms, most of the genes are located within introns of protein-coding and non-protein-coding genes. The intronic C/D box snoRNA genes can either exist individually or as polycistrons, but individual localizations are more common (Weber 2006; Tycowski et al. 2004; Leader et al. 1994; Pelczar and Filipowicz 1998). In general, clusters of several C/D box snoRNAs within an intron can be composed of homologous or heterologous snoRNA genes (Chen et al. 2008). Interestingly, a decent amount of snoRNA genes is located in introns of housekeeping genes that have functions in ribosome biogenesis. This implies a correlation between snoRNA gene expression and the expression of proteins that are involved in the same processes (Dieci et al. 2009; Bratkovic and Rogelj 2011; Maxwell and Fournier 1995).

9.7.1 *Organization of Archaeal C/D Box sRNA Genes*

In archaea, the majority of C/D box sRNAs does not possess independent promoters. In a study that includes more than 300 C/D box sRNAs from six archaea only about 20% of the C/D box sRNA genes were found to possess TATA box promoter elements in the 50 nt upstream region. Instead the C/D box sRNA genes are found to be located in positions that allow for co-transcription with neighboring genes. The genes often overlap with the 3′ or 5′ end of flanking open reading frames

(ORFs) or clusters of several C/D box sRNA genes exist, suggesting polycistronic transcription (Tripp et al. 2016). The archaeon *N. equitans*, which contains a minimal and highly compact genome with several split protein-coding genes, possesses C/D box sRNA genes directly adjacent to the split genes (Waters et al. 2003; Randau et al. 2005; Randau 2012). Other C/D box sRNA gene localizations that ensure the C/D box sRNA transcription are the downstream regions of tRNA genes, in which the C/D box sRNAs are transcribed together with the tRNA. A dicistronic tRNA-C/D box sRNA is reported for the most abundant C/D box sRNA of *N. equitans*. Its gene is located downstream of the gene for $tRNA^{Val}$ (Randau 2012). Furthermore, C/D box sRNA genes exist in tRNA introns. The $pre\text{-}tRNA^{Trp}$ of *H. volcanii* and *Halobacterium salinarum* possess C/D box sRNA genes in their introns. Interestingly, the *H. volcanii* C/D box sRNA guides modification of $pre\text{-}tRNA^{Trp}$ anticodon nucleotides *in cis* and *in trans* (Singh et al. 2004; Clouet d'Orval et al. 2001; Randau 2012). Potential homologs of these C/D box sRNAs encoded in $pre\text{-}tRNA^{Trp}$ introns were identified in additional euryarchaeal genomes (Omer et al. 2000; Singh et al. 2004; Weisel et al. 2010; Clouet d'Orval et al. 2001). The intron of $pre\text{-}tRNA^{Trp}$ in *S. solfataricus* is also highly stable but it does not contain the characteristic boxC and boxD motifs. As it shows sequence similarity to the cognate $tRNA^{Trp}$ it could also play a role in modifying this RNA (Danan et al. 2012). The $pre\text{-}tRNA^{Met}$ of *N. equitans* exhibits a stable intron with boxC and boxD motifs that would be able to form into a k-turn structure but a guide function of this RNA is unlikely (Randau 2012).

C/D box s(no)RNA transcription in polycistronic clusters, in introns or as co-transcripts with flanking ORFs results in the need for maturation of the primary transcripts to obtain functional C/D box s(no)RNAs that can guide the 2′-*O*-methylation of their targets. In eukaryotes, maturation is well understood. Endo- and exoribonucleases are responsible for the processing of polycistronic C/D box snoRNA transcripts. In yeast, polycistronic snoRNAs are flanked by sequences that form short hairpin structures and that are recognized by an RNaseIII-like endoribonuclease (Chanfreau et al. 1998a, b). Further trimming of the 5′ and 3′ ends occurs by 5′-3′ and 3′-5′ exoribonucleases that are not specific for the maturation of snoRNAs (Rat1, Xrn1; exosome) (Allmang et al. 1999; Petfalski et al. 1998; Qu et al. 1999). The dicistronic tRNA-C/D box snoRNA transcripts reported in plants are processed by the tRNA 3′ processing activity of RNaseZ or an alternative pathway which involves cleavage by unknown RNases (Kruszka et al. 2003; Barbezier et al. 2009).

Intron-encoded C/D box snoRNAs are matured by two different pathways of which one is splicing-dependent and the other is splicing-independent (Brown et al. 2008). Predominantly, the introns are spliced out of the pre-mRNA resulting in lariat structures that are debranched. Subsequently, mature C/D box snoRNAs are obtained by exoribonuclease activities (Villa et al. 1998; Kiss and Filipowicz 1995). In the second pathway, introns are not spliced out of the pre-mRNA, but mature C/D box snoRNAs are revealed by the action of endoribonucleases that cleave up- and downstream of the C/D box sRNA and further trimming occurs by exoribonucleases. The second pathway seems to be important for snoRNA processing from introns in plants but also exists in yeast (Leader et al. 1999; Villa et al. 1998).

9.7.2 Plasticity of Archaeal C/D Box sRNA Biogenesis

In archaea, C/D box sRNA maturation is not fully understood. It was shown in *S. acidocaldarius*, that primary precursor transcripts are processed into mature C/D box sRNAs with randomized upstream and downstream sequences. The presence of intact k-turn elements was required. Thus, the C/D box gene contains all the information required for proper C/D box sRNA maturation (Tripp et al. 2016). This suggests that sequence- or structure-specific endoribonucleases targeting primary transcript elements are not involved in the maturation process (Fig. 9.4). An exception are C/D box sRNAs in tRNA introns or located downstream of tRNAs. Here, bulge-helix-bulge motifs at the borders of tRNA introns are recognized by the splicing endonuclease and the 5′ terminus of the C/D box sRNA downstream of tRNAs is created by the tRNA 3′ end processing activity of RNaseZ (Randau 2012; Diener and Moore 1998). For the remaining archaeal C/D box sRNAs, processing by unspecific exoribonucleases is predicted, similar to the final trimming of eukaryotic C/D box snoRNAs. This universal mechanism allows for the maturation of C/D

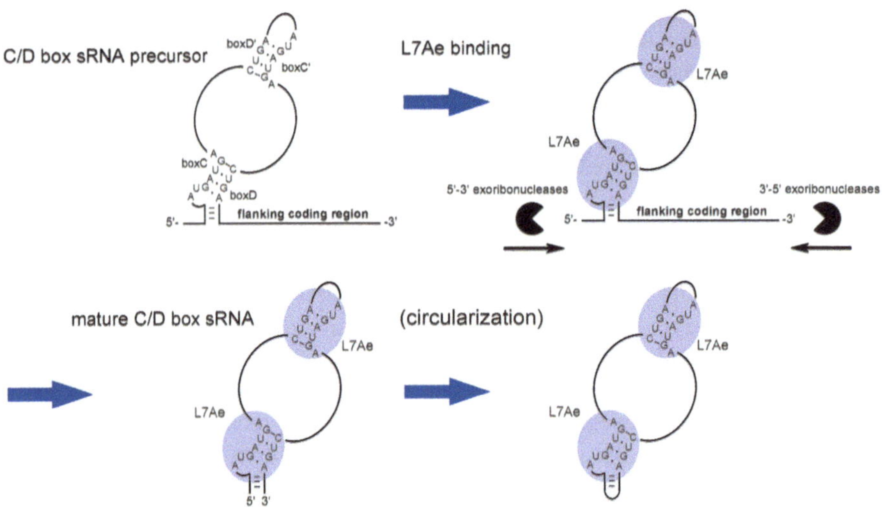

Fig. 9.4 Model for C/D box sRNA maturation. C/D box sRNA co-transcription with flanking ORFs produces primary transcripts in which the 5′ and 3′ ends become trimmed by the action of exoribonucleases. Binding of L7Ae or complete C/D box sRNP assembly (not shown) protects the C/D box sRNA from degradation by exoribonucleases. The 5′ and 3′ ends are defined by steric hindrance of the proteins and can be circularized by RNA ligase activity

box sRNAs that originate from diverse genetic contexts. Homologues of the eukaryotic snoRNA 5′ end processing enzymes Rat1p and Xrn1p do not exist in archaea (Hasenohrl et al. 2011). However, homologues of the bacterial RNaseJ family exist that have 5′-3′ exoribonucleolytic activity and they might be involved in C/D box sRNA processing (Hasenohrl et al. 2011; Clouet-d'Orval et al. 2010, 2015; Martens et al. 2013). In the processing of the 3′ terminus the exosome might be involved as it is also involved in the snoRNA processing of eukaryotes (Koonin et al. 2001; Walter et al. 2006; Lorentzen et al. 2005, 2007).

A protection mechanism for the C/D box sRNA from complete degradation has to exist in the case that sequence-unspecific exoribonucleases are involved in the processing. In eukaryotes, it could be shown that the snoRNP proteins associate with the snoRNA before processing occurs (Caffarelli et al. 1996; Matera et al. 2007). In archaea, the integrity of the k-turn is important for C/D box sRNA stability in *S. acidocaldarius* (Tripp et al. 2016). Mutations in the box motifs likely prevent L7Ae binding and subsequent complete C/D box sRNP assembly. In cases where L7Ae binding or complete C/D box sRNP assembly blocks degradation of the C/D box sRNA, the 5′ and 3′ ends are defined by steric hindrance of the proteins. Interestingly, the 5′ and 3′ ends of archaeal C/D box sRNAs exhibit a length heterogeneity of several nucleotides (Dennis et al. 2015). This observation might be a result of complexes that exhibit a certain degree of flexibility. In agreement, large conformational rearrangements were shown to occur upon substrate binding in the C/D box sRNP of *P. furiosus* (Lapinaite et al. 2013).

An additional mechanism for C/D box sRNA protection is the circularization of C/D box sRNAs that was observed in various thermophilic archaea (Danan et al. 2012; Starostina et al. 2004; Su et al. 2013; Randau 2012). Here, the close proximity of C/D box sRNA termini in assembled C/D box sRNPs might facilitate their ligation via moonlighting RNA ligases, e.g. RtcB, which would provide an added beneficial, but non-essential feature.

9.8 The Emergence of C/D Box sRNA Genes

C/D box s(no)RNAs are found in both archaea and eukaryotes and should have existed in a common ancestor (Omer et al. 2000). Accelerated evolution led to a large difference in the amount of C/D box s(no)RNAs in diverse organisms. In addition, highly variable guide sequences pairs exist and the most conserved sequences were found to be the result of convergent evolution (Dennis et al. 2015). C/D box s(no)RNA genes occur in highly variable genetic contexts and often do not possess their own promoters but are transcribed together with other genes as long precursors (Fig. 9.5).

In archaea, it was observed that the amount of individual C/D box sRNAs differs significantly, which might be related to the strength and regulation of the promoters that were 'hijacked' by C/D box sRNA genes. Interestingly, C/D box sRNA-reporter gene fusions in *S. acidocaldarius* showed that C/D box sRNAs that were

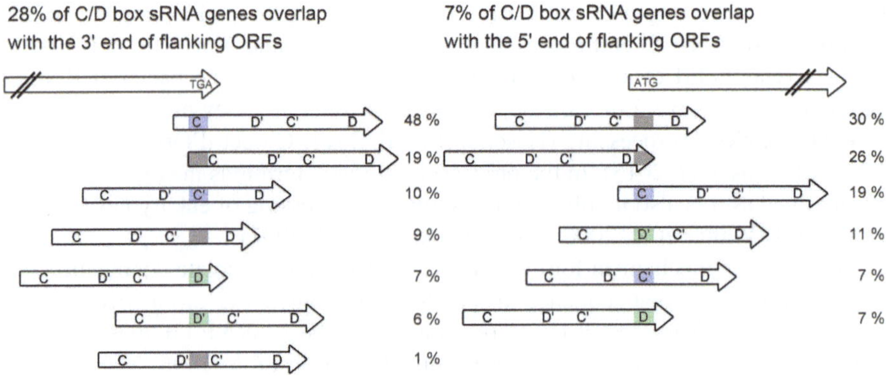

Fig. 9.5 Position of archaeal C/D box sRNAs in relation to adjacent genes. The genetic context of over 300 experimentally confirmed C/D box sRNA genes revealed frequent overlaps with neighboring genes. The percentage of genes with start and stop codons within C/D box sRNA genes is indicated

fused to the ends of a reporter gene have unfavorable effects on its enzyme activity (Tripp et al. 2016). In agreement, C/D box sRNA genes were usually not found to be fused with essential genes. How do these C/D box sRNA gene localizations evolve? Notably, archaeal C/D box sRNA genes that overlap with neighboring genes often possess the sequences of their start (AUG) or stop (e.g. UGA) codon within their boxC/C′ (RUGAUGA) or boxD/D′ (CUGA) motif (Tripp et al. 2016). One possibility for the emergence of C/D box sRNA genes might be that the start or stop codons of a gene promote the formation of k-turn motifs within C/D box sRNA genes. Thus, few mutations in the vicinity of start and stop codons can result in mRNAs with leader and trailer sequences that are recognized and bound by L7Ae. In general, the boxC and boxD motifs, forming k-turns, are the only sequences within the C/D box s(no)RNAs that are required for stability and functionality. It is possible that microhomology between boxC/C′ and boxD/D′ sequences aids in the creation of the overall architecture of a C/D box sRNA gene.

Comparison of C/D box sRNA genes within organisms revealed gene duplications with subsequent circular rearrangements or mutations that led to guide sequence diversification. The accumulation of mutations creates orphan/empty guides that lack targets. These orphan guides can evolve into guides with new targets and novel guide RNA pairs can be selected for (Dennis et al. 2015). In plants, C/D box snoRNA genes were proposed to originate from a repeated series of duplications and the selection of occurring mutations (Brown et al. 2003). Furthermore, mobility of mammalian snoRNA genes was hypothesized. Here, the computational analyses of human snoRNA orthologues in different mammalian genomes revealed several paralogues that display retroposon characteristics: an A-rich tail and an approximately 14 bp long target site duplication (Weber 2006). Evidence for C/D box sRNA gene mobility in archaea is also found in the archaeon *N. equitans*, which contains several split genes that are flanked by C/D box sRNA genes. Here, e.g. the N-terminal gene portion of the reverse gyrase is flanked by two C/D box sRNA genes.

9.9 Conclusions

C/D box sRNAs represent a fast evolving class of RNA molecules whose guide sequences shape the modification landscape of structured target RNAs in eukaryotes and archaea. Hyperthermophilic archaea contain large amounts of C/D box sRNA genes whose presence ensures the potential introduction of beneficial RNA 2′-O-methylations. The alteration of guide sequences allows for the adjustment of the RNA modification profile in changing environmental conditions. The pairing of two guide sequences permits C/D box sRNAs to coordinate two modification events and suggests an RNA chaperone activity. Individual C/D box sRNA genes were not found to be essential for archaea, but the convergent evolution of conserved guide sequences highlight their importance in the cell. The apparent variability of the RNA targeting potential is a result of accelerated C/D box sRNA gene evolution. Archaeal C/D box sRNA genes often hijack promoters of neighboring genes and the formation of k-turns, the L7Ae binding sites, is the sole requirement for proper C/D box sRNA maturation and stabilization. Stop and start codons were found to frequently be incorporated into boxC and boxD motifs. Thus, C/D box sRNA genes shape not only the modification patterns of the ncRNA transcriptomes, but also impact the gene organization of archaeal and eukaryotic genomes.

References

Aittaleb M, Rashid R, Chen Q, Palmer JR, Daniels CJ, Li H (2003) Structure and function of archaeal box C/D sRNP core proteins. Nat Struct Biol 10(4):256–263. doi:10.1038/nsb905

Allmang C, Kufel J, Chanfreau G, Mitchell P, Petfalski E, Tollervey D (1999) Functions of the exosome in rRNA, snoRNA and snRNA synthesis. EMBO J 18(19):5399–5410. doi:10.1093/emboj/18.19.5399

Andersen TE, Porse BT, Kirpekar F (2004) A novel partial modification at C2501 in Escherichia coli 23S ribosomal RNA. RNA 10(6):907–913

Appel CD, Maxwell ES (2007) Structural features of the guide: target RNA duplex required for archaeal box C/D sRNA-guided nucleotide 2′-O-methylation. RNA 13(6):899–911. doi:10.1261/rna.517307

Balakin AG, Smith L, Fournier MJ (1996) The RNA world of the nucleolus: two major families of small RNAs defined by different box elements with related functions. Cell 86(5):823–834

Ban N, Nissen P, Hansen J, Moore PB, Steitz TA (2000) The complete atomic structure of the large ribosomal subunit at 2.4 A resolution. Science 289(5481):905–920

Barbezier N, Canino G, Rodor J, Jobet E, Saez-Vasquez J, Marchfelder A, Echeverria M (2009) Processing of a dicistronic tRNA-snoRNA precursor: combined analysis in vitro and in vivo reveals alternate pathways and coupling to assembly of snoRNP. Plant Physiol 150 (3):1598–1610. doi:10.1104/pp.109.137968

Bleichert F, Baserga SJ (2010) Dissecting the role of conserved box C/D sRNA sequences in di-sRNP assembly and function. Nucleic Acids Res 38(22):8295–8305. doi:10.1093/nar/gkq690

Bleichert F, Gagnon KT, Brown BA 2nd, Maxwell ES, Leschziner AE, Unger VM, Baserga SJ (2009) A dimeric structure for archaeal box C/D small ribonucleoproteins. Science 325 (5946):1384–1387. doi:10.1126/science.1176099

Blouin S, Lafontaine DA (2007) A loop loop interaction and a K-turn motif located in the lysine aptamer domain are important for the riboswitch gene regulation control. RNA 13 (8):1256–1267. doi:10.1261/rna.560307

Bower-Phipps KR, Taylor DW, Wang HW, Baserga SJ (2012) The box C/D sRNP dimeric architecture is conserved across domain Archaea. RNA 18(8):1527–1540. doi:10.1261/rna.033134.112 [pii] rna.033134.112

Bratkovic T, Rogelj B (2011) Biology and applications of small nucleolar RNAs. Cell Mol Life Sci 68(23):3843–3851. doi:10.1007/s00018-011-0762-y

Brown JW, Echeverria M, Qu LH (2003) Plant snoRNAs: functional evolution and new modes of gene expression. Trends Plant Sci 8(1):42–49

Brown JW, Marshall DF, Echeverria M (2008) Intronic noncoding RNAs and splicing. Trends Plant Sci 13(7):335–342. doi:10.1016/j.tplants.2008.04.010

Bruenger E, Kowalak JA, Kuchino Y, McCloskey JA, Mizushima H, Stetter KO, Crain PF (1993) 5S rRNA modification in the hyperthermophilic archaea Sulfolobus solfataricus and Pyrodictium occultum. FASEB J 7(1):196–200

Buchhaupt M, Sharma S, Kellner S, Oswald S, Paetzold M, Peifer C, Watzinger P, Schrader J, Helm M, Entian KD (2014) Partial methylation at Am100 in 18S rRNA of baker's yeast reveals ribosome heterogeneity on the level of eukaryotic rRNA modification. PLoS One 9(2):e89640. doi:10.1371/journal.pone.0089640

Caffarelli E, Fatica A, Prislei S, De Gregorio E, Fragapane P, Bozzoni I (1996) Processing of the intron-encoded U16 and U18 snoRNAs: the conserved C and D boxes control both the processing reaction and the stability of the mature snoRNA. EMBO J 15(5):1121–1131

Caffarelli E, Losito M, Giorgi C, Fatica A, Bozzoni I (1998) In vivo identification of nuclear factors interacting with the conserved elements of box C/D small nucleolar RNAs. Mol Cell Biol 18(2):1023–1028

Caldas T, Binet E, Bouloc P, Richarme G (2000) Translational defects of Escherichia coli mutants deficient in the Um(2552) 23S ribosomal RNA methyltransferase RrmJ/FTSJ. Biochem Biophys Res Commun 271(3):714–718. doi:10.1006/bbrc.2000.2702

Cannone JJ, Subramanian S, Schnare MN, Collett JR, D'Souza LM, Du Y, Feng B, Lin N, Madabusi LV, Muller KM, Pande N, Shang Z, Yu N, Gutell RR (2002) The comparative RNA web (CRW) site: an online database of comparative sequence and structure information for ribosomal, intron, and other RNAs. BMC Bioinformatics 3:2

Cate JH, Yusupov MM, Yusupova GZ, Earnest TN, Noller HF (1999) X-ray crystal structures of 70S ribosome functional complexes. Science 285(5436):2095–2104

Cavaille J, Nicoloso M, Bachellerie JP (1996) Targeted ribose methylation of RNA in vivo directed by tailored antisense RNA guides. Nature 383(6602):732–735. doi:10.1038/383732a0

Chanfreau G, Legrain P, Jacquier A (1998a) Yeast RNase III as a key processing enzyme in small nucleolar RNAs metabolism. J Mol Biol 284(4):975–988. doi:10.1006/jmbi.1998.2237

Chanfreau G, Rotondo G, Legrain P, Jacquier A (1998b) Processing of a dicistronic small nucleolar RNA precursor by the RNA endonuclease Rnt1. EMBO J 17(13):3726–3737. doi:10.1093/emboj/17.13.3726

Chen CL, Chen CJ, Vallon O, Huang ZP, Zhou H, Qu LH (2008) Genomewide analysis of box C/D and box H/ACA snoRNAs in Chlamydomonas reinhardtii reveals an extensive organization into intronic gene clusters. Genetics 179(1):21–30. doi:10.1534/genetics.107.086025

Clouet d'Orval B, Bortolin ML, Gaspin C, Bachellerie JP (2001) Box C/D RNA guides for the ribose methylation of archaeal tRNAs. The tRNATrp intron guides the formation of two ribose-methylated nucleosides in the mature tRNATrp. Nucleic Acids Res 29(22):4518–4529

Clouet-d'Orval B, Rinaldi D, Quentin Y, Carpousis AJ (2010) Euryarchaeal beta-CASP proteins with homology to bacterial RNase J Have 5′- to 3′-exoribonuclease activity. J Biol Chem 285 (23):17574–17583. doi:10.1074/jbc.M109.095117

Clouet-d'Orval B, Phung DK, Langendijk-Genevaux PS, Quentin Y (2015) Universal RNA-degrading enzymes in Archaea: prevalence, activities and functions of beta-CASP ribonucleases. Biochimie 118:278–285. doi:10.1016/j.biochi.2015.05.021

Danan M, Schwartz S, Edelheit S, Sorek R (2012) Transcriptome-wide discovery of circular RNAs in Archaea. Nucleic Acids Res 40(7):3131–3142. doi:10.1093/nar/gkr1009
Decatur WA, Fournier MJ (2002) rRNA modifications and ribosome function. Trends Biochem Sci 27(7):344–351
Dennis PP, Omer A, Lowe T (2001) A guided tour: small RNA function in Archaea. Mol Microbiol 40(3):509–519
Dennis PP, Tripp V, Lui L, Lowe T, Randau L (2015) C/D box sRNA-guided 2′-O-methylation patterns of archaeal rRNA molecules. BMC Genomics 16:632. doi:10.1186/s12864-015-1839-z
Dieci G, Preti M, Montanini B (2009) Eukaryotic snoRNAs: a paradigm for gene expression flexibility. Genomics 94(2):83–88. doi:10.1016/j.ygeno.2009.05.002
Diener JL, Moore PB (1998) Solution structure of a substrate for the archaeal pre-tRNA splicing endonucleases: the bulge-helix-bulge motif. Mol Cell 1(6):883–894
Edelheit S, Schwartz S, Mumbach MR, Wurtzel O, Sorek R (2013) Transcriptome-wide mapping of 5-methylcytidine RNA modifications in bacteria, archaea, and yeast reveals m5C within archaeal mRNAs. PLoS Genet 9(6):e1003602. doi:10.1371/journal.pgen.1003602
Gagnon KT, Zhang X, Qu G, Biswas S, Suryadi J, Brown BA 2nd, Maxwell ES (2010) Signature amino acids enable the archaeal L7Ae box C/D RNP core protein to recognize and bind the K-loop RNA motif. RNA 16(1):79–90. doi:10.1261/rna.1692310
Gagnon KT, Biswas S, Zhang X, Brown BA II, Wollenzien P, Mattos C, Maxwell ES (2012) Structurally conserved Nop56/58 N-terminal domain facilitates archaeal box C/D ribonucleoprotein-guided methyltransferase activity. J Biol Chem 287(23):19418–19428. doi:10.1074/jbc.M111.323253
Galardi S, Fatica A, Bachi A, Scaloni A, Presutti C, Bozzoni I (2002) Purified box C/D snoRNPs are able to reproduce site-specific 2′-O-methylation of target RNA in vitro. Mol Cell Biol 22 (19):6663–6668
Ganot P, Bortolin ML, Kiss T (1997) Site-specific pseudouridine formation in preribosomal RNA is guided by small nucleolar RNAs. Cell 89(5):799–809
Gaspin C, Cavaille J, Erauso G, Bachellerie JP (2000) Archaeal homologs of eukaryotic methylation guide small nucleolar RNAs: lessons from the Pyrococcus genomes. J Mol Biol 297 (4):895–906. doi:10.1006/jmbi.2000.3593
Ge J, Yu YT (2013) RNA pseudouridylation: new insights into an old modification. Trends Biochem Sci 38(4):210–218. doi:10.1016/j.tibs.2013.01.002
Ghalei H, Hsiao HH, Urlaub H, Wahl MC, Watkins NJ (2010) A novel Nop5-sRNA interaction that is required for efficient archaeal box C/D sRNP formation. RNA 16(12):2341–2348. doi:10.1261/rna.2380410
Granneman S, Kudla G, Petfalski E, Tollervey D (2009) Identification of protein binding sites on U3 snoRNA and pre-rRNA by UV cross-linking and high-throughput analysis of cDNAs. Proc Natl Acad Sci U S A 106(24):9613–9618. doi:10.1073/pnas.0901997106
Graziadei A, Masiewicz P, Lapinaite A, Carlomagno T (2016) Archaea box C/D enzymes methylate two distinct substrate rRNA sequences with different efficiency. RNA. doi:10.1261/rna.054320.115
Green R, Noller HF (1996) In vitro complementation analysis localizes 23S rRNA posttranscriptional modifications that are required for Escherichia coli 50S ribosomal subunit assembly and function. RNA 2(10):1011–1021
Green R, Noller HF (1999) Reconstitution of functional 50S ribosomes from in vitro transcripts of Bacillus stearothermophilus 23S rRNA. Biochemistry 38(6):1772–1779. doi:10.1021/bi982246a
Grosjean H, Gaspin C, Marck C, Decatur WA, de Crecy-Lagard V (2008) RNomics and Modomics in the halophilic archaea Haloferax volcanii: identification of RNA modification genes. BMC Genomics 9:470. doi:10.1186/1471-2164-9-470
Hamma T, Ferre-D'Amare AR (2010) The box H/ACA ribonucleoprotein complex: interplay of RNA and protein structures in post-transcriptional RNA modification. J Biol Chem 285 (2):805–809. doi:10.1074/jbc.R109.076893

Hansen JL, Schmeing TM, Moore PB, Steitz TA (2002a) Structural insights into peptide bond formation. Proc Natl Acad Sci U S A 99(18):11670–11675. doi:10.1073/pnas.172404099

Hansen MA, Kirpekar F, Ritterbusch W, Vester B (2002b) Posttranscriptional modifications in the A-loop of 23S rRNAs from selected archaea and eubacteria. RNA 8(2):202–213

Hasenohrl D, Konrat R, Blasi U (2011) Identification of an RNase J ortholog in Sulfolobus solfataricus: implications for 5′-to-3′ directional decay and 5′-end protection of mRNA in Crenarchaeota. RNA 17(1):99–107. doi:10.1261/rna.2418211

Helm M (2006) Post-transcriptional nucleotide modification and alternative folding of RNA. Nucleic Acids Res 34(2):721–733. doi:10.1093/nar/gkj471 [pii] 34/2/721

Heppell B, Lafontaine DA (2008) Folding of the SAM aptamer is determined by the formation of a K-turn-dependent pseudoknot. Biochemistry 47(6):1490–1499. doi:10.1021/bi701164y

Herschlag D, Eckstein F, Cech TR (1993) The importance of being ribose at the cleavage site in the Tetrahymena ribozyme reaction. Biochemistry 32(32):8312–8321

Hury J, Nagaswamy U, Larios-Sanz M, Fox GE (2006) Ribosome origins: the relative age of 23S rRNA Domains. Orig Life Evol Biosph 36(4):421–429. doi:10.1007/s11084-006-9011-z

Huttenhofer A, Kiefmann M, Meier-Ewert S, O'Brien J, Lehrach H, Bachellerie JP, Brosius J (2001) RNomics: an experimental approach that identifies 201 candidates for novel, small, non-messenger RNAs in mouse. EMBO J 20(11):2943–2953. doi:10.1093/emboj/20.11.2943

Khaitovich P, Tenson T, Kloss P, Mankin AS (1999) Reconstitution of functionally active Thermus aquaticus large ribosomal subunits with in vitro-transcribed rRNA. Biochemistry 38(6):1780–1788. doi:10.1021/bi9822473

King TH, Liu B, McCully RR, Fournier MJ (2003) Ribosome structure and activity are altered in cells lacking snoRNPs that form pseudouridines in the peptidyl transferase center. Mol Cell 11 (2):425–435

Kirpekar F, Douthwaite S, Roepstorff P (2000) Mapping posttranscriptional modifications in 5S ribosomal RNA by MALDI mass spectrometry. RNA 6(2):296–306

Kirpekar F, Hansen LH, Rasmussen A, Poehlsgaard J, Vester B (2005) The archaeon Haloarcula marismortui has few modifications in the central parts of its 23S ribosomal RNA. J Mol Biol 348(3):563–573. doi:10.1016/j.jmb.2005.03.009

Kiss T, Filipowicz W (1995) Exonucleolytic processing of small nucleolar RNAs from pre-mRNA introns. Genes Dev 9(11):1411–1424

Kiss-Laszlo Z, Henry Y, Bachellerie JP, Caizergues-Ferrer M, Kiss T (1996) Site-specific ribose methylation of preribosomal RNA: a novel function for small nucleolar RNAs. Cell 85 (7):1077–1088

Kiss-Laszlo Z, Henry Y, Kiss T (1998) Sequence and structural elements of methylation guide snoRNAs essential for site-specific ribose methylation of pre-rRNA. EMBO J 17(3):797–807. doi:10.1093/emboj/17.3.797

Klein DJ, Schmeing TM, Moore PB, Steitz TA (2001) The kink-turn: a new RNA secondary structure motif. EMBO J 20(15):4214–4221. doi:10.1093/emboj/20.15.4214

Koonin EV, Wolf YI, Aravind L (2001) Prediction of the archaeal exosome and its connections with the proteasome and the translation and transcription machineries by a comparative-genomic approach. Genome Res 11(2):240–252. doi:10.1101/gr.162001

Kruszka K, Barneche F, Guyot R, Ailhas J, Meneau I, Schiffer S, Marchfelder A, Echeverria M (2003) Plant dicistronic tRNA-snoRNA genes: a new mode of expression of the small nucleolar RNAs processed by RNase Z. EMBO J 22(3):621–632. doi:10.1093/emboj/cdg040

Kuhn JF, Tran EJ, Maxwell ES (2002) Archaeal ribosomal protein L7 is a functional homolog of the eukaryotic 15.5kD/Snu13p snoRNP core protein. Nucleic Acids Res 30(4):931–941

Lafontaine DL, Tollervey D (1999) Nop58p is a common component of the box C+D snoRNPs that is required for snoRNA stability. RNA 5(3):455–467

Lafontaine DL, Tollervey D (2000) Synthesis and assembly of the box C+D small nucleolar RNPs. Mol Cell Biol 20(8):2650–2659

Lapinaite A, Simon B, Skjaerven L, Rakwalska-Bange M, Gabel F, Carlomagno T (2013) The structure of the box C/D enzyme reveals regulation of RNA methylation. Nature 502 (7472):519–523. doi:10.1038/nature12581

Leader DJ, Sanders JF, Waugh R, Shaw P, Brown JW (1994) Molecular characterisation of plant U14 small nucleolar RNA genes: closely linked genes are transcribed as polycistronic U14 transcripts. Nucleic Acids Res 22(24):5196–5203

Leader DJ, Clark GP, Watters J, Beven AF, Shaw PJ, Brown JW (1997) Clusters of multiple different small nucleolar RNA genes in plants are expressed as and processed from polycistronic pre-snoRNAs. EMBO J 16(18):5742–5751. doi:10.1093/emboj/16.18.5742

Leader DJ, Clark GP, Watters J, Beven AF, Shaw PJ, Brown JW (1999) Splicing-independent processing of plant box C/D and box H/ACA small nucleolar RNAs. Plant Mol Biol 39 (6):1091–1100

Li SG, Zhou H, Luo YP, Zhang P, Qu LH (2005) Identification and functional analysis of 20 Box H/ACA small nucleolar RNAs (snoRNAs) from Schizosaccharomyces pombe. J Biol Chem 280(16):16446–16455. doi:10.1074/jbc.M500326200

Liang D, Zhou H, Zhang P, Chen YQ, Chen X, Chen CL, Qu LH (2002) A novel gene organization: intronic snoRNA gene clusters from Oryza sativa. Nucleic Acids Res 30 (14):3262–3272

Liang XH, Liu Q, Fournier MJ (2009) Loss of rRNA modifications in the decoding center of the ribosome impairs translation and strongly delays pre-rRNA processing. RNA 15 (9):1716–1728. doi:10.1261/rna.1724409

Lilley DM (2014) The K-turn motif in riboswitches and other RNA species. Biochim Biophys Acta 1839(10):995–1004. doi:10.1016/j.bbagrm.2014.04.020

Lin J, Lai S, Jia R, Xu A, Zhang L, Lu J, Ye K (2011) Structural basis for site-specific ribose methylation by box C/D RNA protein complexes. Nature 469(7331):559–563. doi:10.1038/nature09688

Lorentzen E, Walter P, Fribourg S, Evguenieva-Hackenberg E, Klug G, Conti E (2005) The archaeal exosome core is a hexameric ring structure with three catalytic subunits. Nat Struct Mol Biol 12(7):575–581. doi:10.1038/nsmb952

Lorentzen E, Dziembowski A, Lindner D, Seraphin B, Conti E (2007) RNA channelling by the archaeal exosome. EMBO Rep 8(5):470–476. doi:10.1038/sj.embor.7400945

Lovgren JM, Wikstrom PM (2001) The rlmB gene is essential for formation of Gm2251 in 23S rRNA but not for ribosome maturation in Escherichia coli. J Bacteriol 183(23):6957–6960. doi:10.1128/JB.183.23.6957-6960.2001

Maden BE, Corbett ME, Heeney PA, Pugh K, Ajuh PM (1995) Classical and novel approaches to the detection and localization of the numerous modified nucleotides in eukaryotic ribosomal RNA. Biochimie 77(1-2):22–29

Martens B, Amman F, Manoharadas S, Zeichen L, Orell A, Albers SV, Hofacker I, Blasi U (2013) Alterations of the transcriptome of Sulfolobus acidocaldarius by exoribonuclease aCPSF2. PLoS One 8(10):e76569. doi:10.1371/journal.pone.0076569

Matera AG, Terns RM, Terns MP (2007) Non-coding RNAs: lessons from the small nuclear and small nucleolar RNAs. Nat Rev Mol Cell Biol 8(3):209–220. doi:10.1038/nrm2124

Maxwell ES, Fournier MJ (1995) The small nucleolar RNAs. Annu Rev Biochem 64:897–934. doi:10.1146/annurev.bi.64.070195.004341

Mengel-Jorgensen J, Jensen SS, Rasmussen A, Poehlsgaard J, Iversen JJ, Kirpekar F (2006) Modifications in Thermus thermophilus 23 S ribosomal RNA are centered in regions of RNA-RNA contact. J Biol Chem 281(31):22108–22117. doi:10.1074/jbc.M600377200

Newman DR, Kuhn JF, Shanab GM, Maxwell ES (2000) Box C/D snoRNA-associated proteins: two pairs of evolutionarily ancient proteins and possible links to replication and transcription. RNA 6(6):861–879

Ni J, Tien AL, Fournier MJ (1997) Small nucleolar RNAs direct site-specific synthesis of pseudouridine in ribosomal RNA. Cell 89(4):565–573

Nolivos S, Carpousis AJ, Clouet-d'Orval B (2005) The K-loop, a general feature of the Pyrococcus C/D guide RNAs, is an RNA structural motif related to the K-turn. Nucleic Acids Res 33 (20):6507–6514. doi:10.1093/nar/gki962

Noon KR, Bruenger E, McCloskey JA (1998) Posttranscriptional modifications in 16S and 23S rRNAs of the archaeal hyperthermophile Sulfolobus solfataricus. J Bacteriol 180 (11):2883–2888

Ochs RL, Lischwe MA, Spohn WH, Busch H (1985) Fibrillarin: a new protein of the nucleolus identified by autoimmune sera. Biol Cell 54(2):123–133

Omer AD, Lowe TM, Russell AG, Ebhardt H, Eddy SR, Dennis PP (2000) Homologs of small nucleolar RNAs in Archaea. Science 288(5465):517–522

Omer AD, Ziesche S, Ebhardt H, Dennis PP (2002) In vitro reconstitution and activity of a C/D box methylation guide ribonucleoprotein complex. Proc Natl Acad Sci U S A 99 (8):5289–5294. doi:10.1073/pnas.082101999

Pelczar P, Filipowicz W (1998) The host gene for intronic U17 small nucleolar RNAs in mammals has no protein-coding potential and is a member of the 5′-terminal oligopyrimidine gene family. Mol Cell Biol 18(8):4509–4518

Petfalski E, Dandekar T, Henry Y, Tollervey D (1998) Processing of the precursors to small nucleolar RNAs and rRNAs requires common components. Mol Cell Biol 18(3):1181–1189

Petrov AS, Bernier CR, Hershkovits E, Xue Y, Waterbury CC, Hsiao C, Stepanov VG, Gaucher EA, Grover MA, Harvey SC, Hud NV, Wartell RM, Fox GE, Williams LD (2013) Secondary structure and domain architecture of the 23S and 5S rRNAs. Nucleic Acids Res 41 (15):7522–7535. doi:10.1093/nar/gkt513

Polikanov YS, Melnikov SV, Soll D, Steitz TA (2015) Structural insights into the role of rRNA modifications in protein synthesis and ribosome assembly. Nat Struct Mol Biol 22(4):342–344. doi:10.1038/nsmb.2992

Purta E, O'Connor M, Bujnicki JM, Douthwaite S (2009) YgdE is the 2′-O-ribose methyltransferase RlmM specific for nucleotide C2498 in bacterial 23S rRNA. Mol Microbiol 72(5):1147–1158. doi:10.1111/j.1365-2958.2009.06709.x

Qu LH, Henras A, Lu YJ, Zhou H, Zhou WX, Zhu YQ, Zhao J, Henry Y, Caizergues-Ferrer M, Bachellerie JP (1999) Seven novel methylation guide small nucleolar RNAs are processed from a common polycistronic transcript by Rat1p and RNase III in yeast. Mol Cell Biol 19 (2):1144–1158

Randau L (2012) RNA processing in the minimal organism Nanoarchaeum equitans. Genome Biol 13(7):R63. doi:10.1186/gb-2012-13-7-r63

Randau L, Munch R, Hohn MJ, Jahn D, Soll D (2005) Nanoarchaeum equitans creates functional tRNAs from separate genes for their 5′- and 3′-halves. Nature 433(7025):537–541. doi:10.1038/nature03233

Reiter NJ, Osterman A, Torres-Larios A, Swinger KK, Pan T, Mondragon A (2010) Structure of a bacterial ribonuclease P holoenzyme in complex with tRNA. Nature 468(7325):784–789. doi:10.1038/nature09516

Rozhdestvensky TS, Tang TH, Tchirkova IV, Brosius J, Bachellerie JP, Huttenhofer A (2003) Binding of L7Ae protein to the K-turn of archaeal snoRNAs: a shared RNA binding motif for C/D and H/ACA box snoRNAs in Archaea. Nucleic Acids Res 31(3):869–877

Schimmang T, Tollervey D, Kern H, Frank R, Hurt EC (1989) A yeast nucleolar protein related to mammalian fibrillarin is associated with small nucleolar RNA and is essential for viability. EMBO J 8(13):4015–4024

Singh SK, Gurha P, Tran EJ, Maxwell ES, Gupta R (2004) Sequential 2′-O-methylation of archaeal pre-tRNATrp nucleotides is guided by the intron-encoded but trans-acting box C/D ribonucleoprotein of pre-tRNA. J Biol Chem 279(46):47661–47671. doi:10.1074/jbc.M408868200

Starostina NG, Marshburn S, Johnson LS, Eddy SR, Terns RM, Terns MP (2004) Circular box C/D RNAs in Pyrococcus furiosus. Proc Natl Acad Sci U S A 101(39):14097–14101. doi:10.1073/pnas.0403520101

Steitz JA, Tycowski KT (1995) Small RNA chaperones for ribosome biogenesis. Science 270 (5242):1626–1627

Su AA, Tripp V, Randau L (2013) RNA-Seq analyses reveal the order of tRNA processing events and the maturation of C/D box and CRISPR RNAs in the hyperthermophile Methanopyrus kandleri. Nucleic Acids Res 41(12):6250–6258. doi:10.1093/nar/gkt317

Szewczak LB, DeGregorio SJ, Strobel SA, Steitz JA (2002) Exclusive interaction of the 15.5 kD protein with the terminal box C/D motif of a methylation guide snoRNP. Chem Biol 9 (10):1095–1107

Tran E, Zhang X, Lackey L, Maxwell ES (2005) Conserved spacing between the box C/D and C′/D′ RNPs of the archaeal box C/D sRNP complex is required for efficient 2′-O-methylation of target RNAs. RNA 11(3):285–293. doi:10.1261/rna.7223405

Tripp V, Martin R, Orell A, Alkhnbashi OS, Backofen R, Randau L (2016) Plasticity of archaeal C/D box sRNA biogenesis. Mol Microbiol. doi:10.1111/mmi.13549

Tyc K, Steitz JA (1989) U3, U8 and U13 comprise a new class of mammalian snRNPs localized in the cell nucleolus. EMBO J 8(10):3113–3119

Tycowski KT, You ZH, Graham PJ, Steitz JA (1998) Modification of U6 spliceosomal RNA is guided by other small RNAs. Mol Cell 2(5):629–638

Tycowski KT, Aab A, Steitz JA (2004) Guide RNAs with 5′ caps and novel box C/D snoRNA-like domains for modification of snRNAs in metazoa. Curr Biol 14(22):1985–1995. doi:10.1016/j.cub.2004.11.003

van Nues RW, Granneman S, Kudla G, Sloan KE, Chicken M, Tollervey D, Watkins NJ (2011) Box C/D snoRNP catalysed methylation is aided by additional pre-rRNA base-pairing. EMBO J 30(12):2420–2430. doi:10.1038/emboj.2011.148

Villa T, Ceradini F, Presutti C, Bozzoni I (1998) Processing of the intron-encoded U18 small nucleolar RNA in the yeast Saccharomyces cerevisiae relies on both exo- and endonucleolytic activities. Mol Cell Biol 18(6):3376–3383

Walter P, Klein F, Lorentzen E, Ilchmann A, Klug G, Evguenieva-Hackenberg E (2006) Characterization of native and reconstituted exosome complexes from the hyperthermophilic archaeon Sulfolobus solfataricus. Mol Microbiol 62(4):1076–1089. doi:10.1111/j.1365-2958.2006.05393.x

Wang H, Boisvert D, Kim KK, Kim R, Kim SH (2000) Crystal structure of a fibrillarin homologue from Methanococcus jannaschii, a hyperthermophile, at 1.6 A resolution. EMBO J 19 (3):317–323. doi:10.1093/emboj/19.3.317

Waters E, Hohn MJ, Ahel I, Graham DE, Adams MD, Barnstead M, Beeson KY, Bibbs L, Bolanos R, Keller M, Kretz K, Lin X, Mathur E, Ni J, Podar M, Richardson T, Sutton GG, Simon M, Soll D, Stetter KO, Short JM, Noordewier M (2003) The genome of Nanoarchaeum equitans: insights into early archaeal evolution and derived parasitism. Proc Natl Acad Sci U S A 100(22):12984–12988. doi:10.1073/pnas.1735403100

Watkins NJ, Bohnsack MT (2012) The box C/D and H/ACA snoRNPs: key players in the modification, processing and the dynamic folding of ribosomal RNA. Wiley Interdiscip Rev RNA 3(3):397–414. doi:10.1002/wrna.117

Watkins NJ, Segault V, Charpentier B, Nottrott S, Fabrizio P, Bachi A, Wilm M, Rosbash M, Branlant C, Luhrmann R (2000) A common core RNP structure shared between the small nucleoar box C/D RNPs and the spliceosomal U4 snRNP. Cell 103(3):457–466

Watkins NJ, Dickmanns A, Luhrmann R (2002) Conserved stem II of the box C/D motif is essential for nucleolar localization and is required, along with the 15.5K protein, for the hierarchical assembly of the box C/D snoRNP. Mol Cell Biol 22(23):8342–8352

Weber MJ (2006) Mammalian small nucleolar RNAs are mobile genetic elements. PLoS Genet 2 (12):e205. doi:10.1371/journal.pgen.0020205 [pii] 06-PLGE-RA-0372R2

Weisel J, Wagner S, Klug G (2010) The Nop5-L7A-fibrillarin RNP complex and a novel box C/D containing sRNA of Halobacterium salinarum NRC-1. Biochem Biophys Res Commun 394 (3):542–547. doi:10.1016/j.bbrc.2010.03.012

Wimberly BT, Brodersen DE, Clemons WM Jr, Morgan-Warren RJ, Carter AP, Vonrhein C, Hartsch T, Ramakrishnan V (2000) Structure of the 30S ribosomal subunit. Nature 407 (6802):327–339. doi:10.1038/35030006

Xue S, Wang R, Yang F, Terns RM, Terns MP, Zhang X, Maxwell ES, Li H (2010) Structural basis for substrate placement by an archaeal box C/D ribonucleoprotein particle. Mol Cell 39 (6):939–949. doi:10.1016/j.molcel.2010.08.022

Yip WS, Shigematsu H, Taylor DW, Baserga SJ (2016) Box C/D sRNA stem ends act as stabilizing anchors for box C/D di-sRNPs. Nucleic Acids Res. doi:10.1093/nar/gkw576

Yokoyama S, Watanabe K, Miyazawa T (1987) Dynamic structures and functions of transfer ribonucleic acids from extreme thermophiles. Adv Biophys 23:115–147

Zago MA, Dennis PP, Omer AD (2005) The expanding world of small RNAs in the hyperthermophilic archaeon Sulfolobus solfataricus. Mol Microbiol 55(6):1812–1828. doi:10.1111/j.1365-2958.2005.04505.x

Zhang X, Champion EA, Tran EJ, Brown BA II, Baserga SJ, Maxwell ES (2006) The coiled-coil domain of the Nop56/58 core protein is dispensable for sRNP assembly but is critical for archaeal box C/D sRNP-guided nucleotide methylation. RNA 12(6):1092–1103. doi:10.1261/rna.2230106

Ziesche SM, Omer AD, Dennis PP (2004) RNA-guided nucleotide modification of ribosomal and non-ribosomal RNAs in Archaea. Mol Microbiol 54(4):980–993. doi:10.1111/j.1365-2958.2004.04319.x

Chapter 10
Diverse Functions of Small RNAs (sRNAs) in Halophilic Archaea: From Non-coding Regulatory sRNAs to Microprotein-Encoding sRNAs

Jana Kliemt and Jörg Soppa

Abstract The number of experimentally-verified small RNAs (sRNAs) in haloarchaea has increased tremendously since their discovery in 2009. A recent dRNA-Seq study led to the identification of about 2900 non-coding primary transcripts, a number that was much higher than the less than 1900 protein-coding mRNAs that were detected. Intergenic sRNAs have been studied intensively, and it was revealed that they have important regulatory functions under many different conditions. However, the numbers of cis-antisense RNAs and of cis-sense RNAs are much higher, and thus future research will shift to concentrate on their characterization. Other classes of haloarchaeal sRNAs include tRNA-derived fragments, which have been shown to be regulators of translation, CRISPR/CAS defensive sRNAs, C/D box sRNAs that guide methylation, and 7S RNA involved in membrane protein biogenesis. Last, but not least haloarchaea contain hundreds of small mRNAs that encode microproteins, which represent an emerging field of research. Taken together, the network of small RNAs in haloarchaea is much larger than anticipated only a few years ago, and the multitude of sRNAs has changed the view of the function of haloarchaeal genomes.

10.1 Introduction

Small non-coding regulatory RNAs (sRNAs) exist in all three domains of life, archaea, bacteria, and eukaryotes. In recent years an ever increasing number of sRNAs have been discovered, and they were found to be involved in and important for many different biological processes. Several recent reviews summarize various aspects of sRNAs in prokaryotes (Wagner and Romby 2015; Kopf and Hess 2015; Murina and Nikulin 2015; van Puyvelde et al. 2015; Georg and Hess 2011; Waters

J. Kliemt • J. Soppa (✉)
Biocentre, Institute for Molecular Biosciences, Goethe-University, Max-von-Laue-Str. 9, 60438 Frankfurt, Germany
e-mail: kliemt@bio.uni-frankfurt.de; soppa@bio.uni-frankfurt.de

B. Clouet-d'Orval (ed.), *RNA Metabolism and Gene Expression in Archaea*, Nucleic Acids and Molecular Biology 32, DOI 10.1007/978-3-319-65795-0_10

and Storz 2009) and in eukaryotes (Catalanotto et al. 2016; Yang et al. 2016; Borges and Marienssen 2015; Huang et al. 2013). In addition, very recently it has been acknowledged that small RNAs can also contain open reading frames and that the encoded microproteins can have very important functions (reviews: Ramamurthi and Storz 2014; Storz et al. 2014). This chapter will concentrate on the various groups of non-coding and coding sRNAs from halophilic archaea, including e.g. cis antisense RNAs, snoRNAs, and tRNA-derived fragments. On the one hand it updates two earlier reviews (Babski et al. 2014; Schmitz-Streit et al. 2011), on the other hand it gives a broader view and includes additional classes like microprotein-encoding sRNAs. sRNAs from other phylogenetic groups of archaea are discussed in other chapters of this book.

10.2 Identification of Small RNAs and the Changing View of the Haloarchaeal Transcriptome

The first archaeal sRNAs were detected in the euryarchaeon *Archaeoglobus fulgidus* at the beginning of this century (Tang et al. 2002). Shortly thereafter, it was unraveled that sRNAs occur also in crenarchaeota (Tang et al. 2005). The first study with a halophilic archaeon was published a few years later (Straub et al. 2009). A small scale RNomics study with *Haloferax volcanii* led to the identification of 21 intergenic sRNAs and 18 antisense sRNAs (asRNAs). Northern blot analyses revealed that many of the sRNA genes were differentially expressed, indicating that the regulatory roles of the respective sRNAs are confined to specific environmental conditions. The next approach was the bioinformatics comparison of the intergenic regions of *Hfx. volcanii* with those of four other haloarchaea, one crenarchaeon, and one halophilic bacterium (Babski et al. 2011). More than 120 conserved regions that might represent conserved sRNA genes were found. The expression of 61 of these putative sRNA genes was analyzed using a dedicated DNA microarray, and 37 genes were found to be expressed under at least one of the three conditions tested, verifying that the bioinformatics predictions could successfully identify sRNA genes.

High Throughput Sequencing (HTS) of cDNA libraries, which was relatively new at that time, was used to characterize the small transcriptome of sRNAs with lengths between 17 and 500 nt (Heyer et al. 2012). Thereby, the number of sRNAs was increased to 145 intergenic sRNAs and 45 asRNAs. RNAs from cultures grown at six different conditions were used for cDNA library generation, and multiplexing was used to sort the sequences bioinformatically after a single HTS run. Again, it was found that many sRNA genes were differentially expressed. Notably, many sRNAs could only be detected in cultures that were grown under low salt, a condition that represents considerable stress for haloarchaea. Haloarchaea use the so-called “salt in” strategy for osmoadaptation. The salt concentration in the cytoplasm is as high as in the environment. The consequence of this strategy is that all biological processes have to be evolutionary adapted to function in the presence of molar concentrations of salt. Haloarchaeal proteins contain 20% aspartic and glutamic acid residues and have a high negative charge density at their surface. This makes them soluble at high salt concentrations, but on the other hand, this makes them very sensitive to low salt concentrations. Typical haloarchaeal proteins denature at salt concentrations below 1 M NaCl. Most halophilic bacteria apply the so called "salt out" strategy, they have a low salt concentration in the cytoplasm and use organic compatible solutes for osmoadaptation.

Very recently a state of the art differential RNA-Seq (dRNA-Seq) approach was used to characterize the primary transcriptome of *Hfx. volcanii* (Babski et al. 2016). dRNA-Seq makes use of an enzyme that degrades all transcripts without a triphosphate at their 5′-end, while transcripts with a triphosphate remain untouched. Comparison of treated and untreated samples allows the differentiation between primary transcripts and transcripts that were generated by processing or are degradation intermediates. The highly increased sequencing depth led to the identification of nearly 2800 novel non-coding transcripts. Remarkably, the total number of non-coding RNAs was with 2900 considerably higher than the total number of protein-coding RNAs with less than 1900. Taken together, the view of the transcriptome and of the genome function of *Hfx. volcanii* has changed dramatically within the last 6 years. Figure 10.1a schematically shows that according to the original annotation the genome contained nearly exclusively protein-coding genes (Hartman et al. 2010). Figure 10.1b illustrates that small-scale RNomics and the HTS approaches led to the identification of about 200 sRNA genes (Babski et al.

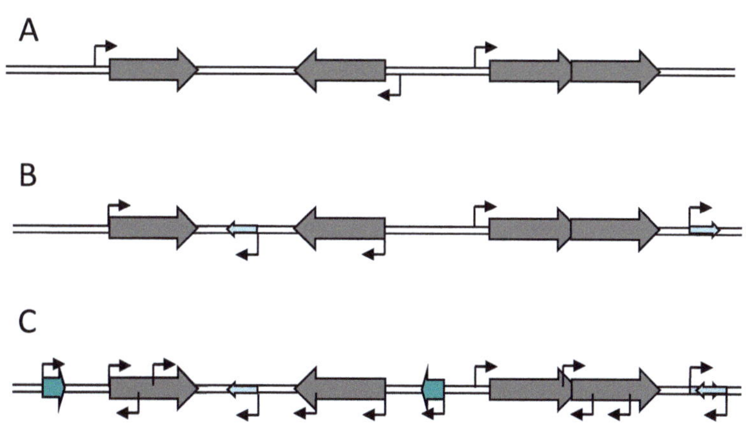

Fig. 10.1 Changing view of the transcriptome and genome function of *Hfx. volcanii* during recent years. (**a**) View based on genome sequencing and annotation in 2010 (Hartmann et al. 2010). (**b**) View based on the identification of sRNAs genes in intergenic regions (Heyer et al. 2012). (**c**) View based on the very recent dRNA-Seq study (Babski et al. 2016). The *arrows* indicate presumed transcription start sites (**a**, **b**) and experimentally verified transcription start sites (**c**)

2011; Heyer et al. 2012). In stark contrast, the dRNA-Seq study uncovered that the number of non-coding RNAs is in fact higher than the number of protein-coding genes (Fig. 10.1c) (Babski et al. 2016). Various different classes of non-coding RNAs were found, which are discussed below. It can be expected that not all sRNAs of *Hfx. volcanii* have been identified yet, because the dRNA-Seq study was performed using cultures grown under optimal conditions. Because it has been shown that sRNA genes can be differentially expressed and can be silent under optimal conditions (see above), it can safely be predicted that further studies with cultures grown under non-optimal conditions will further enhance the number of sRNAs of *Hfx. volcanii*.

dRNA-Seq studies have been performed only for three additional archaeal species, i.e. *Methanolobus psychrophilus* (Li et al. 2015), *Thermococcus kodakarensis* (Jäger et al. 2014), and *Methanosarcina mazei* Gö1 (Jäger et al. 2009). For all three species the number of non-coding RNAs was much smaller than the number of protein-coding RNAs, e.g. only 195 of 2056 transcripts from *M. psychrophilus* were non-coding sRNAs. Therefore, the fraction of non-coding RNAs is not uniformly high in all archaea, and the situation in haloarchaea resemble the situation in higher eukaryotes, which also contain a higher number of non-coding transcripts than protein-coding transcripts (Wan et al. 2014).

Table 10.1 gives on overview of the number of three classes of sRNAs and the number of annotated protein-coding genes for six archaeal species. Only RNA-Seq studies and dRNA-Seq studies since 2009 have been included, because earlier small-scale RNomics studies led to much smaller numbers of identified sRNAs. It should be noted that the numbers should be handled with care, because the number of different culturing conditions, the sequencing depth, and the bioinformatics analysis pipeline can tremendously influence the results. Nevertheless, it can be seen that the numbers especially of intergenic sRNAs and asRNAs (and their ratios) differ considerably in the six investigated species.

Table 10.1 Numbers of protein coding genes (genome annotation) and of three classes of non-coding sRNAs (RNA-Seq or dRNA-Seq) in selected archaeal species

Species	No. of ORFs	Interg. sRNAs	asRNAs	iRNAs	Reference
Haloferax volcanii	4040	395	1244	1153	Babski et al. (2016)
Methanolobus psychr.	3167	195	1110	1440	Li et al. (2015)
Thermococcus kodak.	2306	69	1018	644	Jäger et al. (2014)
Pyrococcus abyssi	1784	107	215	n.r.	Toffano-Nioche et al. (2013)
Haloferax volcanii	4040	145	45	n.r.	Heyer et al. (2012)
Sulfolobus solft.	2994	125	185	n.r.	Wurtzel et al. (2010)
Methanosarcina m.	3371	199	43	n.r.	Jäger et al. (2009)

n.r. not reported

10.3 Various Classes of Small Non-coding Regulatory Haloarchaeal RNAs

10.3.1 Intergenic sRNAs

The intergenic sRNAs were the first sRNAs that have been systematically characterized. Already in the first study two gene deletion mutants have been constructed and phenotypically analyzed (Straub et al. 2009). One of the mutants could not grow at the elevated temperature of 51 °C, the other mutant had a severe growth defect at the low salt concentration of 0.9 M NaCl. Both phenotypes underscored the high importance of sRNAs for the physiology of *H. volanii*. In a subsequent study 27 sRNA gene deletion mutants were generated and characterized (Jaschinski et al. 2014). For 24 of the 27 mutants a phenotypic difference from the wild-type could be detected under at least 1 of the 12 tested conditions. In addition, differential expression of sRNA genes was studied using a variety of different experimental approaches (Northern blot analyses, reporter gene assays, DNA microarray analyses). The results of all approaches revealed that sRNAs are important for the regulation of many biological functions in haloarchaea, which is schematically illustrated in Fig. 10.2. The biological functions include stress adaptation (which is proposed to be the major function of sRNAs in bacteria), but also metabolic regulation, adaptation to the extremes of growth conditions, and, last but not least, regulation of behavior. Remarkably, more than 10 of the 27 deletion mutants exhibited a gain-of-function phenotype. As yet this is unprecedented for any sRNA gene deletion mutant in bacteria. However, gain-of-function phenotypes have also been described for deletion or depletion of miRNAs in higher eukaryotes (Daniel et al. 2014). These results illustrate that regulatory circuits did not evolve to ensure the highest growth rate under one specific (laboratory) condition, but that regulatory networks were favored that had the highest stability and flexibility under the ever changing conditions of natural environments.

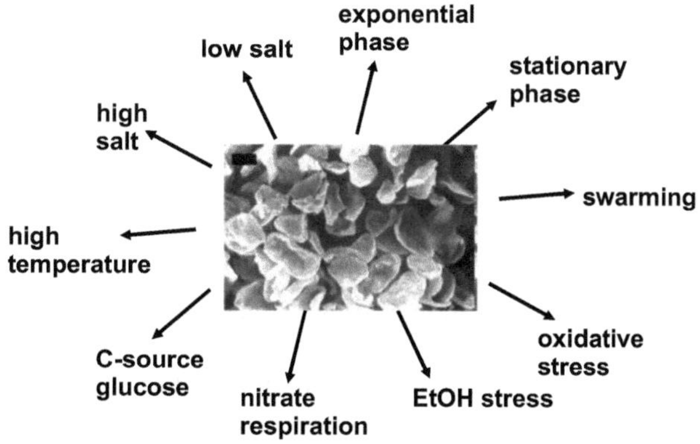

Fig. 10.2 Schematic overview of the diverse biological functions of sRNAs in *Hfx. volcanii*. The functions have been deduced from the phenotypes of sRNA gene deletion mutants and form elevated sRNA levels under specific conditions. The EM picture of *Hfx. volcanii* was supplied by J. Babski, K. Jaschinski, and J. Soppa (unpublished data)

The recent dRNA-Seq study increased the number of intergenic sRNAs to more than 400. Only a small fraction of them have been studied until now, therefore, the already uncovered manifold functions of sRNAs in haloarchaea (Fig. 10.2) represent only the tip of the iceberg. Further work is also needed to identify the molecular targets of sRNAs, which are presumably primarily protein-coding mRNAs, as well as the molecular details of sRNA-target RNA interactions and the molecular mechanisms of regulation. The bioinformatics target prediction algorithms that have been successfully used with bacteria and methanogenic archaea have as yet not been successful with haloarchaea, possibly because the conditions in the high salt cytoplasm are so different from the conditions in mesohalic species. However, comparisons of the transcriptomes of sRNA deletion mutants and the wild-type have already led to the discovery of the target mRNAs for several intergenic sRNAs, and thus it can be expected that experimental approaches will soon shed light on the details of the regulatory functions of intergenic haloarchaeal sRNA. Because 72% of all haloarchaeal protein-coding transcripts are leaderless (Babski et al. 2016), it has been predicted that many sRNAs might bind to the 3′-UTRs of their target mRNAs. This would be analogous to the eukaryotic miRNAs, which also bind to 3′UTRs, and in contrast to bacterial sRNAs, which typically bind to the 5′-region. The sRNA-target mRNA interaction does not seem to be uniform in archaea, first examples include the binding to the 5′-region in *M. mazei* (Prasse et al. 2013; Jäger et al. 2012) as well as the binding to the 3′UTR in *Hfx. volcanii* (Kliemt, Jaschinski, and Soppa, unpublished data) and in *Sulfolobus solfataricus* (Martens et al. 2013).

10.3.2 Cis Sense sRNAs

The analysis of the primary transcriptome of *Hfx. volcanii* led to the identification of more than 1100 sRNAs that were encoded in the same direction and within ORFs of protein-encoding genes (cis sense sRNAs) (Babski et al. 2016). This class of sRNAs had also been found in previous studies, but had not been further discussed, because these RNAs might be meta-stable degradation intermediates of the mRNAs. However, this possibility could be excluded by the experimental design of dRNA-Seq, which enriches for primary transcripts with a triphosphate at the 5′-end. In addition, a high fraction of these internal sRNA genes were preceded by promoter motifs with a high promoter score, also indicating that a high number of ORF-internal promoters exist in *Hfx. volcanii.* Also for *Hbt. salinarum* a large number of internal transcripts have been described, that were preceded by transcription factor binding sites. Therefore, it was concluded that a high number of ORF-internal promoters and cis sense sRNAs exists in *Hbt. salinarum* (Koide et al. 2009). ORF-internal promoters at the 3′-end of the first of two overlapping genes can drive the expression of the down-stream protein-coding gene, and in these cases the transcripts would not be bona fide sRNAs. Such an example has been characterized for the *HVO_2723*/*HVO_2722* gene pair of *Hfx. volcanii*

(Maier et al. 2015a). However, these cases will be only a very minor fraction, because the ORF-internal TSS were not enriched at the 3′-ends of genes, but distributed throughout the ORFs (Babski et al. 2016). The functionality of stand-alone internal sRNA genes was proven by overexpression of two examples, which led to a clear phenotypic difference between the overexpression mutants and the wild-type (Gomez-Filho et al. 2015). Overexpression of two internal sRNAs (VNG_aot0042 and VNG_R0052) resulted in a slight increase in growth rate and an about 50% increase in growth yield, compared with control cultures containing the empty expression vector. The molecular mechanism of action of the cis sense sRNAs is not clear and has to be clarified in the future. One obvious possibility is that the internal sRNAs regulate the protein-coding mRNAs via competition for RNA-binding proteins, in particular RNases. However, as yet there is no experimental evidence that this is the mode of operation of internal sRNAs. The very high number of more than 1100 internal sRNAs indicates that they will have an important influence on the physiology of haloarchaea, and that the understanding of regulatory networks in haloarchaea will remain incomplete without the analysis of internal sRNAs.

A special class of internal sRNAs are Transcription Start Site associated RNAs (TSSaRNAs). They represent transcripts that are initiated at the promoters of protein-coding genes, but are terminated soon after initiation (Zaramela et al. 2014). They have been found to occur and to be ubiquitous in all three domains of life. Most probably they are products of regulatory mechanisms that involve pausing of RNA polymerase to differentially decide about further elongation or termination of transcription. Therefore, these sRNAs do not have a regulatory function themselves, but they are the products of a co-transcriptional regulatory mechanism.

10.3.3 Cis Antisense sRNAs (asRNAs)

The dRNA-Seq study has revealed that *Hfx. volcanii* contains more than 1200 cis antisense RNAs, and thus asRNAs form the largest group of non-coding RNAs (Babski et al. 2016). During exponential growth under optimal conditions asRNAs were present to 30% of all protein-coding genes. Figure 10.3 shows that the levels of the asRNAs and the levels of the cognate sense mRNAs exhibited a very strong negative correlation, i.e. when the levels of the asRNAs were high, the levels of the cognate mRNAs were very low, and vice versa. This is a strong indication that the antisense RNAs are negative regulators of gene expression, and that duplex formation between mRNA and asRNA leads to degradation. This would require the presence of a double-strand specific RNase, which still needs to be identified.

The presence of a high fraction of asRNAs is not confined to haloarchaea, but seems to be widespread in various archaea and bacteria. For example, asRNAs to 26% of all genes have been found in *Methanococcus psychrophilus* (Li et al. 2015) and in *Pyrococcus abyssii* (Toffano-Nioche et al. 2013). Examples for

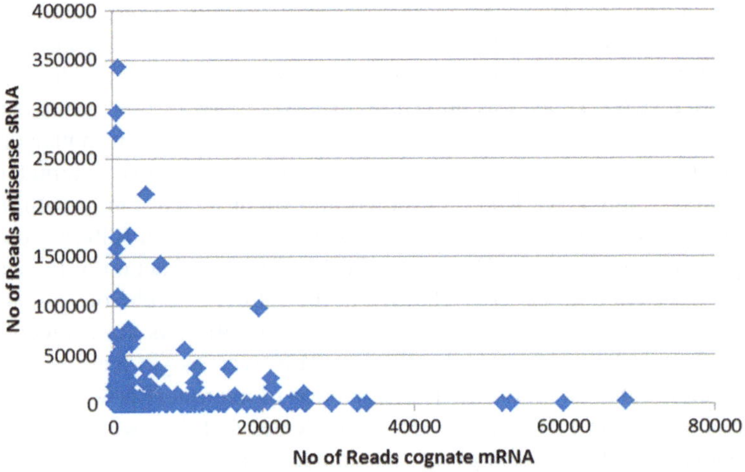

Fig. 10.3 Scatter plot of the levels of antisense sRNAs and the corresponding mRNAs. The scatter plot shows the strong negative correlation between the levels of the asRNAs and the cognate target mRNAs (taken from Babski et al. 2016)

bacteria with a high fraction of asRNAs are *Staphylococcus aureus* with 50% and *Prochlorococcus* strains with up to 75% (Georg and Hess 2011).

One specific function of asRNAs in haloarchaea seems to be the regulation of transposition, which has also been described to be true for bacteria (Brantl 2007). 134 of the 1244 asRNAs of *Hfx. volcanii* were antisense to the genes of transposases, underscoring the model of antisense regulation of transposition. This is not confined to haloarchaea, e.g. asRNAs to transposons have also been described to occur in *T. kodakarensis* (Jäger et al. 2014), *Sulfolobus solfataricus* (Wurtzel et al. 2010), and *M. mazei* (Jäger et al. 2009). Regulation of transposition by asRNAs has also been described to operate in bacteria (review: Ellis and Haniford 2016). While the regulation of transposition by asRNAs seems to be wide-spread in prokaryotes, only a minor fraction of asRNAs target transposons, most are directed against mRNAs encoding proteins of the cellular metabolism. Future work is needed to unravel the molecular regulatory mechanism of asRNAs in *H. volcanii*, irrespective of the identity of their target mRNAs.

10.4 tRNA-Derived Fragments

Recently it was discovered that tRNAs can be cleaved into “tRNA-derived fragments” (tRFs) in all three domains of life. The processing is induced by specific conditions, e.g. stress conditions, and the resulting fragments can have very different half lives and functions (review: Gebetsberger and Polacek 2013). The existence of tRFs in haloarchaea was discovered in the course of a transcriptome analysis via High Throughput Sequencing (Heyer et al. 2012). tRFs were found for 11 of the 51 tRNAs of *Hfx. volcanii*. The tRFs were typically detected under one or two of the six tested conditions, underscoring the differential generation of tRFs. Northern blot analysis was performed to determine the lengths of the tRFs from $tRNA^{Gln}$ (about 40 nt) and from $tRNA^{His}$ (about 65 nt) and to show differential levels under different conditions.

In an independent approach all sRNAs were identified that could be co-purified with ribosomes using density gradients (Gebetsberger et al. 2012). The ribosomes were isolated from cultures that had been exposed to 1 of 11 different stress conditions, respectively, and the co-isolated sRNAs in the range from 20 to 500 nt were identified by HTS. In total, tRFs from 12 tRNAs could be identified, which had a length distribution from 10 to 49 nt. However, 1 tRF of 26 nt dominated the library and generated more than 85% of all reads. It was derived from two paralogous valine tRNAs (GAC) that are encoded adjacently in the genome of *Hfx. volcanii*. Processing of the $tRNA^{Val}$ into the tRF was condition-dependent and occurred nearly exclusively under alkaline stress at a pH of 8.5. In contrast, the tRF was absent under optimal conditions and under various other stress conditions, e.g. a hypoosmotic shock or UV irradiation. The tRF^{Val} was shown to bind to the small subunit of the ribosome, and it could severely inhibit translation in an *in vitro* translation system (Gebetsberger et al. 2012). Furthermore, it could be shown that binding of tRF^{Val} to the ribosome can displace the mRNA, which results in a stress-induced global attenuation of translation *in vitro* and *in vivo* (Gebetsberger et al. 2016). The processing of tRNAs into tRFs that bind to the ribosome and inhibit translation represent an extremely fast response to the onset of stress conditions. In addition, tRNAs are extremely old, and thus it can be speculated that their usage in stress response circuits started early in evolution, in agreement with the occurrence of tRFs in all three domains of life.

10.5 CRISPR/Cas Defence Systems in Haloarchaea

It was not less than a sensation as it was discovered about a decade ago that prokaryotes contain adaptive immune systems that are directed against invading nucleic acids like phages or plasmids (reviews: van der Oost et al. 2014; Westra et al. 2014). The systems are comprised of “Clustered Regularly Interspaced Palindromic Repeats” (CRISPR) and “CRISPR Associated” (Cas) protein genes. In short, when cells survive the attack of a virus or a plasmid, short sequences of the attacking nucleic acid are integrated into CRISPR locus as spacer sequences between repeated motifs. Transcription of the CRISPR locus results in long transcripts, which are processed into small crRNAs that each contain the recognition motif for one invader. Upon a new infection, the crRNAs direct the Cas proteins to the foreign DNA (or RNA) and enables its destruction. About half of all bacteria and nearly all archaea contain such CRISPR/Cas systems. The CRISPR/Cas systems are not identical, based on the inventory of the Cas proteins they have been classified into several groups (Makarova et al. 2011).

The CRISPR/Cas system of *H. volcanii* has been intensely studied in recent years (Maier et al. 2012, 2013, 2015b, c; Marchfelder et al. 2012). *H. volcanii* contains three CRISPR loci and eight Cas genes. All three CRISPR loci are transcribed constitutively and thus the system is active in the absence of any invader. Genetic approaches have been established that allowed the characterization of the importance of the repeat sequences, the spacers, and the Cas proteins. The system has also been modified as a molecular genetic tool to down-regulate the expression of any gene of interest. A whole chapter of this book is devoted to the haloarchaeal CRISPR/Cas system, therefore, it will not be discussed any further in this chapter.

10.6 sRNAs That Are Not Well-Studied in Haloarchaea

In eukaryotes small nucleolar RNAs (snoRNAs) form a large and important class of sRNAs with a variety of functions (Lui and Lowe 2013). Their canonical functions are to be part of RNP complexes and guide enzymes to target sites on ribosomal RNAs, leading either to 2′-O-methylation of ribose (C/D box snoRNAs) or to the formation of pseudouridine (H/ACA snoRNAs). Because archaeal sRNAs fulfill the same functions and interact with archaeal proteins that are homologous to eukaryotic proteins, they were also called "snoRNAs" in spite of the lack of a nucleolus or nucleus in archaea. Recently, it has been proposed to rename them to C/D box sRNAs and H/ACA guide sRNAs. This terminology will be used when one class of these sRNAs is discussed, the term "snoRNAs" will still be used when both classes are summarized. Two recent reviews summarize the knowledge about these classes of archaeal sRNAs (Tripp et al. 2017; Lui and Lowe 2013).

The number of snoRNAs is especially high in thermophilic archaea, e.g. more than 80 C/D box sRNAs have been identified in several species of *Pyrococcus* (Bernick et al. 2012). In contrast, the number of snoRNAs is very low in haloarchaea, and only a single C/D box sRNA is present in the genome annotation of *H. volcanii* (Hartman et al. 2010). In addition, a second C/D box sRNA has been characterized that is encoded in an intron of the tRNATrp, and which was shown to be essential for methylation of the pre-tRNATrp at positions 34 and 39 (Clouet d'Orval et al. 2001). In any case, the number of snoRNAs in haloarchaea is very low. Therefore, it is not surprising that the role of archaeal snoRNAs has not been analyzed in haloarchaea, but in other archaeal groups. A chapter of this book is devoted to the characterization of archaeal snoRNAs.

In eukaryotes, the 7S RNA is part of the signal recognition particle (SRP), which is important for the direction of membrane proteins to the cytoplasmic membrane and their faithful integration. The current model is that the SRP stops translation of mRNAs for membrane proteins after the signal sequence has been translated, and translation is restarted after the interaction between SRP and the SRP receptor in the membrane have ensured the correct localization of the translating ribosome. One very early study showed that the 7S RNA is important for the expression of the gene for the major membrane protein of *Hbt. salinarum*, Bacterioopsin (Gropp et al. 1992). It was concluded that the 7S RNA is probably essential for the expression of membrane protein genes in general. Unfortunately, no study followed to verify this claim for *Hbt. salinarum* or any other haloarchaeal species. However, it is very likely that the 7S RNA is indeed important for membrane protein biosynthesis in haloarchaea in general, because it is conserved also in species that are devoid of bacterioopsin.

In eukaryotes, many circular RNAs (circRNAs) have been described, and they are thought to play important regulatory roles in physiological as well as pathological processes. Recently, the database "circRNADb" was generated, which contains more than 30,000 human exonic circRNAs (Chen et al. 2016). In archaea a single study exists that has used circRNA-seq to systematically identify cicRNAs in *Sulfolobus solfataricus* (Danan et al. 2012). A large number of circRNAs have been found, including expected circRNAs like tRNA introns, but the majority were novel circRNAs of unknown function. Also circular forms of C/D box sRNAs and of RNase P were found. Also for *Pyrococcus furiosus* it has been reported that most, if not all C/D box sRNAs exist not only in linear, but also in circular form (Starostina et al. 2004). In haloarchaea, it is known that the splicing of introns from tRNAs and the processing of pre-rRNA leads to circular RNAs (Salgia et al. 2003). However, these circles are thought to be processing intermediates without further biological function, which are degraded soon after their generation. Based on the wide distribution in other phylogenetic groups it is tempting to speculate that also haloarchaea contain circular RNAs with biological (regulatory) functions, however, no experimental evidence has been presented as yet.

In eukaryotes, studies of non-coding regulatory RNAs have initially focused on very short RNAs of only about 20 nt, e.g. miRNAs, siRNAs, and piRNAs. In recent years the so-called "long non-coding RNAs" (lncRNAs) came into focus, and it was discovered that eukaryotes contain thousands of lncRNAs. However, per definition lncRNAs are longer than 200 nt. Because the definition of sRNAs in archaea covers non-coding RNAs from 20 to 500 nt, non-coding RNAs of 200–500 nt are termed lncRNAs (=long) in eukaryotes and sRNAs (=short) in archaea. The majority of the eukaryotic lncRNAs is longer than 500 nt and lncRNAs can be up to several thousand nt long. Non-coding RNAs of such lengths have not been discovered yet for haloarchaea or any other prokaryote.

10.7 Small RNAs Encoding Microproteins

The first genome sequence was published in 1995, it was the genome sequence of the bacterium *Mycoplasma genitalism*, an intracellular pathogen with a reduced genome size. The first genome sequence of the first archaeon, *Methanococcus jannaschii,* followed soon after in 1996. Today, only 20 years later, more than

60,000 prokaryotic genome sequences are available. For the annotation of open reading frames (ORFs) typically a minimal cutoff of 100 codons was used to avoid the massive annotation of false positive small ORFs, which are not real genes. However, this meant that also real small genes that encode microproteins of less than 100 amino acids escaped annotation. In recent years it became evident that microproteins are prevalent and have very important functions in all three domains of live (Eguen et al. 2015; Ramamurthi and Storz 2014; Storz et al. 2014; Cheng et al. 2011).

Already about 10 years ago a study had focused on the characterization of the "low molecular weight proteome" of *Hbt. salinarum* (Klein et al. 2007). The optimization of several techniques was necessary, because standard experimental approaches usually work well with medium-sized proteins, but do not perform well for microproteins. In total, 380 microproteins of less than 100 amino acids could be identified, which are equivalent to 14% of the annotated proteome. Thus the microproteins make up a non-negligible part of the total proteome. It was noted that 20 of these microproteins contain two CPXCG double cysteine motifs and they were proposed to be one-domain zinc finger proteins. As a proof-of-principle that these putative zinc finger microproteins can have important regulatory functions the gene for one of these proteins was deleted (Tarasov et al. 2008). The mutant was defective in the expression of the bacterioopsin (*bop*) gene and consequently could no longer grow phototrophically. Also the replacement of one of the cysteines by a serine led to a loss *bop* gene expression and the ability to use light to drive the energy metabolism. Furthermore, the mutants were unable to synthesize carotinoids because the transcript level of the phytoene synthase was decreased. These results underscore the importance of one 60 amino acid microprotein for the physiology of *Hbt. salinarum*. Subsequently it was discovered that the transcript was in fact bicistronic and downstream of the zinc finger microprotein another microprotein of 55 amino acids was encoded, which also is involved in regulation of *bop* gene expression (Tarasov et al. 2011). This further enlarged the regulatory network of phototrophy of *H. salinarum*, which was known before to contain several normal-sized proteins.

The experimental analysis of the low molecular weight proteome has aided the annotation of small protein genes in other haloarchaea. For example, the annotation of the genome of *Hfx. volcanii* currently contains 575 genes for microproteins of less than 100 amino acids, 69 of which are putative one-domain zinc finger proteins with two CPXCG motifs. This is equivalent to 14% of all proteins, like in *Hbt. salinarum*, and the fraction is higher than the average fraction in prokaryotes, which is around 11% (Cheng et al. 2011). The vast majority of these microproteins (72%) are annotated as "hypothetical proteins" and do not have known functions. Some examples of microproteins with known functions are several ribosomal proteins, the cold shock proteins, and the Lsm protein. The Lsm (Like Sm) protein belongs to a large family of RNA-binding proteins. In eukaryotes, Sm and Lsm proteins have many different functions, for example, they are components of the splicesosome (review: Wilusz and Wilusz 2013). The Hfq protein also belongs to this protein family, which is important for the function of intergenic sRNAs in bacteria (review:

negative (D, E) and positive (K, R) charges / S,T,Y / N, Q

```
HVO_
0197A MPVCDHCGSHVSERFARVFADKNGQVLACPNCSANAGIAEVARQRARTA
0241  MPAVQCRECGRDVAVHEIETTTKTTPDGFDTRYRCPYCKSEMRDVKTRIV
0325  MSQCNHCDAFVSNNFVRVFGDEDGNVYACPSCSANAGISQVSSERRASSL
0416  MASAPSDDLFDQFLTDRGHETEPARWDRSYNKLQCPDCGALHDMGAATCSVCGWVPEA
0489  MSATTTDDYEFTCSQCGEGFAVNGGMRAAVLERGCPICGSPVSADAFDPCPA
0490  MSDSADYTFVCPECAESMLVNDSMRDAPLENGCVVCSAALTTDAFSAT
0546  MRDAPDPAGPPTVCDDCGARHSFRRRFVTGGGWRTVYRCSECGSRASSDGDGDE
0695  MHTCGNCGEFVSRDFVRVFGNDMDEVVGCPACMNMREVMQGDGAAQTSGRVRWTRA
0758  MKTTRKGLRDGELEKDTYGRLTCSECGESLKKKNDPDEVFSVRICADCGREWKELR
0919  MWCERCGRDTTVRKHAVDEFTRFLCNDCRAVWDRFVSA
0993  MSFEWVSYFTCEECGLERPSPEVPYDRLGYAVCPGCGAETRPAAAAMADASTPVDG
1118  MREFDVTCPECGERYRVNEPMMRTLRETGCVLCTAPLNDAAKSA
1533  MVLHNSVIDDYHPTEGYYECRSCRTRTVSASHLSECPDCGGSVRNIAVARE
1677  MVEAFVRLLCPECGKDWETNPTELPAHRDNYSCQSCGATRRTAEFMRTERDLQTLKQFE
1848A MPTCQNCNSFVTEGYVRVFAPEGMDAPRVCPHCEDLVRDGSQVREARATRH
2400  MSDLEIERECPACGNDTFYLAASMEIHLGTKTKWHCTECDYGYIHITDDIETYAKAEA
2753  MSESEQRHAHQCVSCGINIAGMSAATFKCPDCGQEISRCSKCRKQSNLYECPDCGFMGP
2805A MPECQNCGSFVTPAYARVFTPDGMENPRVCPNCEDMVRDGAQVRAARSPRNH
2901  MAGLQQQRARGRDMLECRGCGAVFPEGRATNDGWTYVCPECEQVEGIGEGLRRL
```

Fig. 10.4 Sequences of 19 arbitrarily chosen one-domain zinc finger microproteins of *H. volcanii*. The two CPXCG motifs are *underlined*. Charged and hydrophilic amino acids are color-coded, as indicated on *top*. The HVO numbers (*left*) are the gene designations in the genome annotation of *H. volcanii* (www.halolex.mpg.de)

Wilusz and Wilusz 2013). The haloarchaeal Lsm protein has also been shown to bind sRNAs and to have important regulatory functions, because a deletion mutant has a very severe growth defect (Fischer et al. 2010).

Also only very few of the 69 *Hfx. volcanii* putative one-domain zinc finger microproteins with CPXCG motifs have annotated functions, e.g. as ribosomal proteins or a small subunit of RNA polymerase. Figure 10.4 shows that proteins of this family have a very high fraction of charged and hydrophilic amino acids, which are indicative of binding to many interaction partners and of posttranslational modifications. It will be interesting to unravel functions of more examples of this interesting family of microproteins.

The characterization of microproteins has also been started in methanogenic archaea. Three microproteins of 23–61 amino acid lengths have been identified by LC-MSMS in cell extracts of *Methanosarcina mazei* (Prasse et al. 2015). Two of them had increased levels during mid-exponential growth phase under nitrogen limitation. Overproduction of the three microproteins resulted in transcript level changes of 40–159 transcripts. However, phenotypic changes between the wild-type and the three microprotein overproducers could not be observed (Prasse et al. 2015). Optimization of experimental approaches, e.g. including gel-free LC-MS, increased the number of experimentally verified microproteins of *M. mazei* to 28 (Cassidy et al. 2016), and it is easy to predict that the number will further increase in the future.

The characterization of microproteins and their biological roles is an emerging field in molecular biology. The German Research Council (DFG) has reacted to this challenge and is currently setting up a Priority Program, which is devoted to the analysis of microproteins in prokaryotes and will operate from August 2017 to July 2023.

10.8 Conclusions and Outlook

The recent improvements of RNA-Seq and derivatives thereof have led to the identification of thousands of non-coding sRNAs, not only in haloarchaea, but also in other phylogenetic groups of archaea and bacteria. However, the prevalence of non-coding sRNAs over protein-coding mRNAs is not universally conserved, but specific for certain species or groups. Most probably not all sRNAs have been discovered yet, because their levels vary substantially in different environmental conditions, and thus further studies under additional conditions will most probably further increase the numbers of haloarchaeal sRNAs.

The most intensely studied group of sRNAs are intergenic sRNAs, which have been shown to be important for many biological functions. Future studies will concentrate on the identification of their target mRNAs, either by experimental approaches or using optimized bioinformatics approaches, and on the analysis of their molecular mechanisms of action. The high fraction of leaderless mRNAs in haloarchaea makes it likely that many sRNAs will be found to interact with the 3′-UTRs of their targets.

Haloarchaea contain a much higher number of cis sense sRNAs than intergenic sRNAs. These cis sense sRNAs have hardly been studied and were long thought to be degradation intermediates. However, the dRNA-Seq approach ensured that all of the listed cis sense sRNAs are primary transcripts and no processing intermediates, and two characterized examples verified that they have a regulatory function *in vivo*. The largest group of haloarchaeal sRNAs are asRNAs. The high negative correlation between their levels and the levels of the cognate mRNAs led to the prediction that most of them will turn out to be negative regulators of gene expression. Additional classes of haloarchaeal sRNAs include the crRNAs from the CRISPR/Cas systems and the tRNA-derived fragments, both of which are being studied intensively, and snoRNAs and circular RNAs, which have not been studied until now.

The last group of sRNAs is formed by small mRNAs that encode microproteins of less than 100 amino acids. The analysis of the roles and mechanisms of microproteins is an emerging field of research, which has been initiated not only with haloarchaea, but also with methanogenic archaea and many groups of bacteria.

In summary, haloarchaea contain a zoo of different small RNAs, most of which have only been identified very recently. It is easy to predict that future work will lead to unprecedented insight into the RNA regulatory networks in haloarchaea—and will yield many surprises. The change of concept has only been started, that will change of the view on haloarchaeal genomes from protein-encoding entities with a few RNA genes to DNA molecules that encode mostly RNA—and additionally contains a minor fraction of protein-encoding genes.

References

Babski J, Tjaden B, Voss B, Jellen-Ritter A, Marchfelder A, Hess WR, Soppa J (2011) Bioinformatic prediction and experimental verification of sRNAs in the haloarchaeon *Haloferax volcanii*. RNA Biol 8:806–816

Babski J, Maier LK, Heyer R, Jaschinski K, Prasse D, Jäger D, Randau L, Schmitz RA, Marchfelder A, Soppa J (2014) Small regulatory RNAs in Archaea. RNA Biol 11:489–493

Babski J, Haas KA, Nather-Schindler D, Pfeiffer F, Forstner KU, Hammelmann M, Hilker R, Becker A, Sharma CM, Marchfelder A, Soppa J (2016) Genome-wide identification of transcriptional start sites in the haloarchaeon *Haloferax volcanii* based on differential RNA-Seq (dRNA-Seq). BMC Genomics 17:629

Bernick DL, Dennis PP, Lui LM, Lowe TM (2012) Diversity of antisense and other non-coding RNAs in archaea revealed by comparative small RNA sequencing in four *Pyrobaculum* species. Front Microbiol 3:231

Borges F, Martienssen RA (2015) The expanding world of small RNAs in plants. Nat Rev Mol Cell Biol 16:727–741

Brantl S (2007) Regulatory mechanims employed by cis-encoded antisense RNAs. Curr Opin Microbiol 10:102–109

Cassidy L, Prasse D, Linke D, Schmitz RA, Tholey A (2016) Combination of bottom-up sD-LC-MS and semi-top-down gelFree-LC-MS enhances coverage of proteome and low molecular weight short open reading frames encoded peptides of the archaeon *Methanosarcina mazei*. J Proteome R 15:3773–3783

Catalanotto C, Cogoni C, Zardo G (2016) MicroRNA in control of gene expression: an overview of nuclear functions. Int J Mol Sci 17:E1712

Chen X, Han P, Zhou T, Guo X, Song X, Li Y (2016) circRNADb: a comprehensive database for human circular RNAs with protein-coding annotations. Sci Rep 6:34985

Cheng H, Chan WS, Li Z, Wang D, Liu S, Zhou Y (2011) Small open reading frames: current prediction techniques and future prospect. Curr Protein Pept Sci 12:503–507

Clouet d'Orval B, Bortolin ML, Gaspin C, Bachellerie JP (2001) Box C/D RNA guides for the ribose methylation or archaeal tRNAs. The tRNATrp intron guides the formation of two ribose-methylated nucleosides in the mature tRNATrp. Nucleic Acids Res 29:4518–4529

Danan M, Schwartz S, Edelheit S, Sorek R (2012) Transcriptome-wide discovery of circular RNAs in Archaea. Nucleic Acids Res 40:3131–3142

Daniel JM, Penzkofer D, Teske R, Dutzmann J, Koch A, Bielenberg W, Bonauer A, Boon RA, Fischer A, Bauersachs J, van Rooij E, Dimmeler S, Sedding DG (2014) Inhibition of miR-92a improves re-endothelialization and prevents neointima formation following vascular injury. Cardiovasc Res 103:564–572

Eguen T, Straub D, Graeff M, Wenkel S (2015) MicroProteins: small size-big impact. Trends Plant Sci 20:477–482

Ellis MJ, Haniford DB (2016) Riboregulation of bacterial and archaeal transposition. WIREs RNA 7:382–398

Fischer S, Benz J, Spath B, Maier LK, Straub J, Granzow M, Raabe M, Urlaub H, Hoffmann J, Brutschy B, Allers T, Soppa J, Marchfelder A (2010) The archaeal Lsm protein binds to small RNAs. J Biol Chem 285:34429–34438

Gebetsberger J, Polacek N (2013) Slicing tRNAs to boost functional ncRNA diversity. RNA Biol 10:1798–1806

Gebetsberger J, Zywicki M, Kunzi A, Polacek N (2012) tRNA-derived fragments target the ribosome and function as regulatory non-coding RNA in *Haloferax volcanii*. Archaea 2012:260909

Gebetsberger J, Wyss L, Mleczko AM, Reuther J, Polacek N (2016) A tRNA-derived fragment competes with mRNA for ribosome binding and regulates translation during stress. RNA Biol. doi:10.1080/15476286.2016.1257470

Georg J, Hess WR (2011) *cis*-antisense RNA, another level of gene regulation in bacteria. Microbiol Mol Biol Rev 75:286–300

Gomes-Filho JV, Zaramela LS, Italiani VC, Baliga NS, Vencio RZ, Koide T (2015) Sense overlapping transcripts in IS*1341*-type transposase genes are functional non-coding RNAs in archaea. RNA Biol 12:490–500

Gropp R, Gropp F, Betlach MC (1992) Association of the haloarchaeal 7S RNA to the polysome correlates with expression of the membrane protein bacterioopsin. Proc Natl Acad Sci U S A 89:1204–1208

Hartman AL, Norais C, Badger JH, Delmas S, Haldenby S et al (2010) The complete genome sequence of *Haloferax volcanii* DS2, a model archaeon. PLoS One 5:e9605

Heyer R, Dorr M, Jellen-Ritter A, Spath B, Babski J, Jaschinski K, Soppa J, Marchfelder A (2012) High throughput sequencing reveals a plethora of small RNAs including tRNA derived fragments in *Haloferax volcanii*. RNA Biol 9:1011–1018

Huang Y, Zhang JL, Yu XL, Xu TS, Wang ZB, Cheng XC (2013) Molecular functions of small regulatory noncoding RNA. Biochemistry (Moscow) 78:221–230

Jäger D, Sharma CM, Thomsen J, Ehlers C, Vogel J, Schmitz RA (2009) Deep sequencing analysis of the *Methanosarcina mazei* Go1 transcriptome in response to nitrogen availability. Proc Natl Acad Sci U S A 106:21878–21882

Jäger D, Pernitzsch SR, Richter AS, Backofen R, Sharma CM, Schmitz RA (2012) An archaeal sRNA targeting *cis*- and *trans*-encoded mRNAs via two distinct domains. Nucleic Acids Res 40:10964–10979

Jäger D, Förstner KU, Sharma CM, Santangelo TJ, Reeve JN (2014) Primary transcriptome map of the hyperthermophilic archaeon *Thermococcus kodakarensis*. BMC Genomics 15:568

Jaschinski K, Babski J, Lehr M, Burmester A, Benz J, Heyer R, Dorr M, Marchfelder A, Soppa J (2014) Generation and phenotyping of a collection of sRNA gene deletion mutants of the haloarchaeon *Haloferax volcanii*. PLoS One 9:e90763

Klein C, Aivaliotis M, Olsen JV, Falb M, Besir H, Scheffer B, Bisle B, Tebbe A, Konstantinidis K, Siedler F, Pfeiffer F, Mann M, Oesterhelt D (2007) The low molecular weight proteome of *Halobacterium salinarum*. J Proteome Res 6:1510–1518

Koide T, Reiss DJ, Bare JC, Pang WL, Facciotti MT, Schmid AK, Pan M, Marzolf B, Van PT, Lo FY, Pratap A, Deutsch EW, Peterson A, Martin D, Baliga NS (2009) Prevalence of transcription promoters within archaeal operons and coding sequences. Mol Syst Biol 5:285

Kopf M, Hess WR (2015) Regulatory RNAs in photosynthetic cyanobacteria. FEMS Microbiol Rev 39:301–315

Li J, Guo Y, Yue L, Li Y, Ge W et al (2015) Global mapping of transcriptional start sites revealed both transcriptional and post-transcriptional regulation of cold adaptation in the methanogenic archaeon *Methanolobus psychrophilus*. Sci Rep 5:9209

Lui L, Lowe T (2013) Small nucleolar RNAs and RNA-guided post-transcriptional modifictation. Biochem Soc Essays Biochem 54:53–77

Maier LK, Fischer S, Stoll B, Brendel J, Pfeiffer F, Dyall-Smith M, Marchfelder A (2012) The immune system of halophilic archaea. Mob Genet Elements 2:228–232

Maier LK, Lange SJ, Stoll B, Haas KA, Fischer S, Fischer E, Duchardt-Ferner E, Wohnert J, Backofen R, Marchfelder A (2013) Essential requirements for the detection and degradation of invaders by the *Haloferax volcanii* CRISPR/Cas system I-B. RNA Biol 10:865–874

Maier LK, Benz J, Fischer S, Alstetter M, Jaschinski K, Hilker R, Becker A, Allers T, Soppa J, Marchfelder A (2015a) Deletion oft he SM1 encoding motif in the *lsm* gene results in distinct changes in the transcriptome and enhanced swarming activity of *Haloferax* cells. Biochimie 117:129–137

Maier LK, Dyall-Smith M, Marchfelder A (2015b) The adaptive immune system of *Haloferax volcanii*. Life (Basel) 5:521–537

Maier LK, Stachler AE, Saunders SJ, Backofen R, Marchfelder A (2015c) An active immune defense with a minimal CRISPR (clustered regularly interspaced short palindromic repeats) RNA and without the Cas6 protein. J Biol Chem 290:4192–4201

Makarova KS, Haft DH, Barrangou R, Brouns SJ, Charpentier E, Horvath P, Moineau S, Mojica FJ, Wolf YI, Yakunin AF, van der Oost J, Koonin EV (2011) Evolution and classification of the CRISPR-Cas systems. Nat Rev Microbiol 9:467–477

Marchfelder A, Fischer S, Brendel J, Stoll B, Maier LK, Jager D, Prasse D, Plagens A, Schmitz RA, Randau L (2012) Small RNAs for defence and regulation in archaea. Extremophiles 16:685–696

Martens B, Manoharadas S, Hasenohrl D, Manica A, Blasi U (2013) Antisense regulation by transposon-derived RNAs in the hyperthermophilic archaeon *Sulfolobus solfataricus*. EMBO Rep 14:527–533

Murina VN, Nikulin AD (2015) Bacterial small regulatory RNAs and Hfq protein. Biochemistry (Moscow) 80:1647–1654

Prasse D, Ehlers C, Backofen R, Schmitz RA (2013) Regulatory RNAs in archaea: first target identification in methanoarchaea. Biochem Soc Trans 41:344–349

Prasse D, Thomsen J, De Santis R, Muntel J, Becher D, Schmitz RA (2015) First description of small proteins encoded by spRNAs in *Methanosarcina mazei* strain Go1. Biochimie 117:138–148

Ramamurthi KS, Storz G (2014) The small protein floodgates are opening; now the functional analysis begins. BMC Biol 12:96

Salgia SR, Singh SK, Gurha P, Gupta R (2003) Two reactions of *Haloferax volcanii* RNA splicing enzymes: joining of exons and circularization of introns. RNA 9:319–330

Schmitz-Streit R, Jäger D, Jellen-Ritter A, Babski J, Soppa J, Marchfelder A (2011) Archaea employ small RNAs as regulators. In: Hess M (ed) Regulatory RNAs in prokaryotes. Springer, Wien, pp 131–145

Starostina NG, Marshburn S, Johnson LS, Eddy SR, Terns RM, Terns MP (2004) Circular box C/D RNAs in *Pyrococcus furiosus*. Proc Natl Acad Sci U S A 101:14097–14101

Storz G, Wolf YI, Ramamurthi KS (2014) Small proteins can no longer be ignored. Annu Rev Biochem 83:753–777

Straub J, Brenneis M, Jellen-Ritter A, Heyer R, Soppa J, Marchfelder A (2009) Small RNAs in haloarchaea: identification, differential expression and biological function. RNA Biol 6 (3):281–292

Tang TH, Bachellerie JP, Rozhdestvensky T, Bortolin ML, Huber H, Drungowski M, Elge T, Brosius J, Huttenhofer A (2002) Identification of 86 candidates for small non-messenger RNAs from the archaeon *Archaeoglobus fulgidus*. Proc Natl Acad Sci U S A 99:7536–7541

Tang TH, Polacek N, Zywicki M, Huber H, Brugger K, Garrett R, Bachellerie JP, Huttenhofer A (2005) Identification of novel non-coding RNAs as potential antisense regulators in the archaeon *Sulfolobus solfataricus*. Mol Microbiol 55:469–481

Tarasov VY, Besir H, Schwaiger R, Klee K, Furtwangler K, Pfeiffer F, Oesterhelt D (2008) A small protein from the *bop-brp* intergenic region of *Halobacterium salinarum* contains a zinc finger motif and regulates *bop* and *crtB1* transcription. Mol Microbiol 67:772–780

Tarasov V, Schwaiger R, Furtwängler K, Dyall-Smith M, Oesterhelt D (2011) A small basic protein from the *brz-brp* operon is involved in the regulation of *bop* transcription in *Halobacterium salinarum*. BMC Mol Biol 12:42

Toffano-Nioche C, Ott A, Crozat E, Nguyen AN, Zytnicke M, Leclerc F, Forterre P, Bouloc P, Gautheret D (2013) RNA at 92°C: the non-coding transcriptome of the hyperthermophilic archaeon *Pyrococcus abyssi*. RNA Biol 10:1211–1220

Tripp V, Martin R, Orell A, Alkhnbashi OS, Backofen R, Randau L (2017) Plasticity of archaeal C/D box sRNA biogenesis. Mol Microbiol 103:151–164

Van der Oost J, Westra ER, Jackson RN, Wiedenheft B (2014) Unravelling the structural and mechanistic basis of CRIPR-Cas systems. Nat Rev Microbiol 12:479–492

Van Puyvelde S, Vanderleyden J, De Keersmaecker SC (2015) Experimental approaches to identify small RNAs and their diverse roles in bacteria–what we have learnt in one decade of MicA research. Microbiologyopen 4:699–711

Wagner EG, Romby P (2015) Small RNAs in bacteria and archaea: who they are, what they do, and how they do it. Adv Genet 90:133–208

Wan G, Liu Y, Han C, Zhang X, Lu X (2014) Noncoding RNAs in DNA repair and genome integrity. Antioxid Redox Signal 20:655–677

Waters LS, Storz G (2009) Regulatory RNAs in bacteria. Cell 136:615–628

Westra ER, Buckling A, Fineran PC (2014) CRISPR-Cas systems: beyond adaptive immunity. Nat Rev Microbiol 12:317–326

Wilusz CJ, Wilusz J (2013) Lsm proteins an Hfq: life at the 3'-end. RNA Biol 10:592–601

Wurtzel O, Sapra R, Chen F, Zhu Y, Simmons BA, Sorek R (2010) A single-base resolution map of an archaeal transcriptome. Genome Res 20:133–141

Yang JX, Rastetter RH, Wilhelm D (2016) Non-coding RNAs: an introduction. Adv Exp Med Biol 886:13–32

Zaramela LS, Vencio RZN, ten Caten F, Baligy N, Koide T (2014) Transcription start site associated RNAs (TSSaRNAs) are ubiquitous in all domains of life. PLoS One 9:e107680

Chapter 11
CRISPR and Salty: CRISPR-Cas Systems in Haloarchaea

Lisa-Katharina Maier, Omer S. Alkhnbashi, Rolf Backofen, and Anita Marchfelder

Abstract CRISPR-Cas (CRISPR: Clustered Regularly Interspaced Short Palindromic Repeats and Cas: CRISPR associated) systems are unique defence mechanisms since they are able to adapt to new invaders and are heritable. CRISPR-Cas systems facilitate the sequence-specific elimination of invading genetic elements in prokaryotes, they are found in 45% of bacteria and 85% of archaea. Their general features have been studied in detail, but subtype- and species-specific variations await investigation. Haloarchaea is one of few archaeal classes in which CRISPR-Cas systems have been investigated in more than one genus. Here, we summarize the available information on CRISPR-Cas defence in three Haloarchaea: *Haloferax volcanii*, *Haloferax mediterranei* and *Haloarcula hispanica*. Haloarchaea share type I CRISPR-Cas systems, with subtype I-B being dominant. Type I-B systems rely on Cas proteins Cas5, Cas7, and Cas8b for the interference reaction and these proteins have been shown to form a Cascade (CRISPR-associated complex for antiviral defence) -like complex in *Hfx* (*Haloferax*). *volcanii*. Cas6b is the endonuclease for crRNA (CRISPR RNA) maturation in type I-B systems but the protein is dispensable for interference in *Hfx. volcanii*. Haloarchaea share a common repeat sequence and crRNA-processing pattern. A prerequisite for successful invader recognition in *Hfx. volcanii* is base pairing over a ten-nucleotide-long non-contiguous seed sequence. Moreover, *Hfx. volcanii* and *Har* (*Haloarcula*). *hispanica* rely each on certain specific PAM (protospacer adjacent motif) sequences to elicit interference, but they share only one PAM sequence. Primed adaptation in *Har. hispanica* relies on another set of PAM sequences.

Keywords CRISPR-Cas • crRNA • PAM • Cas6 • Cascade • Archaea • *Haloferax volcanii* • *Haloferax mediterranei* • *Haloarcula hispanica* • Type I-B • Adaptation • Interference

L.-K. Maier • A. Marchfelder (✉)
Biology II, University of Ulm, Ulm, Germany
e-mail: anita.marchfelder@uni-ulm.de

O.S. Alkhnbashi • R. Backofen
Bioinformatics Group, Department of Computer Science and Centre for Biological Signalling Studies (BIOSS), University of Freiburg, Freiburg, Germany

B. Clouet-d'Orval (ed.), *RNA Metabolism and Gene Expression in Archaea*,
Nucleic Acids and Molecular Biology 32, DOI 10.1007/978-3-319-65795-0_11

11.1 The Prokaryotic Immune System CRISPR-Cas

Prokaryotic organisms, especially archaea, thrive in nature's most hostile habitats. Apart from the environmental stressors imposed by the abiotic nature of their ecological niche, prokaryotes also face a constant threat by the virome, which exceeds their number by a power of 10 (Suttle 2007). Moreover, prokaryotes face a multitude of invasive entities, including plasmids, transposons and other mobile genetic elements. To balance the integration of beneficial elements and the elimination of detrimental invaders, prokaryotes apply a range of defence strategies (Labrie et al. 2010). The RNA-mediated CRISPR-Cas system has recently attracted increasing attention because it confers adaptive, specific and hereditary immunity against viruses and mobile genetic elements (for recent reviews see (Barrangou 2015; Hille and Charpentier 2016; Mohanraju et al. 2016; Mojica and Rodriguez-Valera 2016)).

Although CRISPR-Cas systems come in different versions (Makarova et al. 2015; Mohanraju et al. 2016; Shmakov et al. 2017), they share one common functional principle. A small RNA guide sequence, called the crRNA, specifically recognizes together with Cas proteins an invading nucleic acid and mediates target degradation. The nature of the effector defines the CRISPR-Cas system as class 1 if a multiprotein complex is present (termed Cascade for type I systems and Csm- or Cmr-complex for type III systems) and class 2 if only a single effector protein is required (e.g. Cas9 for type II, Cas12a[1] for type V) (Makarova et al. 2015; Mohanraju et al. 2016; Shmakov et al. 2017). The *cas* genes include a great variety of nucleic acid binding and processing activities that are crucial for CRISPR-Cas immunity (Jansen et al. 2002a; Makarova et al. 2011). In addition to the mechanistic details of the defence reaction, the presence of these proteins and their characteristic arrangement within the *cas* gene loci give rise to a multilayer classification that currently encompasses 2 classes, 6 types and more than 20 subtypes (Burstein et al. 2017; Makarova et al. 2015; Mohanraju et al. 2016; Shmakov et al. 2015, 2017; Vestergaard et al. 2014).

In contrast the nature of the small RNA guide is relatively uniform. Organisms with an active CRISPR-Cas system encode arrays of recurring repeat sequences that are interspaced by short sequence stretches (spacers) captured from foreign genetic elements in close proximity to the aforementioned *cas* gene cassettes (Bolotin et al. 2005; Jansen et al. 2002a, b; Mojica et al. 2005). The adaptability and expandability of the CRISPR loci through the integration of new spacers of foreign origin upon infection is the basis of the immunogenic power of the CRISPR-Cas system (Barrangou et al. 2007; Brouns et al. 2008; Deveau et al. 2008; Garneau et al. 2010; Pourcel et al. 2005). The cell maintains an ongoing record of previously encountered pathogens or mobile genetic elements that confers specific immunity upon reinfection. Adaptation to an invading genetic element through the integration of new spacers is one of the three stages of CRISPR-Cas immunity (Fig. 11.1)

[1]Cas12a was formerly termed Cpf1.

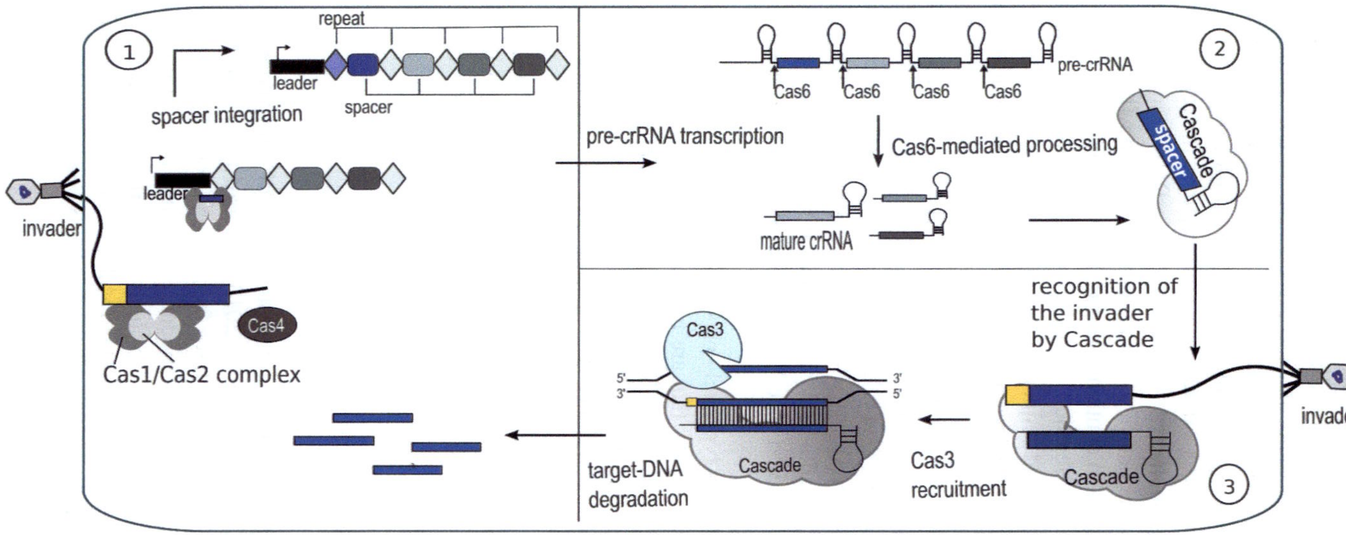

Fig. 11.1 The three stages of type I interference. The first stage (1) ensures the recognition of a foreign invader via a Cas1/Cas2 complex (adaptation step). Part of the invading DNA, the protospacer sequence, is selected by identifying the PAM sequence. Cas1/Cas2-mediated integration into the CRISPR locus at the leader end expands the array by one repeat spacer unit. All type I-B systems contain also a Cas4 protein, that has been shown to be involved in adaptation in *Har. hispanica* (Li et al. 2014b). However, details on how Cas4 is involved in the adaptation step are not known yet. The transcription of the CRISPR locus into a long pre-crRNA initiates the second stage (2) (expression stage). After being processed into mature crRNAs by the Cas6 endonuclease, the crRNAs are integrated into the multiprotein effector complex termed Cascade. The crRNA-loaded Cascade complexes can detect the foreign invader during the final interference stage (3). Upon reinfection with a cognate invader, Cascade scans the foreign nucleic acid, and upon identification of the PAM sequence and base pairing of the crRNA spacer region and the protospacer sequence, the nuclease Cas3 is recruited and the invader DNA is degraded

(detailed reviews can be found in (Amitai and Sorek 2016; Mohanraju et al. 2016; Sternberg et al. 2016)). To use the genetic information stored within the spacer sequences, the CRISPR array is transcribed in the second stage of the defence reaction into a long precursor molecule, the pre-crRNA, which is subsequently processed into the mature crRNAs. This reaction is catalysed in type I systems by Cas proteins and in most type II systems by RNase III in conjunction with tracrRNA and Cas9. Every crRNA comprises a unique spacer flanked by the remainder of the repeat sequence. Depending on the type of system, each crRNA is joined by one or more Cas proteins to form the active effector complex, in type II and type V-B systems the effector complex also contains the tracrRNA (Shmakov et al. 2017). In the third and final step, the interference, the crRNA-loaded complex mediates the recognition of foreign nucleic acid sequences through the base pairing between the crRNA and the invader. This interaction leads to the degradation and subsequent elimination of the targeted nucleic acid: in type I systems the degrading nuclease Cas3 is recruited to the effector complex whereas in type III systems Cas10 and Cmr/Csm subunits of the effector complex mediate target degradation; in class 2 systems degradation is achieved via the activity of the single effector protein. Type I, II, and V systems target DNA, whereas the activity of type III systems is transcription-dependent and results in degradation of RNA and DNA (Mohanraju et al. 2016), the recently discovered type VI systems target RNA (Mohanraju et al. 2016; Shmakov et al. 2017).

Moreover, some CRISPR-Cas types rely on short sequence motifs, termed PAM (type I, II and V) or PFS[2] (type VI), that are located on the invading nucleic acid (Deveau et al. 2008; Jinek et al. 2012; Mojica et al. 2009; Shmakov et al. 2017; Westra et al. 2013; Zetsche et al. 2015). These motifs direct protospacer selection during adaptation as well as interference by ensuring self/non-self discrimination (Amitai and Sorek 2016; Shah et al. 2013).

Despite the rapid pace in scientific activity regarding CRISPR-Cas systems, much has yet to be elucidated with regard to the protein and RNA machinery that execute CRISPR-Cas function as well as the regulatory circuits that orchestrate it.

11.2 CRISPR-Cas Systems in Haloarchaea

CRISPR-Cas systems are present in approximately 45% of bacteria and 85% of archaea (Alkhnbashi et al. 2014; Lange et al. 2013; Makarova et al. 2015). Despite their prevalence in archaea, most studies have focused on CRISPR-Cas systems in bacteria, whereas only a few archaeal model organisms have been analysed. Archaeal CRISPR-Cas systems are almost exclusively restricted to class 1 systems, which rely on a multisubunit effector complex (Makarova et al. 2015; Vestergaard et al. 2014). Just recently a few class 2 systems have been found in archaeal

[2]PFS is the abbreviation for protospacer flanking site.

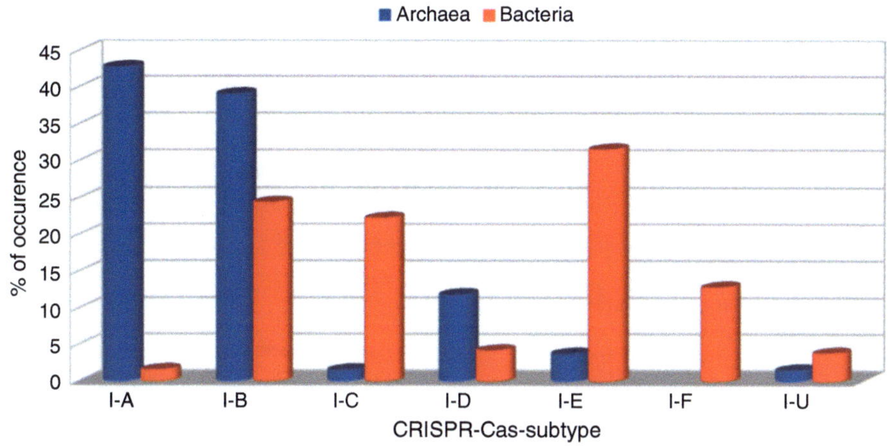

Fig. 11.2 Distribution of type I subtypes in Archaea. Subtypes I-A and I-B are the dominant subtypes in Archaea, subtypes I-C, I-D, I-E and I-U are only present in few archaeal systems, whereas subtype I-F systems are completely absent. In Bacteria however, all subtypes are present with subtypes I-B, I-C and I-E predominant

genomes (Burstein et al. 2017; Makarova et al. 2015; Mohanraju et al. 2016): Cas9 protein genes were identified in a metagenomic analysis in two uncultured nanoarchaeal genomes: *Candidatus* Micrarchaeum acidiphilum ARMAN-1 and *Candidatus* Parvarchaeum acidiphilum ARMAN-4 (Burstein et al. 2017) and a putative type V system could be identified in *Candidatus* Methanomethylophilus alvus (Makarova et al. 2015).

Archaea encode more type III systems than bacteria but the most prevalent systems in archaea are the type I systems with type I-A and I-B being the most abundant (Fig. 11.2). Interestingly, type I-F is completely absent from archaeal genomes (Fig. 11.2) (Makarova et al. 2015; Staals and Brouns 2013). The distribution of CRISPR-Cas types in the archaeal domain is not uniform: crenarchaeota encode mostly type I-A and III-B systems, whereas in euryarchaeota a greater diversity is found with examples of type I-A, -B, -D and type III-A (Makarova et al. 2015; Vestergaard et al. 2014). The type I-B systems are overrepresented within the Euryarchaeota and are most abundant in Haloarchaea (Makarova et al. 2015; Vestergaard et al. 2014). Haloarchaea thrive in the most saline habitats found on earth, e.g., salterns, salt lakes, tidal evaporation ponds, deep-sea salt domes, salt mines, salty soils and anthropogenic salt-dominated environments, such as salted-fish-fermented foods (Oren 2006). They can tolerate salt concentrations up to saturation but also depend on a species-specific minimal salinity within their environment (Oren 2006).

Regarding their CRISPR-Cas content, haloarchaea also form a coherent group; 12 of the 24 publicly available genomes possess complete CRISPR-Cas systems, all of subtype I-B (Fig. 11.3). Most haloarchaea contain only one CRISPR-Cas subtype, and only two strains (*Haloquadratum walsbyi* C23 and *Halorubrum*

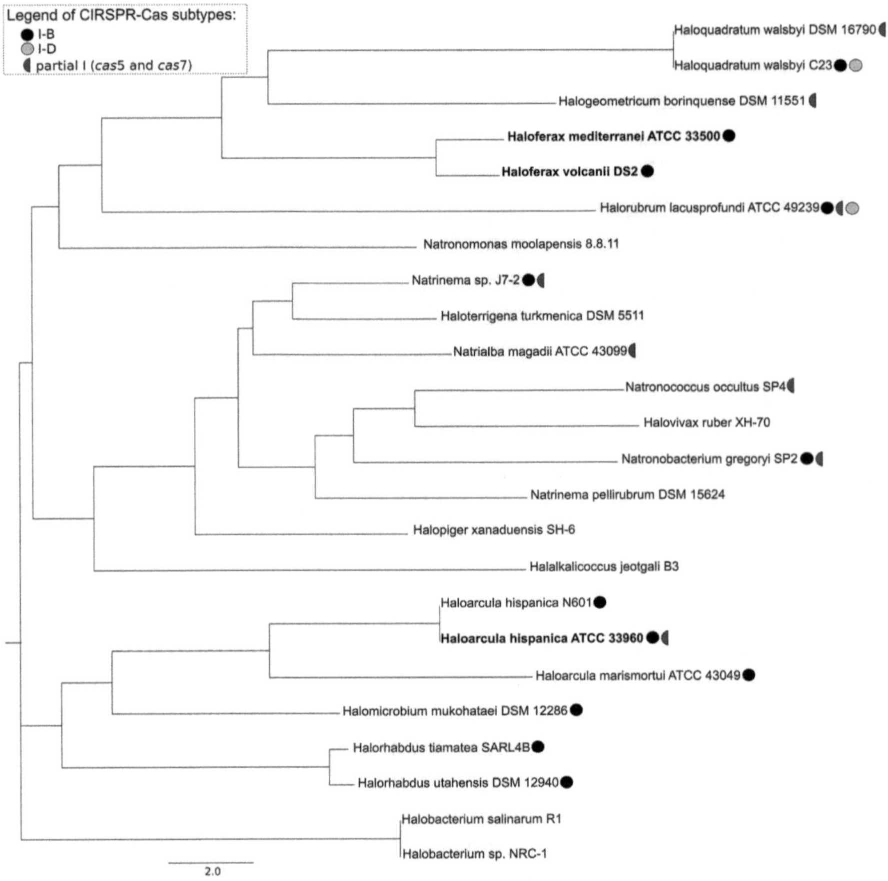

Fig. 11.3 Phylogenetic distribution of CRISPR-Cas systems in Haloarchaea. This phylogenetic tree was constructed using all of the haloarchaeal genomes that are available in public databases (as of October 2016). The presence and type of CRISPR-Cas system found in each species are given. The distribution of *cas* gene cassettes does not show a pattern that correlates with the phylogenetic relationship of the haloarchaea depicted. All of the CRISPR-Cas-positive species exclusively encode type I systems, and most of them only possess a single *cas* gene cassette of subtype I-B (*black dot*). Dual CRISPR-Cas systems are rare and represent a combination of subtypes I-B (*black dot*) and I-D (*grey dot*). Moreover, partial *cas* gene cassettes comprising only cas5 and cas7 genes (*grey halfmoon*) are the sole trace of a CRISPR-Cas system in some species but also co-occur with complete systems of both subtypes. The strains referred to in the text are given in bold

lacusprofundi ATCC49239) encode two different complete CRISPR-Cas systems of subtypes I-B and I-D. Interestingly, some haloarchaeal genomes contain isolated *cas* genes that represent partial effector modules comprising *cas* genes *cas5* and *cas7*. These partial *cas* gene clusters are found in isolation in four species (*Haloquadratum walsbyi* DSM16790, *Halogeometricum borinquense* DSM11551, *Natrialba magadii* ATCC43099, and *Natronococcus occultus* SP4). In *Natrinema sp.* J7-2, *Natronobacterium gregoryi* SP2 and *Haloarcula hispanica* ATCC 33960, orphan *cas* genes accompany a subtype I-B *cas* gene cluster, whereas in *Hrr*[3]. *lacusprofundi* ATCC49239, a partial cluster is present together with type I-B and I-D systems. Five representatives of haloarchaea were completely devoid of both CRISPR loci and *cas* genes, whereas four strains (*Halogeometricum borinquense* DSM 11551, *Haloquadratum walsbyi* DSM 16790, *Haloterrigena turkmenica* DSM 5511, *Natrinema pellirubrum* DSM 15624) were missing *cas* genes but possessed CRISPR loci, so called orphan CRISPR loci.

The presence and absence, as well as the distribution of CRISPR-Cas types and combinations of subtypes, do not reflect the phylogenetic relationships among species. This is not only true in haloarchaea but was observed throughout all classification efforts (Garrett et al. 2011; Haft et al. 2005; Makarova et al. 2011, 2015; Mohanraju et al. 2016; Vestergaard et al. 2014). An uneven distribution pattern reflects the dynamic nature and rapid pace of evolution of the CRISPR-Cas components as well as the system's propensity to be transmitted by mobile genetic elements or to be lost due to self-targeting or selective pressure, favouring the uptake of mobile DNA elements (Makarova et al. 2015; Shah and Garrett 2011). Thus, the distribution of CRISPR-Cas activity may reflect the balance between the costs and benefits of maintaining a CRISPR-Cas system in the individual habitat of the respective species (Jiang et al. 2013).

A growing body of information on haloarchaeal CRISPR-Cas systems is available and has, until now, been concentrated on subtype I-B and focussed on *Haloferax volcanii* (Brendel et al. 2014; Cass et al. 2015; Fischer et al. 2012; Maier et al. 2012, 2013a, b, 2015b; Stachler and Marchfelder 2016; Stoll et al. 2013), *Haloferax mediterranei* (Li et al. 2013) and *Haloarcula hispanica* (Li et al. 2014a, b; Wang et al. 2016). The signature gene of this subtype is *cas8b*, and the *cas* gene clusters show a conserved arrangement (Makarova et al. 2015), whereas the number and location of the associated CRISPR loci differ widely (Fig. 11.4). However, the nature of the repeats within these CRISPR loci is strictly conserved within the Haloarchaea, resulting in a near-identical repeat sequence and length over the phylogenetic tree (Fig. 11.5). Aspects of all stages of CRISPR-Cas activity have been studied in different haloarchaeal systems, and the resulting picture is summarized in the following paragraphs.

11.3 The Adaptation Process in *Haloarcula hispanica*

The most striking feature of the CRISPR-Cas defence is its capacity to adapt to previously unknown invaders (as reviewed in (Amitai and Sorek 2016; Sternberg et al. 2016)). Until a decade ago, adaptive immunity was exclusively assigned to eukaryotic organisms, a paradigm that was swept aside by the characterization of

[3]*Halorubrum.*

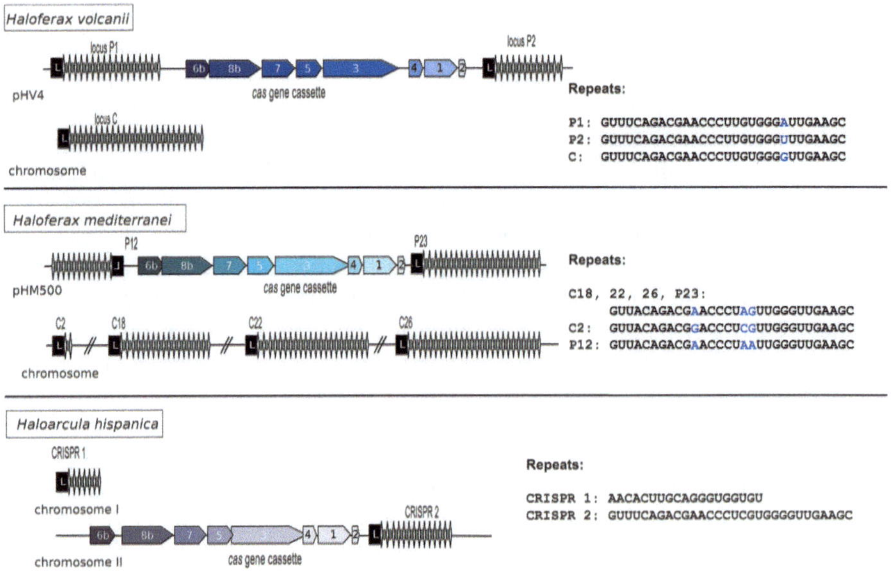

Fig. 11.4 The CRISPR-Cas type I-B systems of the haloarchaeal species discussed in this review. All three species possess a *cas* gene cassette of subtype I-B comprising eight *cas* genes. Gene synteny is conserved, whereas intergenic spacing as well as individual gene sequences are not. A characteristic of type I systems is the presence of the nuclease Cas3, whereas the Cas8b protein further characterizes a subtype I-B system. The *cas* genes are accompanied by a species-specific number of associated CRISPR loci. The repeat sequences within each locus are the same, whereas the sequences of different loci found within one genome vary in few positions (bold, *blue*). In *Haloferax volcanii*, the *cas* gene cassette is encoded on the pHV4 plasmid and is flanked by two CRISPR loci. A third locus is located on the primary chromosome. Laboratory strain H119 shows a deletion of 23 spacers within locus P1 with respect to the genome sequence published for the type strain *Hfx. volcanii* DS2 (Fischer et al. 2012). The repeat sequences of each locus are identical except for position 23. *Haloferax mediterranei* possesses six CRISPR loci. There are two loci flanking the *cas* gene cassette on plasmid pHM500 and four loci distributed on the primary chromosome. The repeat sequences in the four loci are identical, whereas loci C2 and P12 deviate at two positions (bold, *blue*). The *Haloarcula hispanica* CRISPR-Cas system is found on chromosome II. Downstream of the *cas* gene cassette is a single CRISPR locus (CRISPR2). CRISPR2 is also the only locus that encodes a full-length repeat sequence (with respect to the conserved repeat sequence given in Fig. 11.5). CRISPR1 on chromosome I comprises only repeats of 19 nucleotides, which widely deviate from the conserved haloarchaeal repeat (Fig. 11.5). Whether this locus actually results in mature crRNAs is unknown

CRISPR-Cas systems. Early on, the spacer content of CRISPR loci was linked to environmental sequences, such as phages or transposable elements (Bolotin et al. 2005; Mojica et al. 2005; Pourcel et al. 2005), conferring adaptive phage resistance (Barrangou et al. 2007; Brouns et al. 2008; Hale et al. 2009; Marraffini and Sontheimer 2008). The adaptation step expands the CRISPR locus by one repeat-spacer unit, and new spacers are preferentially added to the leader end of the locus, resulting in a near-chronological record of past encounters. First, the intruding nucleic acid has to be identified, and a small portion of its sequence known as the

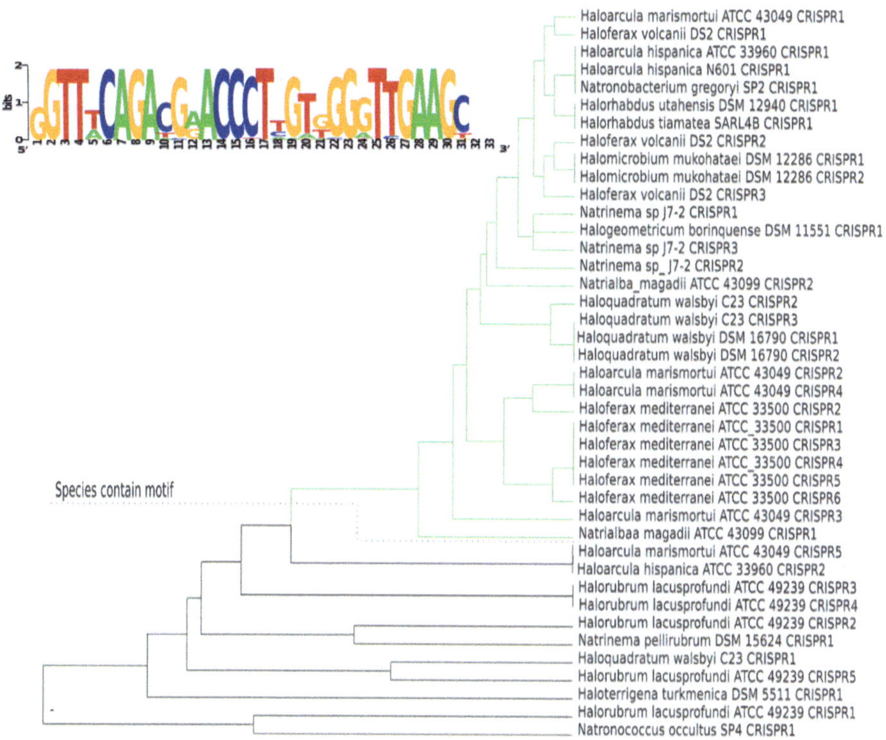

Fig. 11.5 The haloarchaeal repeat sequence is highly conserved and nearly identical throughout Haloarchaea. The repeat sequences found in the haloarchaeal-encoded CRISPR loci (listed on the *right*) have been combined into a sequence logo using the software WebLogo (Crooks et al. 2004), and the corresponding phylogenetic tree with repeat conservation is provided on the *right* (Lange et al. 2013; Alkhnbashi et al. 2014). The overall sequence conservation of the haloarchaeal repeat is very high, and only a few positions show interspecies differences. This trend, together with the conservation of the processing site utilized by Cas6 endonuclease during crRNA maturation, gives rise to highly uniform crRNA populations

protospacer has to be selected for integration. Crucial for this step in type I systems is a conserved series of a few nucleotides upstream of the protospacer, which is the PAM (Deveau et al. 2008; Erdmann and Garrett 2012; Mojica et al. 2009; Shmakov et al. 2015; Swarts et al. 2012; Wang et al. 2015). The PAM not only allows for the selection of a sequence as a protospacer during adaptation but also ensures discrimination between endogenous CRISPR loci that encode the spacer and the invader carrying the protospacer during the interference stage (Shah et al. 2013). The Cas proteins that are essential for the adaptation step are Cas1 and Cas2, which form a complex (Nunez et al. 2015; Plagens et al. 2012; Yosef et al. 2012). The concerted activity of the Cas1/Cas2 complex leads to the integration of the new spacer at the leader-repeat junction via a transposase/integrase-like mechanism, as shown in studies of the *E. coli* I-E system (Arslan et al. 2014; Nunez et al. 2015; Yosef et al. 2012). In *E. coli* (type I-E) DNA polymerase I and presumably other not yet identifies factors are involved (Ivancic-Bace et al. 2015) and adaptation depends on RecBCD activity occurring at sites of double strand breaks found e.g. at replication forks (Ivancic-Bace et al. 2015; Levy et al. 2015). In type I-B systems the Cas4 protein is also required for adaptation but details on its involvement are not unravelled yet.

In addition to this *de novo* capture of spacers known as naïve adaptation, a second form of spacer acquisition called primed adaptation was shown (Datsenko et al. 2012; Fineran et al. 2014; Künne et al. 2016; Li et al. 2014b; Richter et al. 2014; Semenova et al. 2016; Swarts et al. 2012; Vorontsova et al. 2015). During primed adaptation, a pre-existing spacer induces a positive-feedback loop, which leads to enhanced spacer acquisition from the targeted genetic element. A non-perfect match between a pre-existing crRNA and an invader DNA results in a defective interference reaction, and the elimination of the invader is not achieved; however, enhanced acquisition activity is induced (Datsenko et al. 2012; Fineran et al. 2014). Accordingly, in addition to the key acquisition proteins Cas1 and Cas2, primed adaptation also requires the presence of the Cascade interference complex as well as the Cas3 nuclease (Datsenko et al. 2012; Künne et al. 2016; Li et al. 2014b; Swarts et al. 2012). The co-occurrence of naïve and primed adaptation has so far been demonstrated in the *E. coli* type I-E system (Datsenko et al. 2012; Fineran et al. 2014; Swarts et al. 2012; Yosef et al. 2012), and the type I-F systems of *Pectobacterium atrosepticum* and *Pseudomonas aeruginosa* (subtype I-F, (Richter et al. 2014; Staals et al. 2016; Vorontsova et al. 2015)), whereas in the *Har. hispanica* subtype I-B system, only primed adaptation seems to be employed (Li et al. 2014b). *Har. hispanica* is the only haloarchaeal system for which adaptation could be shown to date. The deletion of the adaptation genes *cas1*, *cas2* and *cas4*, as well as the deletion of the interference module (*cas5–8*) in its entirety or the effector nuclease Cas3, will render *Har. hispanica* cells incapable of acquiring new spacers (Li et al. 2014b). In accordance with the necessity of the interference machinery, this process is strictly limited to primed adaptation, as the deletion of the priming spacer, with limited complementarity to the invader sequence, likewise hinders the integration of new spacers (Li et al. 2014b). During the priming process, a Cascade loaded with the imperfectly matched crRNA binds to the protospacer region of the invader in low-fidelity binding mode, triggering the priming process, as shown by FRET (Förster resonance energy transfer) analysis in *E. coli* (Blosser et al. 2015). Cas3 was speculated to be involved in the provision of acquisition substrates (Ivancic-Bace et al. 2015; Swarts et al. 2012). This speculation was supported by the observation that mutation of conserved residues within *Har. hispanica* Cas3 clearly show the active involvement of both the HD nuclease and the DxD/H-helicase domain in spacer acquisition (Li et al. 2014b). Recent work in *E. coli* further confirmed it: Cas3 degradation products were bound by the Cas1-Cas2 complex and integrated as new spacers (Künne et al. 2016). In *P. atrosepticum* (I-F system) interference promotes a targeted spacer acquisition process similar to priming (Staals et al. 2016).

Primed adaptation results in a biased sampling of new spacers with respect to the location of the priming protospacer. In *E. coli*, the DNA strand from which the new spacers are derived from is the same as that of the priming protospacer (Datsenko et al. 2012; Fineran et al. 2014). However, in *Har. hispanica*, the pattern is more diverse. Upstream of the imperfect match, spacers stem from the non-target strand, and their orientation matches the priming protospacer, whereas downstream, the target strand is the preferred source of spacers with the opposite directionality (Li et al. 2014b). A similar distorted acquisition pattern from both strands has also been observed in type I-F (Richter et al. 2014). The spacers acquired by *Har. hispanica* over the course of these first experiments were all sampled exclusively from the infecting viral particles, and the cognate protospacers were always preceded by the PAM TTC (Li et al. 2014b). Li and co-workers further used a mutational approach, presenting *Har. hispanica* cells with all possible three-nucleotide PAM combinations within an invader plasmid to study the motif's impact on the initiation of primed adaptation as well as its variability (Li et al. 2014a). Twenty-three of the 64 tested PAM sequences induced primed adaptation when the 5′ end of the protospacer was targeted by the priming spacer, these PAMs were termed priming permissive (Fig. 11.6) (Li et al. 2014a). Further investigation

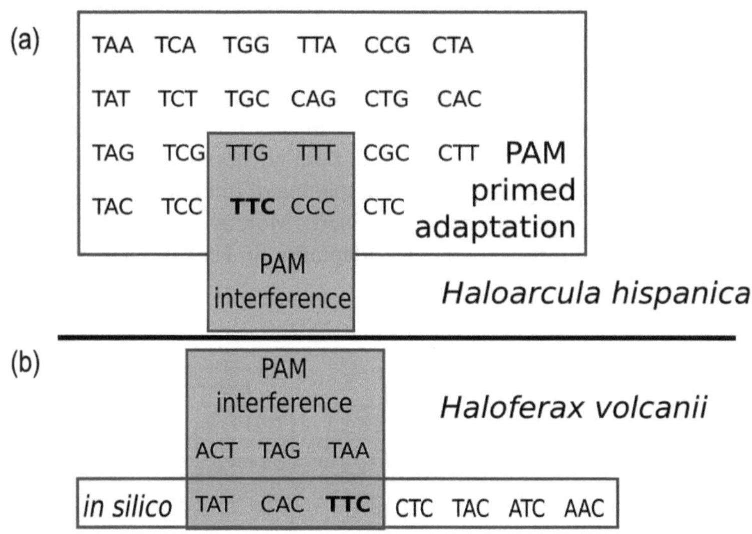

Fig. 11.6 Overview of PAM requirements during primed adaptation and interference. (**a**) *Haloarcula hispanica* is the first haloarchaeon for which primed adaptation was studied. 23 out of 64 possible PAM sequences triggered primed adaptation (Li et al. 2014a, b). In contrast, interference was triggered only with four PAM sequences (highlighted in *grey*) (Li et al. 2014a). (**b**) For *Hfx. volcanii,* the PAM requirements have only been determined for the interference stage. Here, six sequences trigger successful elimination of an invader DNA (Fischer et al. 2012). Additional *in silico* analyses revealed seven motifs found upstream of sequences matching spacers of the *Hfx. volcanii* CRISPR loci (Maier et al. 2015a). Three of them are identical to the identified interference PAM sequences. Despite their close phylogenetic relationship, both haloarchaea only share one PAM motif: TTC (marked in bold)

also revealed that these sequences are not sensed by a base-pairing mechanism but rather through the authentication of the PAM sequence (Li et al. 2014a). In addition, repeat sequences flanking the targeted protospacer do not impair priming as long as a cognate priming-permissive PAM is present (Li et al. 2014a). Similar findings were made in *E. coli*, here some PAM sequences can trigger both, interference and priming, but more PAM sequences were permissive for primed adaptation than for direct interference (Fineran et al. 2014; Li et al. 2014a).

Requirements for the integration of new spacers were also studied in detail in *Har. hispanica* (Wang et al. 2016). The sequences surrounding the leader-repeat junction are highly conserved within Haloarchaea and the conserved leader sequence plays a critical role during spacer integration (Wang et al. 2016). The leader-proximal cut occurs consistently at the leader-repeat junction (Fig. 11.7, integration site 1). However, the leader distal cut site does not have specific sequence requirements but is located at a constant distance to the second conserved repeat motif GTGGG (Fig. 11.7, integration site 2). A mutational analysis of the repeat sequence revealed that two conserved motifs in the repeat sequence are required for integration of new spacers (Fig. 11.7). The first motif (AACCC) needs to be 10 base pairs downstream of the leader-repeat junction and presumably serves as docking site for the integrase complex. The second motif (GTGGG) seems to be the anchor for a molecular ruler to direct the second cut 10 base pairs downstream thereby determining the size of the repeat duplication. Analysis of adaptation in *E. coli* confirms the presence of a ruler mechanism to define repeat length (Goren et al. 2016). Here, two rulers are employed, both anchored in the repeat sequence. Whether this model on the governing of the spacing of integration events is also true for other systems has yet to be determined. Moreover, an analysis of other haloarchaeal species will reveal whether adaptation in Haloarchaea is truly limited to a priming process.

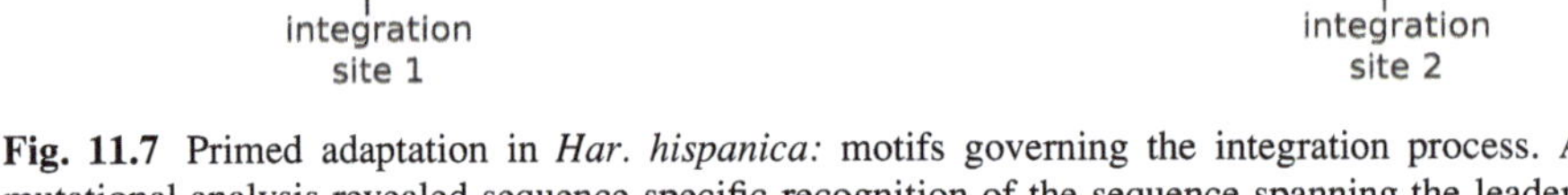

Fig. 11.7 Primed adaptation in *Har. hispanica:* motifs governing the integration process. A mutational analysis revealed sequence-specific recognition of the sequence spanning the leader-repeat-junction as well as two important motifs in the middle of the repeat sequence. The first motif (motif 1 depicted in *green*) has to be located ten nucleotides downstream of the leader-repeat-junction, whereas motif 2 (shown in *blue*) serves as an anchor-point to direct the second cleavage

11.4 Expression of the crRNA and Assembly of the Cascade Complex

The key player in CRISPR-Cas interference is the crRNA. The small guide that confers invader specificity to the Cascade complex is allocated during the expression stage of CRISPR-Cas activity. The crRNA is bound to and positioned within the Cascade by a subset of Cas proteins. Each complex is loaded with an individual guide that allows for the sequence-specific identification of an invader, equipping the cell with a multitude of effector complexes (Brouns et al. 2008; Jore et al. 2011; Künne et al. 2014).

The transcription of CRISPR loci is driven by a promoter region within the leader sequence (Pul et al. 2010). In *Hfx. volcanii* as well as in *Hfx. mediterranei*, the expression of the long precursor, pre-crRNA, is constitutive (Fischer et al. 2012; Li et al. 2013). The release of the crRNAs follows through the processing of the repeat region, catalysed by the Cas6b protein. This was confirmed in *Hfx. volcanii* and *Hfx. mediterranei*, where crRNA production is lost upon deletion of the *cas6b* gene (Brendel et al. 2014; Li et al. 2013). Apart from Cas6b, multiple other Cas proteins are involved in the maintenance of a stable crRNA population within the cell. In *Hfx. volcanii*, a deletion of the *cas5* or *cas7* gene does not impair but rather severely lessens the steady-state level of crRNA, which indicates that there is a Cas5- and Cas7-mediated protection against degradation (Brendel et al. 2014). The protective effect of Cas5 and Cas7 is even more pronounced in the *Hfx. mediterranei* system, in their absence, no mature crRNA is detectable (Li et al. 2013).

By contrast to *cas6b* deletion, the loss of *cas5/7* clearly leaves endonucleolytic processing intact because a leader-first repeat product still accumulates in *Hfx. mediterranei* (Li et al. 2013). As revealed by structural studies with the subtype I-E Cascade in *E. coli*, Cas5 binds to the repeat-derived 5′-handle, whereas Cas7 covers the spacer sequence, thereby enclosing the crRNA within the Cascade and making it less accessible for the degradation machinery (Jackson et al. 2014; Mulepati et al. 2014; Zhao et al. 2014). Both Cas proteins are also integral parts of the *Hfx. volcanii* Cascade, which in addition includes Cas6b (Fig. 11.8) (Brendel et al. 2014). The Cas8b protein was only occasionally obtained and therefore seems to be only loosely associated with the *Haloferax* Cascade complex. This finding is also mirrored by the minor stabilizing effect of Cas8b on the crRNA population in both *Haloferax* species (Brendel et al. 2014; Li et al. 2013).

The crRNA itself also affects the structure and composition of the effector complex. The Cascade complex of the type I-B system in *Haloferax* has not yet been structurally characterized, but a combination of mass spectrometry and intensity-based absolute quantification (iBAQ) identified the core complex as being composed of Cas5, Cas6b and Cas7 in a ratio of 1.7:1:8.5 (Brendel et al. 2014). This complex differs from the composition of the *E. coli* type I-E Cascade, for which the stoichiometry was also determined as Cas5, Cas6, Cas7, Cas8, and Cse2: 1:1:6:1:2 (Jore et al. 2011; Wiedenheft et al. 2011). The small subunit Cse2

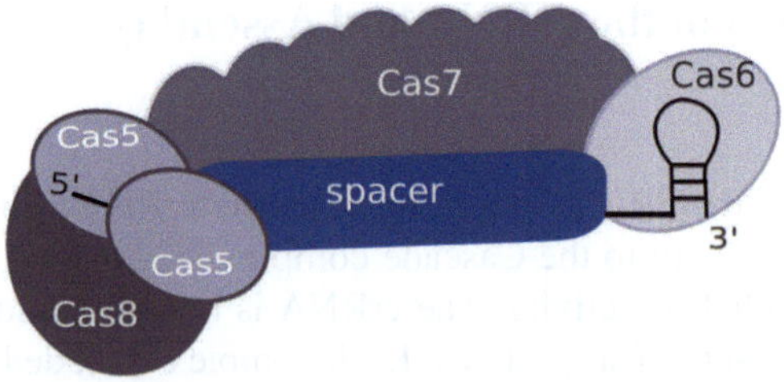

Fig. 11.8 Potential structures of Cascade complexes in *Hfx. volcanii*. Since structural data are not available for the *Haloferax* Cascade or any I-B Cascade complex, a schematic representation based on the published structure of the *E. coli* type I-E complex is given. Co-purification approaches combined with quantitative mass spectrometry identified a Cascade complex composed of Cas5, Cas6, Cas7 in a stoichiometry of 1.7:1:8.5 (Brendel et al. 2014). Cas8b seems to be loosely associated and could only be occasionally co-purified (Brendel et al. 2014). A minimal stable complex might be formed that includes only Cas5 and the Cas7 backbone alongside the crRNA, as Cas6b is dispensable for interference (Brendel et al. 2014; Maier et al. 2015b). The minimal crRNA that elicits an interference reaction only comprises the 5′-handle and the spacer sequence; thus, the minimal Cascade may well be further reduced by omitting the crRNA 3′ handle (Maier et al. 2015b)

and the Cas8 protein are integral parts of the *E. coli* I-E Cascade, which is the most striking difference, but the composition of the core *Hfx. volcanii* I-B Cascade also shows two additional copies of Cas7 (Brendel et al. 2014). This composition might reflect differences in the length of the crRNA because the spacer length in *Haloferax* is 34–39 nt, as opposed to the 32 nt in *E. coli*. Given that Cas7 forms the backbone of Cascade receiving the spacer portion of the crRNA, additional subunits are needed to cover the entire 2- to 7-nt-longer spacer sequence found in *Haloferax*. The elongation of the Cascade backbone to accommodate an elongated crRNA has also been observed in a study that analysed the subunit composition of *Shewanella putrefaciens* I-F Cascade. Upon the extension of the spacer portion of the crRNA, more Cas7 subunits are incorporated (Gleditzsch et al. 2016).

Further analysis of *cas* gene deletion mutants in *Hfx. mediterranei* revealed a negative effect of Cas1, 3 and 4 on the crRNA level (Li et al. 2013), however, it is not clear how these proteins might contribute to crRNA stabilization or influence pre-crRNA expression.

11.5 The crRNA Populations of Haloarchaeal CRISPR-Cas Systems

The position of the Cas6 cleavage site within the repeat regions of the pre-crRNA is highly conserved in type I systems and shows the tight evolutionary and phylogenetic link between the repeat sequence and Cas6 protein family (Kunin et al. 2007; Wang et al. 2012). However, individual Cas6 proteins share neither a common mode for substrate binding nor conserved catalytic residues (Brendel et al. 2014).

Their amino acid sequences show limited conservation with only two common motifs: the ferredoxin fold and a glycine-rich motif (Li 2015). The crRNAs of type I systems are consistently composed of the spacer sequence accompanied by an eight-nucleotide 5′-handle and the remainder of the downstream repeat as a 3′-handle (Charpentier et al. 2015). The extraordinary conservation of the repeat sequences present in haloarchaeal CRISPR-Cas systems results in near-identical crRNA flanking sequences, and the conserved repeat length results in 22 nt-long 3′-handles in almost all haloarchaeal species.

Interestingly, the three CRISPR loci of *Hfx. volcanii* each differ in their repeat sequence at position 23, resulting in a mixed population of mature crRNAs starting with either U, A or G as the first nucleotide of the 5′-handle (Fig. 11.4) (Fischer et al. 2012). The genome of *Hfx. mediterranei* encodes six CRISPR loci, also sharing a common repeat sequence with slight variations in the repeat of locus C2 at positions 11 (A to G) and 17 (A to C), and the repeat sequence of locus P12 deviates at position 18 (G to A) (Fig. 11.4). Those variants also result in a population of crRNAs with varying 3′-handle sequences (Li et al. 2013).

Despite the close phylogenetic relationship of both *Haloferax* species, they differ in the size distribution of the crRNA population that is detectable *in vivo* (Fig. 11.9). The analysis of crRNA sequences in *Hfx. mediterranei* by CR-RT-PCR (circularized-RNA RT-PCR) revealed one population of mature crRNAs with a size range from 64 to 68 nt and with differences accounted for by the varying spacer lengths (approximately 34–39 nt, as expected) (Li et al. 2013). However, an RNA-Seq approach in *Hfx. volcanii* identified a second group of crRNAs that is stably maintained separate from the dominant crRNA population of 64–69 nt due to the spacer length (Maier et al. 2015b). These crRNAs are substantially shorter due to having a 3′-handle of only five nucleotides. A similar trimming of mature crRNAs has been reported for other type I-B systems of *Clostridium thermocellum* and *Methanococcus maripaludis* and is assumed to be characteristic of type I-B as well as I-A and I-D systems in contrast to the type I-C, I-E and I-F systems featuring non-trimmed crRNAs (Charpentier et al. 2015). This variety illustrates the diversity of CRISPR-Cas mechanisms, reaching beyond the subtype-level and making it even more important to study a wide variety of CRISPR-Cas systems in different species to complete the picture of this most elaborate defence system.

Moreover, the crRNA population of *Hfx. volcanii* revealed that crRNAs originating from the same CRISPR locus are not present in equal amounts (Maier et al. 2013a), an observation confirmed in several other organisms of different subtypes (Deng et al. 2012; Hale et al. 2012; Nickel et al. 2013; Richter et al. 2012; Scholz et al. 2013; Zhang et al. 2012). This finding might reflect an imminent technical problem in the currently available RNA-Seq approach but it might also be of biological relevance. Furthermore, as shown for *Hfx. volcanii*, the different crRNAs diverged in their ability to fend off the plasmid invader (Maier et al. 2013a). The different crRNAs present in the cell vary in their spacer sequence, which may not only contain signals that trigger a faster degradation of some of the molecules but may also influence the effectivity of Cascade binding. In addition, they may

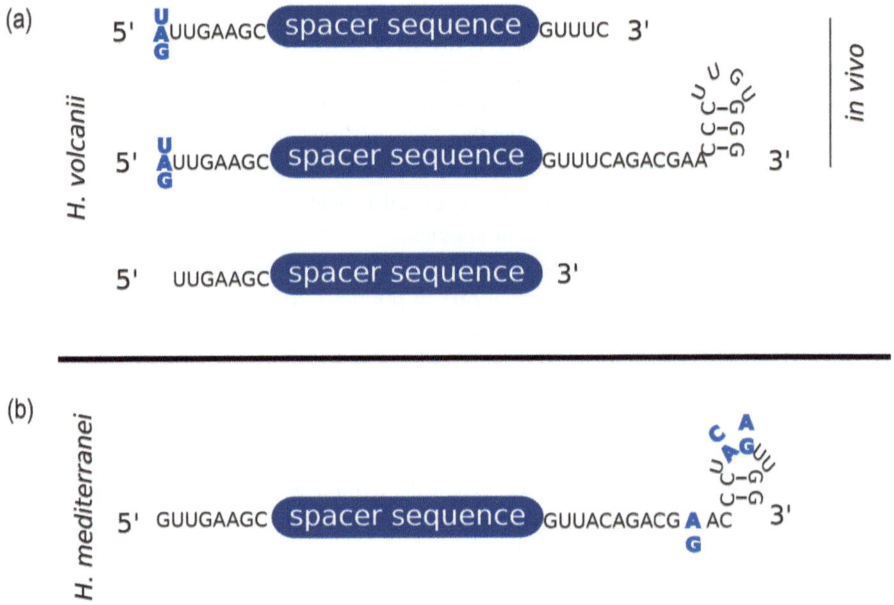

Fig. 11.9 Different crRNA molecules *in Hfx. volcanii* and *Hfx. mediterranei*. (**a**) In addition to the spacer sequence, the long form of the crRNA identified in vivo in both *Haloferax* strains possesses an 8-nucleotide 5′ handle and a 22 nucleotide long 3′ handle (Li et al. 2013; Maier et al. 2013a). In *Hfx. volcanii,* the first nucleotide of the 5′ handle differs due to sequence variation within the three CRISPR loci. The varying repeat sequences in *Hfx. mediterranei* result in a mixed population of crRNAs with variable 3′ handles. The RNA-Seq analysis of the crRNA pool in *Hfx. volcanii* also revealed a shortened crRNA variant with only five nucleotides as the 3′ handle and seven instead of eight nucleotides at the 5′ handle (Maier et al. 2015b). Moreover, mutational analysis demonstrated that the crRNA is still active when the 3′ handle is completely removed (Maier et al. 2015b)

influence the microarchitecture and topology of the Cascade complex and thereby the efficiency of the interaction with the target.

The crRNA structure has so far only been investigated in *Hfx. volcanii*, and although the 22-nucleotide 3′-handle encodes a set of inverted repeats, offering the possibility of forming a hair-pin structure at the very 3′ end, no such structure has been detected in *in vitro* studies (Fischer et al. 2012). Nevertheless, a hairpin structure might be stabilized upon interaction with the Cas6b protein during processing, as seen in *Thermus thermophilus* and *Sulfolobus solfataricus* (Niewoehner et al. 2014; Shao and Li 2013).

The characteristics of the crRNA with importance beyond processing were studied in *Hfx. volcanii* using a system for the Cas6-independent generation of crRNAs based on the tRNA-maturation machinery (Maier et al. 2015b). The independent biogenesis pathway results in a crRNA with a 5′-phosphate and 3′-hydroxyl group, in contrast to a crRNA processed by Cas6b possessing a 5′-hydroxyl and 2′-3′-cyclic phosphate group. This independently generated crRNA (termed icrRNA) was active, therefore, neither the loading of a crRNA into Cascade nor the interference reaction depends on the chemical nature of the end groups (Maier et al. 2015b). A mutational analysis of independently generated crRNAs revealed that the 3′-handle of the crRNA was completely dispensable for the *in vivo* interference reaction (Maier et al. 2015b). A comparison with the structure of the type I Cascade complexes of *E. coli* showed that the 3′-handle would come into contact with Cas6b (Jackson et al. 2014; Mulepati et al. 2014; Zhao et al. 2014). However, in a *Hfx. volcanii* strain capable of Cas6b-independent crRNA maturation, *cas6b* could be deleted without affecting the interference step (Maier et al. 2015b). This finding implies that Cas6b is not an essential part of the type I-B Cascade complex for the interference step (Fig. 11.8). The crRNA 5′-handle, by contrast, is indispensable for crRNA function, only the first nucleotide can be removed without loss of activity (Maier et al. 2015b). Inference from the atomic structures of type I-E Cascade showed that in *E. coli* is bound by the Cas5 subunit (Jackson et al. 2014; Mulepati et al. 2014; Zhao et al. 2014), which was also shown to be an integral part of the *Haloferax* type I-B Cascade (Brendel et al. 2014).

11.6 Determinants for a Successful Defence Reaction in Subtype I-B

CRISPR-Cas systems have recently attracted attention as a molecular biological tool that out-competes all the available nucleic-acid targeting proteins because its targeting activity is based on an easily interchangeable module: crRNA. Through the embodied spacer sequence, different crRNAs guide the Cascade complex to a defined targeting site within an invading nucleic acid. This identification depends on base pairing between the spacer part of the crRNA and the target molecule (Künne et al. 2014). A mutational analysis of the protospacer sequence within a plasmid invader was used to determine how strictly defined this interaction is regarding the *Hfx. volcanii* type I-B system (Maier et al. 2013a). The first ten nucleotides of the spacer sequence were identified as being critical for triggering a defence reaction. Within this sequence, which is denoted as the seed region, only a mismatch at position 6 is tolerated (Fig. 11.10). A similar seed sequence was also determined for *E. coli* and *P. aeruginosa* (Künne et al. 2014; Semenova et al. 2011; Wiedenheft et al. 2011). For the *E. coli* system, every sixth position within the crRNA is not involved in a base-pairing activity (Semenova et al. 2011). As structural data show, the thumb domain of the Cas7 proteins in the *E. coli* Cascade extrude every sixth nucleotide from the axis that runs down the Cas7 backbone, rendering it inaccessible for base pairing with the protospacer region (Jackson et al. 2014; Mulepati et al. 2014; Zhao et al. 2014). Increments of six could not be confirmed in the *Hfx. volcanii* analysis, but this difference might very well reflect the aforementioned differences in the Cascade composition (Maier et al. 2013a).

An important prerequisite for having and keeping a CRISPR-Cas defence system is, that self-targeting is excluded, since this can be fatal for the cells. Such an

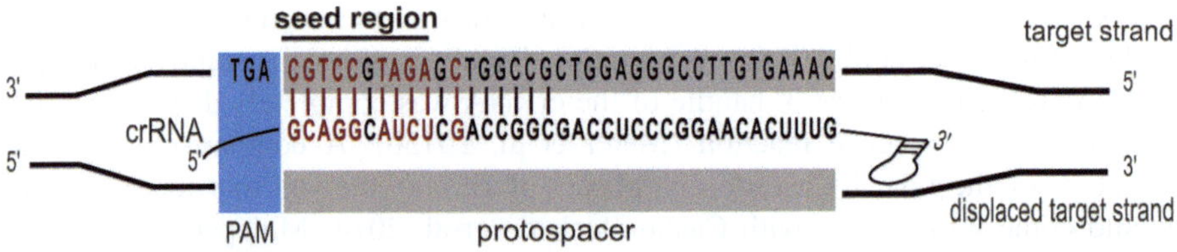

Fig. 11.10 A seed sequence is required for interference in *Hfx. volcanii*. During invader recognition, the crRNA base pairs with the protospacer region of the invading DNA. Base pairing over a ten-nucleotide-long non-contiguous seed sequence elicits the interference reaction. Essential base pairs are shown in *red*. Pairing at position six is not required, but *Hfx. volcanii* does not exhibit a six nucleotide increment as seen in the spacer-protospacer interactions in *E. coli* (Maier et al. 2013a)

auto-immune reaction is prevented by the absence of the PAM in the host DNA. The role of PAM sequences in adaptation has already been discussed, but these sequences also play an important role during the interference stage (Deveau et al. 2008; Mojica et al. 2009). PAM sequences can be found in type I, type II and type V CRISPR-Cas systems and are unique features of the protospacer. Despite perfect base pairing within the seed sequence, interference takes place only if a cognate PAM is present at the 5′ end of the protospacer sequence in type I and type V systems and at the 3′ end in type II systems (Shah et al. 2013; Zetsche et al. 2015). PAM sequences of haloarchaeal species could not be directly inferred by comparing the spacer contents and publicly available sequences of mobile genetic elements because haloarchaeal viruses are grossly underrepresented in public databases (Fischer et al. 2012). Moreover, the population of mobile genetic elements present today likely differs substantially from the one that was present on the isolation date of the laboratory strains under investigation. Therefore, the PAM sequences of the *Hfx. volcanii* type I-B system have been identified *in vivo* using a mutational approach based on a plasmid invader (Fischer et al. 2012). A systematic analysis of all possible three nucleotide sequences preceding a protospacer revealed the following six PAM sequences: ACT, CAC, TTC, TAT, TAG, and TAA (Fig. 11.6).

The stimulation of an interference reaction by more than one sequence motif is a strategy for coping with the divergence of invader populations, rendering escape via individual PAM mutations less likely. Moreover, this stimulation increases the possibility that closely related foreign elements are also susceptible to CRISPR-Cas interference. The authentication of the PAM sequence is a crucial step in the transition of Cascade from a DNA-sensing to a DNA-degrading complex. Studies of *E. coli* type I-E Cascade show that upon detection of a cognate PAM sequence, the conformation of the Cascade is changed and the processing endonuclease Cas3 is recruited (Hochstrasser et al. 2014). The subunit responsible for determining the PAM identity in type I-E systems is the large subunit Cas8e (Sashital et al. 2012). Accordingly, upon the deletion of Cas8b in *Haloferax*, the interference was lost without affecting the crRNA level or stability (Cass et al. 2015). More importantly, the response of Cas8b variants with mutated conserved residues varied with regard to the PAM sequence presented by the invader (Cass et al. 2015). However, the exact mechanism for the read-out of PAM sequences is still under investigation. Although *Har. hispanica* shares the same subtype and a near-identical repeat sequence with *Hfx. volcanii*, the *Har. hispanica* CRISPR-Cas system responds to only four PAM sequences, namely TTT, TTC, TTG, and CCC, and besides TTC, no other PAM is shared between both organisms (Fig. 11.6) (Li et al. 2014a). The Cas8b proteins found in both species only share 22.6% sequence identity, and given the likely role of Cas8b as the PAM-sensing Cascade subunit, this low similarity might account for the low conservation of PAM sequences (Li et al. 2014a). Moreover, findings from a bioinformatics analysis of the spacer content of *Hqr*[4]. *walsbyi* provide support that a certain degree of PAM sequence conservation is present (Fischer et al. 2012; Garcia-Heredia et al. 2012). Several matches to viral contigs from the metavirome data of the isolation sites reveal protospacers preceded by the PAM TTC. A recent bioinformatics analysis of PAM sequences in *Hfx. volcanii* could match eight of the *Hfx. volcanii* spacers to sequences in the database (Maier et al. 2015a). These target sequences are flanked by seven different PAM sequences: TAT, CAC, CTC, TTC, TAC, ATC and AAC at the protospacer 5′-end. Three of them are identical to the experimentally determined PAMs: CAC, TTC and TAT. However, the motifs inferred from *in silico* analysis were obtained by comparing sequences that were not necessarily derived from the same biological context.

Together with the adaptation analysis in *Har. hispanica* (see paragraph above), this evidence illustrates that the requirements for PAM sequences during the adaptation and interference stages are not identical but can overlap. As this trend is also seen in other systems, PAMs have been subdivided into motifs important for adaptation, termed spacer acquisition motif (SAMs) and motifs essential for interference (target interference motif: TIMs) (Shah et al. 2013). These processes rely on different protein machineries: the Cas1/Cas2 complex is interacting with the SAM during naïve adaptation, while the Cascade complex is interacting with TIM during interference. Different binding partners or different conformations of the binding subunit within the complex might result in different PAM demands (Shah et al. 2013).

An analysis of the *Hfx. volcanii* interference reaction revealed an interesting detail that influenced the success of the defence reaction. Whether a plasmid invader effectively triggers an interference reaction depends on the origin of the replication (Maier et al. 2013a). Only the plasmid with a pHV1 origin, replicated by a mechanism depending on an origin recognition complex (ORC), was successfully eliminated (Delmas et al. 2009; Maier et al. 2013a; Norais et al. 2007). By contrast, the type I-B system did not overcome a plasmid based on a pHV2 origin whose replication presumably depended on the Rep protein (Charlebois et al. 1987; Maier et al. 2013a; Woods and Dyall-Smith 1997). The experimental design of the studied plasmids places the targeted protospacer next to the origin of replication. Whether these differences are solely due to steric constraints or reflect a functional interaction requires further analysis.

[4]*Haloquadratum.*

11.7 Using CRISPR-Cas as Tool to Regulate Gene Expression in *Hfx. volcanii*

CRISPR-Cas systems have been developed into a plethora of different tools and have been exploited for numerous scientific analyses outside their natural function in both prokaryotes and eukaryotes (Cong et al. 2013; Fineran and Dy 2014; Sampson and Weiss 2014; Sternberg and Doudna 2015). Given the presence of a cognate PAM sequence, any region of interest can be targeted. The CRISPR-Cas system that is most extensively used in genetic studies is the type II system, which only requires a single effector protein: Cas9. One of the applications is targeted gene regulation, which is performed via a mechanism called CRISPRi (CRISPR interference) (Qi et al. 2013). Here, the expression of a gene specified by the targeting guide RNA is repressed through the binding of a catalytically inactive Cas9 (dCas9) (Qi et al. 2013). The protein is engineered to locate and bind the target sequence as defined by the incorporated crRNA but not to cleave it, thereby preventing or severely hampering transcription initiation or elongation. *Streptococcus pyogenes* dCas9 has successfully been used in eukaryotes as well as in bacteria (Bikard et al. 2013; Qi et al. 2013). The endogenous type I system was likewise repurposed as tool for CRISPRi in *E. coli* by deleting the *cas*3 gene (Luo et al. 2015; Rath et al. 2015). Upon the loss of the targeting endonuclease, Cascade still binds the target region specified by the crRNA but does not cleave it (Luo et al. 2015; Rath et al. 2015).

Molecular biology studies in Archaea become more and more widespread, but tools for transcriptional repression are not available. Due to their extremophilic nature, most archaea pose a challenge to the heterologous expression of proteins, such as Cas9, which is predominantly found in mesophilic bacteria. Therefore, the most straightforward approach is to repurpose the endogenous CRISPR-Cas systems, circumventing the need for the heterologous expression of the Cas9 protein. Similar to the approach used in *E. coli*, the type I-B system in *Hfx. volcanii* has successfully been modified and converted into the first tool for transcriptional repression in archaea (Stachler and Marchfelder 2016). To eliminate the DNA cleavage activity, the *cas*3 gene is deleted and to achieve the efficient downregulation of the targeted gene, the endogenous crRNA population has to be depleted (Stachler and Marchfelder 2016). The latter can be achieved via the deletion of *cas6b* or the deletion of the endogenous CRISPR loci. If Cas6b was no longer present, crRNAs have been provided by the aforementioned icrRNA system (see above) (Maier et al. 2015b; Stachler and Marchfelder 2016). Using this system a plasmid-borne reporter gene, a chromosomal gene, a gene cluster as well as an essential gene were successfully knocked down. The strongest repression effect observed was a down regulation to 8% of the transcript level, highlighting the potential of CRISPRi for archaeal systems (Stachler and Marchfelder 2016).

Various possible targeting regions within the promoter and coding regions of the genes of interest were explored. In general, the crRNAs targeting the promoter region and, more precisely, its template strand are the most efficient, whereas those directed towards the coding strand or the open reading frame had little or no effect on gene expression (Stachler and Marchfelder 2016). Currently, a clear connection between successful targeting and the characteristics of the crRNA could not be inferred, more experimental data on this topic are required, which might then allow to implement a tool for designing efficient crRNAs.

11.8 Conclusions

Haloarchaea form a coherent group with respect to their CRISPR-Cas content, they all encode type I systems, with the subtype I-B being the most dominant. General characteristics of the archaeal type I-B systems can be drawn on the basis of detailed studies of three haloarchaeal species.

The Cascade-like effector complex analysed in *Hfx. volcanii* closely resembles other type I complexes in terms of its Cas protein composition. The same *cas* genes are consistently indispensable for the *Hfx. mediterranei* CRISPR-Cas activity. *Hfx. volcanii* is the only species that generates two types of crRNAs that differ in the lengths of their 3′ ends. This second processing event has also been described in other subtype I-B systems (Richter et al. 2012) but has not been found in the other haloarchaeal type I-B systems. The shortened crRNA variant of *Hfx. volcanii* lacks part of the 3′-handle responsible for Cas6b binding, probably resulting in a Cascade-like complex lacking this subunit. The minimal requirements for a successful defence reaction in *Hfx. volcanii* support this interpretation because the Cas6b protein has been shown to be dispensable during interference.

The adaptation step was hitherto only investigated in two haloarchaea, namely *Hfx. mediterranei* and *Har. hispanica*. Here, interestingly, only the primed adaptation triggered by the presence of a pre-existing spacer could be shown. The PAM sequences for the adaptation reaction have been determined for *Har. hispanica* revealing 23 PAM sequences that allow primed adaptation. They only partially overlap with PAM sequences, that trigger a defence reaction.

The PAM motifs required for effective interference reactions were systematically analysed in *Hfx. volcanii* revealing that *Hfx. volcanii* responds to six PAM sequences. The *Har. hispanica* and *Hfx. volcanii* PAM requirements overlap in only one motif, illustrating that even closely related haloarchaea differ in their defence requirements. PAM sequences in haloarchaea are situated 5′ to the protospacer sequence on the invading DNA, and similar to other type I systems, interference relies on the presence of a seed sequence as shown in *Hfx. volcanii*.

The information summarized here, along with that of other known type I systems, highlights the subtype-specific and inter-subtype strain-specific peculiarities and differences of type I systems. Further exploration will complete our knowledge on CRISPR-Cas immunity, and subsequently allow the application of CRISPR-Cas systems and their components. The details gathered on CRISPR-Cas immunity in *Hfx. volcanii* enabled the first application of an endogenous archaeal CRISPR-Cas system for targeted gene regulation, meeting a long-standing need for gene regulatory tools in archaea that will further promote archaeal research in many fields.

References

Alkhnbashi OS, Costa F, Shah SA, Garrett RA, Saunders SJ, Backofen R (2014) CRISPRstrand: predicting repeat orientations to determine the crRNA-encoding strand at CRISPR loci. Bioinformatics 30:i489–i496. doi:10.1093/bioinformatics/btu1459

Amitai G, Sorek R (2016) CRISPR-Cas adaptation: insights into the mechanism of action. Nat Rev Microbiol 14:67–76. doi:10.1038/nrmicro.2015.1014. Epub 2016 Jan 11

Arslan Z, Hermanns V, Wurm R, Wagner R, Pul U (2014) Detection and characterization of spacer integration intermediates in type I-E CRISPR-Cas system. Nucleic Acids Res 42:7884–7893. doi:10.1093/nar/gku7510. Epub 2014 Jun 11

Barrangou R (2015) Diversity of CRISPR-Cas immune systems and molecular machines. Genome Biol 16:247. doi:10.1186/s13059-13015-10816-13059

Barrangou R, Fremaux C, Deveau H, Richards M, Boyaval P, Moineau S, Romero DA, Horvath P (2007) CRISPR provides acquired resistance against viruses in prokaryotes. Science 315:1709–1712

Bikard D, Jiang W, Samai P, Hochschild A, Zhang F, Marraffini LA (2013) Programmable repression and activation of bacterial gene expression using an engineered CRISPR-Cas system. Nucleic Acids Res 41:7429–7437. doi: 10.1093/nar/gkt7520. Epub 2013 Jun 12

Blosser TR, Loeff L, Westra ER, Vlot M, Kunne T, Sobota M, Dekker C, Brouns SJ, Joo C (2015) Two distinct DNA binding modes guide dual roles of a CRISPR-Cas protein complex. Mol Cell 58:60–70. doi:10.1016/jmolcel201510011028. Epub 2015 Mar 15

Bolotin A, Quinquis B, Sorokin A, Ehrlich SD (2005) Clustered regularly interspaced short palindrome repeats (CRISPRs) have spacers of extrachromosomal origin. Microbiology 151:2551–2561

Brendel J, Stoll B, Lange SJ, Sharma K, Lenz C, Stachler AE, Maier LK, Richter H, Nickel L, Schmitz RA, Randau L, Allers T, Urlaub H, Backofen R, Marchfelder A (2014) A complex of Cas proteins 5, 6, and 7 is required for the biogenesis and stability of clustered regularly interspaced short palindromic repeats (crispr)-derived rnas (crrnas) in Haloferax volcanii. J Biol Chem 289:7164–7177. doi: 10.1074/jbc.M7113.508184. Epub 2014 Jan 23

Brouns SJ, Jore MM, Lundgren M, Westra ER, Slijkhuis RJ, Snijders AP, Dickman MJ, Makarova KS, Koonin EV, van der Oost J (2008) Small CRISPR RNAs guide antiviral defense in prokaryotes. Science 321:960–964

Burstein D, Harrington LB, Strutt SC, Probst AJ, Anantharaman K, Thomas BC, Doudna JA, Banfield JF (2017) New CRISPR-Cas systems from uncultivated microbes. Nature 542:237–241. doi:10.1038/nature21059. Epub 2016 Dec 22

Cass SD, Haas KA, Stoll B, Alkhnbashi O, Sharma K, Urlaub H, Backofen R, Marchfelder A, Bolt EL (2015) The role of Cas8 in type I CRISPR interference. Biosci Rep 5:5

Charlebois RL, Lam WL, Cline SW, Doolittle WF (1987) Characterization of pHV2 from Halobacterium volcanii and its use in demonstrating transformation of an archaebacterium. Proc Natl Acad Sci U S A 84:8530–8534

Charpentier E, Richter H, van der Oost J, White MF (2015) Biogenesis pathways of RNA guides in archaeal and bacterial CRISPR-Cas adaptive immunity. FEMS Microbiol Rev 39:428–441. Epub 2015 May 19

Cong L, Ran FA, Cox D, Lin S, Barretto R, Habib N, Hsu PD, Wu X, Jiang W, Marraffini LA, Zhang F (2013) Multiplex genome engineering using CRISPR/Cas systems. Science 339:819–823 doi:10.1126/science1231143

Crooks GE, Hon G, Chandonia JM, Brenner SE (2004) WebLogo: a sequence logogenerator. Genome Res 14:1188–90

Datsenko KA, Pougach K, Tikhonov A, Wanner BL, Severinov K, Semenova E (2012) Molecular memory of prior infections activates the CRISPR/Cas adaptive bacterial immunity system. Nat Commun 3:945. doi:10.1038/ncomms1937

Delmas S, Shunburne L, Ngo HP, Allers T (2009) Mre11-Rad50 promotes rapid repair of DNA damage in the polyploid archaeon Haloferax volcanii by restraining homologous recombination. PLoS Genet 5:e1000552. Epub 2009 Jul 10

Deng L, Kenchappa CS, Peng X, She Q, Garrett RA (2012) Modulation of CRISPR locus transcription by the repeat-binding protein Cbp1 in Sulfolobus. Nucleic Acids Res 40:2470–2480. Epub 2011 Dec 1

Deveau H, Barrangou R, Garneau JE, Labonte J, Fremaux C, Boyaval P, Romero DA, Horvath P, Moineau S (2008) Phage response to CRISPR-encoded resistance in Streptococcus thermophilus. J Bacteriol 190:1390–1400. Epub 2007 Dec 13

Erdmann S, Garrett RA (2012) Selective and hyperactive uptake of foreign DNA by adaptive immune systems of an archaeon via two distinct mechanisms. Mol Microbiol 85:1044–1056. doi:10.1111/j1365-2958201208171x. Epub 2012 Jul 27

Fineran PC, Dy RL (2014) Gene regulation by engineered CRISPR-Cas systems. Curr Opin Microbiol 18:83–89. doi: 10.1016/j.mib.2014.1002.1007. Epub 2014 Mar 15

Fineran PC, Gerritzen MJ, Suarez-Diez M, Kunne T, Boekhorst J, van Hijum SA, Staals RH, Brouns SJ (2014) Degenerate target sites mediate rapid primed CRISPR adaptation. Proc Natl Acad Sci U S A 111:E1629–E1638. doi:10.1073/pnas.1400071111. Epub 2014 Apr 17

Fischer S, Maier LK, Stoll B, Brendel J, Fischer E, Pfeiffer F, Dyall-Smith M, Marchfelder A (2012) An archaeal immune system can detect multiple protospacer adjacent motifs (PAMs) to target invader DNA. J Biol Chem 287:33351–33365

Garcia-Heredia I, Martin-Cuadrado AB, Mojica FJ, Santos F, Mira A, Anton J, Rodriguez-Valera F (2012) Reconstructing viral genomes from the environment using fosmid clones: the case of haloviruses. PLoS One 7:e33802. doi:10.31371/journalpone0033802. Epub 2012 Mar 30

Garneau JE, Dupuis ME, Villion M, Romero DA, Barrangou R, Boyaval P, Fremaux C, Horvath P, Magadan AH, Moineau S (2010) The CRISPR/Cas bacterial immune system cleaves bacteriophage and plasmid DNA. Nature 468:67–71

Garrett RA, Vestergaard G, Shah SA (2011) Archaeal CRISPR-based immune systems: exchangeable functional modules. Trends Microbiol 19:549–556. Epub 2011 Sep 22

Gleditzsch D, Muller-Esparza H, Pausch P, Sharma K, Dwarakanath S, Urlaub H, Bange G, Randau L (2016) Modulating the Cascade architecture of a minimal Type I-F CRISPR-Cas system. Nucleic Acids Res 44:5872–5882. doi:10.1093/nar/gkw5469. Epub 2016 May 23

Goren MG, Doron S, Globus R, Amitai G, Sorek R, Qimron U (2016) Repeat size determination by two molecular rulers in the Type I-E CRISPR array. Cell Rep 16:2811–2818. doi:10.1016/j.celrep.2016.2808.2043

Haft DH, Selengut J, Mongodin EF, Nelson KE (2005) A guild of 45 CRISPR-associated (Cas) protein families and multiple CRISPR/Cas subtypes exist in prokaryotic genomes. PLoS Comput Biol 1:e60. Epub 2005 Nov 11

Hale CR, Zhao P, Olson S, Duff MO, Graveley BR, Wells L, Terns RM, Terns MP (2009) RNA-guided RNA cleavage by a CRISPR RNA-Cas protein complex. Cell 139:945–956. doi:10.1016/j.cell.2009.1007.1040

Hale CR, Majumdar S, Elmore J, Pfister N, Compton M, Olson S, Resch AM, Glover CV, III, Graveley BR, Terns RM, Terns MP (2012) Essential features and rational design of CRISPR RNAs that function with the Cas RAMP module complex to cleave RNAs. Mol Cell 45:292–302. Epub 2012 Jan 15

Hille F, Charpentier E (2016) CRISPR-Cas: biology, mechanisms and relevance. Philos Trans R Soc Lond B Biol Sci 371(1707):20150496. doi:10.20151098/rstb.20152015.20150496

Hochstrasser ML, Taylor DW, Bhat P, Guegler CK, Sternberg SH, Nogales E, Doudna JA (2014) CasA mediates Cas3-catalyzed target degradation during CRISPR RNA-guided interference. Proc Natl Acad Sci U S A 111:6618–6623. doi:10.1073/pnas.1405079111. Epub 2014 Apr 18

Ivancic-Bace I, Cass SD, Wearne SJ, Bolt EL (2015) Different genome stability proteins underpin primed and naive adaptation in E. coli CRISPR-Cas immunity. Nucleic Acids Res 43:10821–10830. doi:10.11093/nar/gkv11213. Epub 2015 Nov 17

Jackson RN, Golden SM, van Erp PB, Carter J, Westra ER, Brouns SJ, van der Oost J, Terwilliger TC, Read RJ, Wiedenheft B (2014) Crystal structure of the CRISPR RNA-guided surveillance complex from Escherichia coli. Science 345:1473–1479

Jansen R, Embden JD, Gaastra W, Schouls LM (2002a) Identification of genes that are associated with DNA repeats in prokaryotes. Mol Microbiol 43:1565–1575
Jansen R, van Embden JD, Gaastra W, Schouls LM (2002b) Identification of a novel family of sequence repeats among prokaryotes. OMICS 6:23–33
Jiang W, Maniv I, Arain F, Wang Y, Levin BR, Marraffini LA (2013) Dealing with the evolutionary downside of CRISPR immunity: bacteria and beneficial plasmids. PLoS Genet 9: e1003844. doi:10.1001371/journalpgen1003844. Epub 2013 Sep 26
Jinek M, Chylinski K, Fonfara I, Hauer M, Doudna JA, Charpentier E (2012) A programmable dual-RNA-guided DNA endonuclease in adaptive bacterial immunity. Science 337:816–821. doi:10.1126/science.1225829. Epub 2012 Jun 28
Jore MM, Lundgren M, van Duijn E, Bultema JB, Westra ER, Waghmare SP, Wiedenheft B, Pul Ü, Wurm R, Wagner R, Beijer MR, Barendregt A, Zhou K, Snijders AP, Dickman MJ, Doudna JA, Boekema EJ, Heck AJ, van der Oost J, Brouns SJ (2011) Structural basis for CRISPR RNA-guided DNA recognition by Cascade. Nat Struct Mol Biol 18:529–536
Kunin V, Sorek R, Hugenholtz P (2007) Evolutionary conservation of sequence and secondary structures in CRISPR repeats. Genome Biol 8:R61
Künne T, Swarts DC, Brouns SJ (2014) Planting the seed: target recognition of short guide RNAs. Trends Microbiol 22:74–83. doi:10.1016/j.tim.2013.1012.1003. Epub 2014 Jan 14
Künne T, Kieper SN, Bannenberg JW, Vogel AI, Miellet WR, Klein M, Depken M, Suarez-Diez M, Brouns SJ (2016) Cas3-derived target DNA degradation fragments fuel primed CRISPR adaptation. Mol Cell 63:852–864. doi:10.1016/j.molcel.2016.1007.1011. Epub 2016 Aug 18
Labrie SJ, Samson JE, Moineau S (2010) Bacteriophage resistance mechanisms. Nat Rev Microbiol 8:317–327. Epub 2010 Mar 29
Lange SJ, Alkhnbashi OS, Rose D, Will S, Backofen R (2013) CRISPRmap: an automated classification of repeat conservation in prokaryotic adaptive immune systems. Nucleic Acids Res 41:8034–8044
Levy A, Goren MG, Yosef I, Auster O, Manor M, Amitai G, Edgar R, Qimron U, Sorek R (2015) CRISPR adaptation biases explain preference for acquisition of foreign DNA. Nature 520:505–510. doi: 10.1038/nature14302. Epub 2015 Apr 13
Li H (2015) Structural Principles of CRISPR RNA Processing. Structure 23:13–20. doi:10.1016/j.str.2014.1010.1006. Epub 2014 Nov 26
Li M, Liu H, Han J, Liu J, Wang R, Zhao D, Zhou J, Xiang H (2013) Characterization of CRISPR RNA biogenesis and Cas6 cleavage-mediated inhibition of a provirus in the haloarchaeon Haloferax mediterranei. J Bacteriol 195:867–875. doi:10.1128/JB.01688-01612. Epub 2012 Dec 14
Li M, Wang R, Xiang H (2014a) Haloarcula hispanica CRISPR authenticates PAM of a target sequence to prime discriminative adaptation. Nucleic Acids Res 42:7226–7235. doi:10.1093/nar/gku7389. Epub 2014 May 26
Li M, Wang R, Zhao D, Xiang H (2014b) Adaptation of the Haloarcula hispanica CRISPR-Cas system to a purified virus strictly requires a priming process. Nucleic Acids Res 42:2483–2492. doi:10.1093/nar/gkt1154. Epub 2013 Nov 21
Luo ML, Mullis AS, Leenay RT, Beisel CL (2015) Repurposing endogenous type I CRISPR-Cas systems for programmable gene repression. Nucleic Acids Res 43:674–681. doi:10.1093/nar/gku1971. Epub 2014 Oct 17
Maier LK, Fischer S, Stoll B, Brendel J, Pfeiffer F, Dyall-Smith M, Marchfelder A (2012) The immune system of halophilic archaea. Mob Genet Elements 2:228–232
Maier LK, Lange SJ, Stoll B, Haas KA, Fischer S, Fischer E, Duchardt-Ferner E, Wohnert J, Backofen R, Marchfelder A (2013a) Essential requirements for the detection and degradation of invaders by the Haloferax volcanii CRISPR/Cas system I-B. RNA Biol 10:865–874
Maier LK, Stoll B, Brendel J, Fischer S, Pfeiffer F, Dyall-Smith M, Marchfelder A (2013b) The ring of confidence: a haloarchaeal CRISPR/Cas system. Biochem Soc Trans 41:374–378. doi:10.1042/BST20120263

Maier LK, Dyall-Smith M, Marchfelder A (2015a) The adaptive immune system of Haloferax volcanii. Life (Basel) 5:521–537. doi:10.3390/life5010521

Maier LK, Stachler AE, Saunders SJ, Backofen R, Marchfelder A (2015b) An active immune defense with a minimal CRISPR (clustered regularly interspaced short palindromic repeats) RNA and without the Cas6 protein. J Biol Chem 290:4192–4201. doi:10.1074/jbc.M4114.617506. Epub 2014 Dec 15

Makarova KS, Aravind L, Wolf YI, Koonin EV (2011) Unification of Cas protein families and a simple scenario for the origin and evolution of CRISPR-Cas systems. Biol Direct 6:38. doi:10.1186/1745-6150-1186-1138

Makarova KS, Wolf YI, Alkhnbashi OS, Costa F, Shah SA, Saunders SJ, Barrangou R, Brouns SJ, Charpentier E, Haft DH, Horvath P, Moineau S, Mojica FJ, Terns RM, Terns MP, White MF, Yakunin AF, Garrett RA, van der Oost J, Backofen R, Koonin EV (2015) An updated evolutionary classification of CRISPR-Cas systems. Nat Rev Microbiol 13:722–736. doi:10.1038/nrmicro3569. Epub 2015 Sep 28

Marraffini LA, Sontheimer EJ (2008) CRISPR interference limits horizontal gene transfer in staphylococci by targeting DNA. Science 322:1843–1845

Mohanraju P, Makarova KS, Zetsche B, Zhang F, Koonin EV, van der Oost J (2016) Diverse evolutionary roots and mechanistic variations of the CRISPR-Cas systems. Science 353: aad5147. doi:10.1126/science.aad5147

Mojica FJ, Rodriguez-Valera F (2016) The discovery of CRISPR in archaea and bacteria. FEBS J 283:3162–3169. doi:10.1111/febs.13766. Epub 2016 Jun 15

Mojica FJ, Diez-Villasenor C, Garcia-Martinez J, Soria E (2005) Intervening sequences of regularly spaced prokaryotic repeats derive from foreign genetic elements. J Mol Evol 60:174–182

Mojica FJ, Diez-Villasenor C, Garcia-Martinez J, Almendros C (2009) Short motif sequences determine the targets of the prokaryotic CRISPR defence system. Microbiology 155:733–740

Mulepati S, Heroux A, Bailey S (2014) Crystal structure of a CRISPR RNA-guided surveillance complex bound to a ssDNA target. Science 345:1479–1484

Nickel L, Weidenbach K, Jäger D, Backofen R, Lange SJ, Heidrich N, Schmitz RA (2013) Two CRISPR-Cas systems in Methanosarcina mazei strain Göo1 display common processing features despite belonging to different types I and III. RNA Biol 10:779–791

Niewoehner O, Jinek M, Doudna JA (2014) Evolution of CRISPR RNA recognition and processing by Cas6 endonucleases. Nucleic Acids Res 42:1341–1353. doi:10.1093/nar/gkt1922. Epub 2013 Oct 22

Norais C, Hawkins M, Hartman AL, Eisen JA, Myllykallio H, Allers T (2007) Genetic and physical mapping of DNA replication origins in Haloferax volcanii. PLoS Genet 3:e77. Epub 2007 Apr 5

Nunez JK, Harrington LB, Kranzusch PJ, Engelman AN, Doudna JA (2015) Foreign DNA capture during CRISPR-Cas adaptive immunity. Nature 527:535–538. doi:10.1038/nature15760. Epub 2015 Oct 21

Oren A (2006) The order halobacteriales. In: Dworkin M, DeLong EF, Stackebrandt E, Lory S, Thompson F (eds) The prokaryotes, vol 3. Springer, New York, pp 113–164

Plagens A, Tjaden B, Hagemann A, Randau L, Hensel R (2012) Characterization of the CRISPR/Cas subtype I-A system of the hyperthermophilic crenarchaeon Thermoproteus tenax. J Bacteriol 194:2491–2500. doi:10.1128/JB.00206-00212. Epub 2012 Mar 9

Pourcel C, Salvignol G, Vergnaud G (2005) CRISPR elements in Yersinia pestis acquire new repeats by preferential uptake of bacteriophage DNA, and provide additional tools for evolutionary studies. Microbiology 151:653–663

Pul Ü, Wurm R, Arslan Z, Geissen R, Hofmann N, Wagner R (2010) Identification and characterization of E. coli CRISPR-cas promoters and their silencing by H-NS. Mol Microbiol 75:1495–1512

Qi LS, Larson MH, Gilbert LA, Doudna JA, Weissman JS, Arkin AP, Lim WA (2013) Repurposing CRISPR as an RNA-guided platform for sequence-specific control of gene expression. Cell 152:1173–1183. doi:10.1016/j.cell.2013.1102.1022

Rath D, Amlinger L, Hoekzema M, Devulapally PR, Lundgren M (2015) Efficient programmable gene silencing by Cascade. Nucleic Acids Res 43:237–246. doi:10.1093/nar/gku1257. Epub 2014 Nov 30

Richter H, Zoephel J, Schermuly J, Maticzka D, Backofen R, Randau L (2012) Characterization of CRISPR RNA processing in Clostridium thermocellum and Methanococcus maripaludis. Nucleic Acids Res 40:9887–9896. doi:10.1093/nar/gks9737. Epub 2012 Aug 8

Richter C, Dy RL, McKenzie RE, Watson BN, Taylor C, Chang JT, McNeil MB, Staals RH, Fineran PC (2014) Priming in the Type I-F CRISPR-Cas system triggers strand-independent spacer acquisition, bi-directionally from the primed protospacer. Nucleic Acids Res 42:8516–8526. doi:10.1093/nar/gku8527. Epub 2014 Jul 12

Sampson TR, Weiss DS (2014) Exploiting CRISPR/Cas systems for biotechnology. Bioessays 36:34–38. doi:10.1002/bies.201300135

Sashital DG, Wiedenheft B, Doudna JA (2012) Mechanism of foreign DNA selection in a bacterial adaptive immune system. Mol Cell 46:606–615. doi:101016/jmolcel201210031020. Epub 2012 Apr 19

Scholz I, Lange SJ, Hein S, Hess WR, Backofen R (2013) CRISPR-Cas systems in the cyanobacterium Synechocystis sp. PCC6803 exhibit distinct processing pathways involving at least two Cas6 and a Cmr2 protein. PLoS One 8:e56470. doi:10.51371/journalpone0056470. Epub 2013 Feb 18

Semenova E, Jore MM, Datsenko KA, Semenova A, Westra ER, Wanner B, van der Oost J, Brouns SJ, Severinov K (2011) Interference by clustered regularly interspaced short palindromic repeat (CRISPR) RNA is governed by a seed sequence. Proc Natl Acad Sci U S A 108:10098–10103. Epub 2011 Jun 6

Semenova E, Savitskaya E, Musharova O, Strotskaya A, Vorontsova D, Datsenko KA, Logacheva MD, Severinov K (2016) Highly efficient primed spacer acquisition from targets destroyed by the Escherichia coli type I-E CRISPR-Cas interfering complex. Proc Natl Acad Sci U S A 113:7626–7631. doi:10.1073/pnas.1602639113. Epub 2016 Jun 20

Shah SA, Garrett RA (2011) CRISPR/Cas and Cmr modules, mobility and evolution of adaptive immune systems. Res Microbiol 162:27–38. Epub 2010 Sep 21

Shah SA, Erdmann S, Mojica FJ, Garrett RA (2013) Protospacer recognition motifs: mixed identities and functional diversity. RNA Biol 10:5

Shao Y, Li H (2013) Recognition and cleavage of a nonstructured CRISPR RNA by its processing endoribonuclease Cas6. Structure 21:385–393

Shmakov S, Abudayyeh OO, Makarova KS, Wolf YI, Gootenberg JS, Semenova E, Minakhin L, Joung J, Konermann S, Severinov K, Zhang F, Koonin EV (2015) Discovery and functional characterization of diverse class 2 CRISPR-Cas systems. Mol Cell 60:385–397. doi:101016/jmolcel201510101008. Epub 2015 Oct 22

Shmakov S, Smargon A, Scott D, Cox D, Pyzocha N, Yan W, Abudayyeh OO, Gootenberg JS, Makarova KS, Wolf YI, Severinov K, Zhang F, Koonin EV (2017) Diversity and evolution of class 2 CRISPR-Cas systems. Nat Rev Microbiol 15:169–182. doi:10.1038/nrmicro.2016.1184. Epub 2017 Jan 23

Staals RH, Brouns SJ (2013) Distribution and mechanism of the type I CRISPR-Cas systems. In: Barrangou R, van der Oost J (eds) CRISPR-Cas systems. Springer, Berlin, pp 145–169

Staals RH, Jackson SA, Biswas A, Brouns SJ, Brown CM, Fineran PC (2016) Interference-driven spacer acquisition is dominant over naive and primed adaptation in a native CRISPR-Cas system. Nat Commun 7:12853. doi:10.1038/ncomms12853

Stachler AE, Marchfelder A (2016) Gene repression in Haloarchaea using the CRISPR (clustered regularly interspaced short palindromic repeats) – Cas I-B system. J Biol Chem 291:15226–15242

Sternberg SH, Doudna JA (2015) Expanding the biologist's toolkit with CRISPR-Cas9. Mol Cell 58:568–574. doi:10.1016/j.molcel.2015.1002.1032

Sternberg SH, Richter H, Charpentier E, Qimron U (2016) Adaptation in CRISPR-Cas systems. Mol Cell 61:797–808. doi:10.1016/jmolcel201610011030. Epub 2016 Mar 13

Stoll B, Maier LK, Lange SJ, Brendel J, Fischer S, Backofen R, Marchfelder A (2013) Requirements for a successful defence reaction by the CRISPR-Cas subtype I-B system. Biochem Soc Trans 41:1444–1448. doi:10.1042/BST20130098

Suttle CA (2007) Marine viruses–major players in the global ecosystem. Nat Rev Microbiol 5:801–812

Swarts DC, Mosterd C, van Passel MW, Brouns SJ (2012) CRISPR interference directs strand specific spacer acquisition. PLoS One 7:e35888. doi:10.31371/journalpone0035888. Epub 2012 Apr 27

Vestergaard G, Garrett RA, Shah SA (2014) CRISPR adaptive immune systems of Archaea. RNA Biol 11:156–167. doi:10.4161/rna.27990. Epub 2014 Feb 7

Vorontsova D, Datsenko KA, Medvedeva S, Bondy-Denomy J, Savitskaya EE, Pougach K, Logacheva M, Wiedenheft B, Davidson AR, Severinov K, Semenova E (2015) Foreign DNA acquisition by the I-F CRISPR-Cas system requires all components of the interference machinery. Nucleic Acids Res 43:10848–10860. doi:10.11093/nar/gkv11261. Epub 2015 Nov 19

Wang R, Zheng H, Preamplume G, Shao Y, Li H (2012) The impact of CRISPR repeat sequence on structures of a Cas6 protein-RNA complex. Protein Sci 21:405–417. doi:10.1002/pro.2028. Epub 2012 Feb 9

Wang J, Li J, Zhao H, Sheng G, Wang M, Yin M, Wang Y (2015) Structural and mechanistic basis of PAM-dependent spacer acquisition in CRISPR-Cas systems. Cell 163:840–853. doi:10.1016/jcell201510101008. Epub 2015 Oct 17.

Wang R, Li M, Gong L, Hu S, Xiang H (2016) DNA motifs determining the accuracy of repeat duplication during CRISPR adaptation in Haloarcula hispanica. Nucleic Acids Res 44:4266–4277. doi:10.1093/nar/gkw4260. Epub 2016 Apr 16

Westra ER, Semenova E, Datsenko KA, Jackson RN, Wiedenheft B, Severinov K, Brouns SJ (2013) Type I-E CRISPR-cas systems discriminate target from non-target DNA through base pairing-independent PAM recognition. PLoS Genet 9:e1003742. doi:10.1001371/journalpgen1003742. Epub 2013 Sep 5

Wiedenheft B, van Duijn E, Bultema JB, Waghmare SP, Zhou K, Barendregt A, Westphal W, Heck AJ, Boekema EJ, Dickman MJ, Doudna JA (2011) RNA-guided complex from a bacterial immune system enhances target recognition through seed sequence interactions. Proc Natl Acad Sci U S A 108:10092–10097. Epub 2011 May 2

Woods WG, Dyall-Smith ML (1997) Construction and analysis of a recombination-deficient (radA) mutant of Haloferax volcanii. Mol Microbiol 23:791–797

Yosef I, Goren MG, Qimron U (2012) Proteins and DNA elements essential for the CRISPR adaptation process in Escherichia coli. Nucleic Acids Res 40:5569–5576. doi:10.1093/nar/gks5216. Epub 2012 Mar 8

Zetsche B, Gootenberg JS, Abudayyeh OO, Slaymaker IM, Makarova KS, Essletzbichler P, Volz SE, Joung J, van der Oost J, Regev A, Koonin EV, Zhang F (2015) Cpf1 is a single RNA-guided endonuclease of a class 2 CRISPR-Cas system. Cell 163:759–771. doi:10.1016/j.cell.2015.1009.1038. Epub 2015 Sep 25

Zhang J, Rouillon C, Kerou M, Reeks J, Brugger K, Graham S, Reimann J, Cannone G, Liu H, Albers SV, Naismith JH, Spagnolo L, White MF (2012) Structure and mechanism of the CMR complex for CRISPR-mediated antiviral immunity. Mol Cell 45:303–313. doi:10.1016/j.molcel.2011.1012.1013. Epub 2012 Jan 15

Zhao H, Sheng G, Wang J, Wang M, Bunkoczi G, Gong W, Wei Z, Wang Y (2014) Crystal structure of the RNA-guided immune surveillance Cascade complex in Escherichia coli. Nature 515(7525):147–150

MIX
Papier aus verantwortungsvollen Quellen
Paper from responsible sources
FSC® C105338

If you have any concerns about our products,
you can contact us on
ProductSafety@springernature.com

In case Publisher is established outside the EU,
the EU authorized representative is:
Springer Nature Customer Service Center GmbH
Europaplatz 3, 69115 Heidelberg, Germany

Printed by Libri Plureos GmbH
in Hamburg, Germany